# WIFI REFUGEE

## Plight of the Modern-day Canary

SHANNON ROWAN

**WiFi Refugee; Plight of the Modern-day Canary**

**Shannon Rowan**

Paperback Edition First Published in the United States in 2022 by Shannon Rowan

Hardcover Edition First Published in the United States in 2022 by Shannon Rowan

eBook Edition First Published in the United States in 2022 by Shannon Rowan



ISBN 979-8-9866706-0-7

# Praise For WiFi Refugee

*As an EMF consultant, and a provider of EMF remediation products, I get calls and emails from people around the world, all suffering the effects of exposure to various types of EMF. Most often, the people I speak to have been told they are too sensitive, they are crazy, or their symptoms are all in their heads, etc. People always feel better speaking with me as I completely understand the challenges they are experiencing and I know those psychological and biological challenges are real.*

*With this said, I highly suggest Shannon's excellent book! It speaks directly to the heart of those of us who can feel and are affected by the toxic exposure to various types of EMF. Her book lets us know that we are not crazy or suffering from psychosomatic illness. We are not 'too sensitive' nor are our symptoms just in 'our heads'. We are normal, however, the present global proliferation of toxic, man-made, non-native EMF exposure is not normal! Shannon provides relevant and validating stories from many people, along with excellent resources and holistic suggestions for living a healthier life. Highly recommended reading for our present times!*

**—Neil Steven Cohen**
EMF Consultant, www.emfsafetyzone.com
EMF Safety Zone YouTube Channel.

WiFi Refuge *is not only a wake-up call; it is a testament to the determination and resourcefulness required of today's WiFi refugees. It is much more than an informative glimpse into the challenges of WiFi-free living and the realities of today's poisonous world. It is a voice for those who have been pushed into the corners and crevasses of the United States and abroad, sharing their stories in commonality and caution. It is a manual for clean and natural living, with many helpful and alternative medical and health options. It is a warning, as the canaries of the mines also were, in foretelling of coming danger, which has, unfortunately, already spread across most of the country and planet. You will be scared, you will be educated, and you will be thankful for Shannon's commitment to tell this story as no one has before.*

—**Katherine Cottle**, Ph.D., author of *Baltimore Sideshow, The Hidden Heart of Charm City: Baltimore Letters and Lives, Halfway: A Journal through Pregnancy,* and others

*This is probably one of the most important, if not the most important and comprehensive books ever written on this topic by one of the so-called 'canaries.' I have known Shannon for several years and am proud to be able to call her an extremely good friend—as well as a collaborator on several other geopolitical writing projects, but I write this review not from any misplaced sense of loyalty or bias, but simply as an independent observation of the merits of this truly seminal work.*

*Within the pages of her semi-autobiographical account, Shannon has somehow managed to concisely capture and disseminate the entire, vast subject in a sometimes humorous, sometimes moving, yet often sad, poignant, and evocative way. Beautifully written and immaculately researched, this book is a must read for anyone who wishes to understand exactly how the insidious EMF industries operate and the desperately sad truth of the devastating impact their products have upon us all—and also upon this Earth, the only home we humans will ever have. The over-riding message is that we are under attack—and deliberately so—and Shannon pulls no punches whatsoever in providing her readers with all the often disturbing and unsettling truths surrounding this increasingly prescient and important topic.*

—**John Hamer**, geopolitical researcher, public speaker, and author of several books including *The Falsification of History, The Falsification of Science* plus others and latterly (jointly with Shannon), *Welcome to the Masquerade.*

*Having been driven from life as a world traveller into an off-grid existence, Shannon Rowan writes about life as a person sensitive to electric and magnetic fields (EMFs) with the authority, sensitivity, and insight that can come only from lived experience. A canary in the coalmine of modernity, Rowan's shocking and absorbing account of her own story, along with those of her fellow 'canaries,' is told in such a compelling manner that it allows the reader to see the world through their eyes and walk in their shoes. This important window into a world that for most of us has either been completely unknown or shrouded in mystery cannot help but inspire greater compassion, empathy, and understanding for those who suffer from EMF and other sensitivities, while raising important questions about what we as a society can do to protect and care for those among us who are most vulnerable to the impacts of our own creations.*

--**Anise Eden**, MSW, LCSW, author of the *Healing Edge Series* and the *Things Unseen Series*

*Shannon Rowan had my complete attention from the very beginning with presenting the worldview of Chief Seattle, one of my heroes. She gives us an intimate view into her personal journey with electro- and chemical sensitivity, as well as that of many others who have crossed her path. I identify and relate with these life experiences, as I am EHS as well. Shannon condenses many facts, research, and facets of electro-sensitivity and presents it in such a way that makes it easily accessible and relatable. She connects the dots in ways that show how our civilization got here, where it's going, and what we can do about it, offering practical solutions. Reading pure technical data and jargon gets tedious very quickly in some other sources, but reading the experiences of others helps me empathize and connect with others sharing the same journey. It's a human journey. Sometimes this journey can make you feel so alone and isolated. This is a book I can recommend to my friends and family who don't understand what I and others are going through and what's at stake. They may not be canaries, but they are equally affected by encroaching EMF and life out of balance. Great work Shannon. Thank you!*

—**Tim H.**, Oregon

*I went to the woods because I wished to live deliberately, to front only the essential facts of life, and see if I could not learn what it had to teach and not, when I came to die, discover that I had not lived.*

*I did not wish to live what was not life, living is do dear; nor did I wish to practice resignation, unless it was quite necessary. I wanted to live deep and suck out all of the marrow of life, to live sturdily, and Spartan–like as to rout all that was not life, to cut a broad swatch and shave close, to drive life into a corner, and reduce it to its lowest terms.*

—David Henry Thoreau, *Walden*

# *Dedication*

*To all of the silenced canaries, may your plight at last be heard.*

*And to my dear departed friend, Adam, who taught me to face life's adventures fearlessly and with loud laughter.*

# CONTENTS

# GLOSSARY

The following is a short glossary of terms and abbreviations used in this book, which may be unfamiliar to the reader.

**EMF**: *Electromagnetic Field.* Refers to manmade sources of artificial electromagnetic fields, which result from electronics of all kinds. Electromagnetic fields emitted from electronic devices and motorized vehicles, machines, power lines, power transfer stations, radar and mobile communications installations, etc., and can be in either the low or high frequency range.

**EMR**: *Electromagnetic Radiation.* This term is basically interchangeable with EMF, but often used to refer specifically to the radiation produced by electromagnetic fields. All EMFs/electromagnetic fields no matter at what power level or frequency give off electromagnetic radiation.

**RF**: *Radiofrequency*. This refers directly to frequencies emitted by radio-transmitting technologies—any wireless device, and the infrastructure powering wireless/mobile device usage. Radiofrequencies encompass microwave and millimeter-wave frequencies, high on the electromagnetic spectrum, but lower than x-rays and gamma rays.

**ELF**: *Extremely Low Frequency.* ELFs are EMFs produced by low-frequency producing electronic motors/appliances or power stations and power lines.

**EHS**: *Electro-hypersensitivity/Electro-hypersensitive.* (Also what I jokingly call *Exploding Head Syndrome* as it accurately describes one of the symptoms I personally experience from having EHS.) The current most widely used label given to those who can directly feel the harmful effects of all of the above (EMFs, EMR, RFs, ELFs). This condition when first documented in the 1930s was called

"Microwave Sickness" or "Microwave Syndrome" as it was understood to have been caused by excessive exposures to microwave radiation.

**ES**: Electro-sensitive/electro-sensitivity. Used interchangeably with EHS. Also sometimes used to reference a less severe form of sensitivity than EHS.

**EMS**: Electromagnetic Sensitive/Sensitivity. Also used interchangeably with EHS/ES, but mostly reserved for legal purposes, as currently in the US, it is the only legally accepted label for those suffering with EHS/ES.

**SMART:** Referring to *wireless, remotely operable technology*, that is generally equipped with radiofrequency-emitting antennae with the intention of tracking and monitoring people alongside everything else within our environments, including energy usage. SMART is an acronym. It does *not* mean that these technologies are intelligent, although some may make use of AI (artificial intelligence) software. SMART, when speaking of consumer products, stands for "Self-Monitoring And Reporting Technology," (Note: dozens of other acronyms are listed under SMART on acryonms.com, mostly used by the US Military, some of which include, "Secret Militarized Arms in Residential Technologies," "Satellite Monitoring And Remote Tracking," and "Simulation Modeling Anchored in Real-World Testing.")

**White Zone**: A radio-quiet area, free from radiofrequency pollution.

**Dead Zone**: An area without cellphone coverage, thus similar to a White Zone, but possibly not free from other radiofrequency or other EMF interference.

# Foreword

## By Claire Edwards

Why should you read this book? Is the topic of electrosensitivity relevant to you and your family? Quite simply, yes, it is. It is a matter of health, life and death.

Even before the advent of 5G, scientists estimated that 100 million people worldwide were suffering from electrosensitivity.[1] But these are just the people who are all too painfully aware of it. You are suffering from it, too, but you probably have no idea because you are either not yet aware of your injuries or you attribute your symptoms to other causes. Medical doctors receive zero training on the effects of exposure to electromagnetic radiation (EMR) so they fail to recognise the signs and diagnose diseases without realising that they were caused by exposure to EMR. Yet an extensive bibliography of Soviet research on the biological effects of EMR exposure compiled in the 1970s by the US military documented the vast range of effects on the body.[2] A World Health Organization symposium held in 1973 was actually entitled "*Biologic* Effects and Hazards of Electromagnetic Radiation".[3] The EMF Portal hosted by the University of Aachen provides a database of over 36 thousand studies on these biological effects.[4]

Why, then, have you never heard anything about the biological effects of exposure to electromagnetic radiation, otherwise known as "electrosensitivity"? Ah, thereby hangs a tale! Before being secretly brought to the United States under Operation Paperclip to work for the US Navy post-WWII, Nazi scientist Herman P. Schwan had worked for the Kaiser Wilhelm Institute for Anthropology, Human Heredity and *Eugenics*. Is it any surprise, then, that when the US Navy asked him to advise on guidelines for exposure to electromagnetic radiation, he withheld his full knowledge of the biological effects and proposed guidelines that insisted on only thermal effects? This implies that, if you feel no heat when exposed, you have nothing to worry about.

Schwan's 1953 guidelines effectively gave the US military carte blanche to expose their personnel to excessive levels of EMR at a time when the Cold War was used to justify vast arrays of radar around the Arctic circle and on the West's border with East Germany. To this day, the so-called "thermal hypothesis" holds sway across the world and the biological effects are denied by corporate- and military-controlled national and international regulatory and health agencies and the relevant agencies of the UN, while defenceless and unsuspecting populations are attacked with ever-increasing levels of power and numbers of frequencies from cell phones, antennas, so-called "smart" devices, satellites, "smart" meters and much more.[5]

EMR exposure limits across the world are set so high that it is almost impossible to exceed them.[6] It's akin to claiming that you care about road safety while setting car speed limits at a million miles per hour. However, the comparison is a poor one, for speed limits are enforceable by law, while EMR exposure *guidelines* are not. You may be prosecuted for exceeding the road speed limit, while powerful telecommunications corporations can and do exceed the so-called guidelines as much as they like with no penalty. However, the problem is not just with high power levels: given the delicacy of the electrical functions of our bodies, even the tiniest power levels cause biological effects.[7]

Horrifyingly, an EU Briefing for parliamentarians of February 2020 discussing the Effects of 5G Wireless Communication on Human Health stated, inter alia, that *no one knows* what 5G actually is and whether it has impacts on human health and the environment, that *no research* has addressed the constant exposure that 5G would introduce, that the European Commission has conducted *no studies* on the potential health risks of 5G, and that *it is not currently possible* to accurately simulate or measure 5G emissions in the real world (emphasis added). Populations are not being warned about these unquantified, existential risks.[8] The US military explicitly stated in 1976 in a since declassified document that populations should not be warned of the dangers of EMR because this would have "unfavourable effects on industrial output and military functions". This is code for the message that warning the public would impede the production of EMR weapons and limit the profit to be derived from them![9]

In 2019, when a 20-year international health study was published, its lead author posited "a hidden epidemic". Hidden indeed, and deliberately so. The current driving force for hiding the horrendous injuries and damage from ubiquitous EMR exposure is the 17-trillion-dollar projected profit from the rollout of 5G. The United Nations literature on 5G repeatedly states the intention of "blanketing" the entire world with 5G, including the oceans and rainforests. As of March 2022, permissions had been sought from the US Federal Communications Commission alone for 440 thousand 5G satellites.[10] To give you an idea of your massively increased exposure, just four years earlier there were only about 2 thousand functioning satellites in space. Governments used their "covid" measures of 2020 to accelerate the rollout of 5G on the false premise that the emergency services and hospitals would need it. While populations were locked down, 5G antennas were installed everywhere, and especially in, on or adjacent to hospitals, schools and kindergartens.

The campaign to stop 5G leads people to believe that we would be safe if we just stopped at 4G but this is a misunderstanding. *All* wireless technology injures and kills, but 5G simply does so faster than previous generations.[11] You could think of the generations up to 4G as creating a soup of electromagnetic radiation, whereas 5G is beam-forming; in other words, the radiation is concentrated into a beam something like a laser. The telecoms industry indulges in some intellectual acrobatics to convince you that this means that your exposure will be lower with 5G. As far as I can make out, the argument goes something like this: if one particular beam is pointing in one particular direction, then it is not pointing in the other 359-degree possible directions. Hey presto! The average exposure is much lower! The truth is, though, that 5G will be beaming in all directions all the time rather like a searchlight rotating at high speed, except that this 5G "searchlight" is made up of singular beams shooting out ceaselessly.

You may ask how EMR could affect us when our bodies are solid. But they are not! We all learned in school about molecules and atoms, and lately scientists have asserted that there is no matter at all in atoms.[12] Every function in our bodies happens electrically. Each of our brain waves is attuned to the harmonics of the Schumann Resonance of the Earth.[13] Each of our organs has a frequency.[14] Our cell membranes operate electrically.[15] Nobel-prize-winning scientist Luc Montagnier posited in 2011 that DNA reconstitutes itself apparently from nothing when subject to the electromagnetic signal emanating from another string of DNA.[16]

When public WiFi access points were installed in December 2015 at the UN in Vienna, I was ill for months alternately with flu and colds. Public access points are way more powerful than your home Internet router, and schoolchildren are routinely exposed to dangerous levels of EMR in classrooms equipped with them. Children's bodies are wetter than those of adults, and water is a conductor so they are at greater risk.[17] What's more, their brains are still developing and should never be put at risk of injury by exposure to EMR. Military EMR expert Barrie Trower has warned that children exposed to EMR in schools are being made sterile.[18] Tablets used by children are typically held directly over their genitals.[19] Children's school performance is falling because it is motor skills that develop the brain and not tapping at keyboards.[20] Teenagers become addicted to their cell phones because, according to Trower, cell phones are 17 times as addictive as heroin.[21]

The warning signs are there for us all to see if we but recognise them. Birds' wings are EMR antennas and this is why we so frequently hear reports from around the world of birds falling from the sky.[22] Beekeepers are well aware that bees become aggressive and hives die off when they are sited between antennas and bee loss is not due to the specious "colony collapse disorder".[23] We have lost 80 percent of our insects in the last 20 years.[24] Look around your neighbourhood at trees growing next to the new digital streetlights. You will see how the irradiated leaves have died on the side exposed to the radiation from the streetlights.[25] Whales are beaching themselves and other marine life is washing up dead on beaches – are they being affected by the extremely low frequency (ELF) waves used for submarine communications and/or 5G being emitted onto or under the oceans?[26]

Rates of autism in the US have gone from 1 in 150 in 2000, to a predicted 1 in 2 by 2025, according to a Massachusetts Institute of Technology scientist.[27] Brain tumour incidence and rates of depression and suicide have skyrocketed in recent years.[28] Rates of burnout are off the scale.[29] With brain fog being a common result of EMR exposure, is it any wonder that society appears to have become so illogical and emotionally unstable in recent years?

I have never believed in "covid". From the very first moment that it was announced, I believed people's symptoms to be the result of massive exposure to exceptionally intense electromagnetic radiation. When "covid" was alleged to have broken out in Wuhan, China in October 2019, 10 thousand 5G antennas had just been switched on and the world's biggest ELF antenna had just been completed in Hubei Province, where Wuhan is situated.[30] The symptoms of "covid" are identical to those of exposure to EMR and doctors keep adding to the list because EMR affects every system of the body.[31]

There is a long history of sickness associated with the expansion of the use of electromagnetic radiation. Electric lighting was first installed in the late 1800s, and the first people to complain of strange symptoms were wealthy women, the very people who were the first to be able to afford domestic electric lighting. Berlin's electrification around 1880 coincided with Bismarck's introduction of compulsory health insurance and ordinary people started flooding doctors' offices with complaints whose cause could not be identified and which were diagnosed under the umbrella term "neurasthenia". This universal malaise was called "the modern disease" and attributed to the increased pace of life under electrification. Almost no one realised that it was the electrification itself that was making everyone sick.[32]

At the end of the nineteenth century and the beginning of the twentieth, half of the telegraph operators became sick and later half of the telephone operators.[33] During WWI, runners laid out telephone lines through the trenches so that the generals could communicate with the front lines. Many soldiers who became sick were said to be suffering from shell shock. This was medically labelled neurasthenia, which is an umbrella term for a set of symptoms whose cause is not identifiable, and it is notable that many soldiers suffering from neurasthenia had never served at the front and never come into contact with any shells! Most languished for months in convalescent homes, although one doctor became famous for successfully treating his neurasthenia patients in a single session. What did he do? He sent them out to dig in the fields – in other words, he grounded them![34]

It is also notable that six months before the so-called Spanish Flu broke out, the US had joined WWI and brought powerful radio antennas. Despite all the research efforts of doctors, they were never able to prove that Spanish Flu was transmissible.[35]

And the lamentable story goes on. The early radar operators became sick. Rates of disease increased with the advent of AM radio and television, then with the first launches of satellites. Hong Kong flu, avian flu, swine flu and now "covid" all correlate with the expansion in the use of EMR.[36]

You have been lied to, but it is a lie of omission. No one has warned you of the danger from electromagnetic radiation. The military and telecoms companies refute campaigners' claims that EMR will harm and kill you, but notice that they never tell you that it is safe because they cannot provide a single study that has found it to be safe and because 5G enables a novel kind of covert warfare.[37] Rather, the telcos are "war gaming" the science, exactly as they did with the tobacco scandal.[38] You surround yourselves and your children with increasing numbers of lethal EMR-emitting devices: baby alarms, routers, cell phones, "smart" devices and meters, DECT phones, "wearables", and much else. Outside your door, EMR is coming at you from everywhere. No studies have proven the safety of any of it, especially your cell phone.

While we are discussing the lies of the 5G and wireless technology apologists, let us deal with their three constantly repeated canards[39] [40]:

1. *Non-ionising radiation is not strong enough to break chemical bonds and damage DNA.*

The University of Aachen EMF Portal contains 623 articles citing DNA damage, and telco Swisscom filed a patent in 2004 on a method and system for reducing electrosmog in wireless local networks that states clearly that WiFi damages DNA.

2. *The Sun is irradiating you every day with higher frequency and more intense EMR than you are going to get from 5G.*

Current public exposure to wireless radiation (*before* the addition of 5G) is approximately a quintillion times (1,000,000,000,000,000,000) greater than natural, background radiation. And radiation from the Sun is neither pulsed nor modulated, whereas wireless signals are. Harm is caused by both the high-frequency carrier wave and the low-frequency pulsations. Even that corporate shill, the self-proclaimed International Commission on Non-Ionizing Radiation Protection (ICNIRP), admitted in its 1998 *Guidelines* that, "Compared with continuous-wave radiation, pulsed microwave fields with the same average rate of energy deposition in tissues are generally more effective in producing a biological response".

3. *Provided you are not being heated, you have nothing to worry about.*

See the discussion above on the thermal hypothesis promulgated by a Nazi eugenicist.

To conclude, let us nail the lie that all of this is being done in the name of communications. The real reason for the ubiquitous exposure of populations is the desire by the elites to impose a new technocratic system in which the resources of the world can be inventoried and controlled by them, and rationed.[41] 5G (and 4G LTE) is the technology that makes this possible. Everything — including our bodies if they have their way — will be connected to the Internet and thereby inventoried so that everything — and everyone — can be monitored and controlled. Your behaviour will be controlled by social crediting; if you do not follow their diktats, you will be blacklisted, as at least 13 million people in China already have been.[42] Digital money will be time-limited and you will be told how you may spend it and within what time frame, while dissidents will simply be made non-persons in the digital dystopia.

Doctors in at least 12 countries have confirmed that the "covid" injections contain graphene oxide[43], which multiplies the frequency of incoming EMR signals from gigahertz to terahertz.[44] Dr. Robert O. Young has described those who have been injected as walking time bombs who will be internally incinerated when they are targeted by a 5G signal.[45] I agree with fellow campaigners such as EMR weapons expert Mark Steele who assert that 5G is a weapon system designed to depopulate.[46]

The stories in this book are from the canaries in the coal mine. We don't know why some people suffer earlier than others; all we know is that the body of each individual is unique and reacts differently to exposure. When people become tortured by EMR, their spouses and/or family members often don't believe that they are suffering because they themselves are feeling no effects. Doctors have often placed the EMR-tortured in psychiatric hospitals, believing them to be suffering from delusions. The fact is that "electrosensitivity" is a political term. It conveniently points the finger at the sufferer, as if he or she has some weakness that makes him or her particularly vulnerable and it reassures us that the problem is unique to them and does not affect us because we are not "electrosensitive" so we have no need to worry or take any action to protect ourselves. The truth is that we are all electrical beings and we are all being horribly injured by ubiquitous EMR. If we continue to allow 5G to be deployed, we will all wind up being horribly tortured.

Exposure is cumulative. When we suffer a visible injury such as a broken leg, we expect it to heal within a short time, but exposure to EMR is cumulative — you never return to the way you were before exposure.

People have claimed that they became "electrosensitive" when a 5G antenna was installed outside their home but this is not so. The 5G installation was simply the straw that broke the camel's back, for the injury had been oncoming through years of careless EMR exposure. Professor Emeritus Martin Pall believes that everyone in society will lose cognitive function within months of 5G being fully activated[47], which implies that 25 years of cell phone use has already caused a loss of cognitive function across society. I believe that if 5G is ever fully switched on, people exposed to it will lose cognitive function immediately, and death will quickly follow.

It is way beyond time for human beings to wake up to the EMR war that has been waged against us for many decades already. I am not crying wolf here, as the mainstream media constantly does about invented threats such as anthropogenic climate change.[48] So let me quote some experts. Describing electromagnetic weapons in 1960, Russian President Khrushchev said that "*these new weapons could wipe out all life on earth if unrestrainedly used*". His successor as president, Leonid Brezhnev, described EMR weapons in 1976 as "*more frightful than the mind of man has ever imagined*". Whistleblower physicist US Lt. Col. Tom Bearden wrote in 1991 that "*This technology will give us the capability to engineer reality itself — both physical reality, life and mind — And we will be able to engineer it to be either a heaven or a hell; the choice is up to us.*" ELF radiation is amplified in the magnetosphere of the Earth up to 100 thousand-fold, changing the Earth's electromagnetic environment. Medical doctor Robert O. Becker warned in 1985 that the continued expansion of this effect "*threatens the viability of all life on Earth*".[49]

Without any exaggeration, the blanketing of our planet with electromagnetic radiation is the single greatest threat that humanity has ever faced. Failure to stop wireless technology and dismantle all of it immediately, before the cognitive, emotional and physiological toll on humanity gets any worse than it is already, spells the extermination of our species. Listen to the testimony of the people who tell their stories in this book. Don't let their torture and injuries be yours or your children's. Ditch your cell phone yesterday, cable your computer, get a landline and fight to have all the EMR infrastructure in your locality dismantled, as well as all satellites and their ground stations, and to stop the EMR that is torturing marine life in our oceans and obliterating the insects, in particular the pollinators, without whose help we would have no food. And do it now. Your voice and your action have never been more needed than right now, this moment, before we pass the point of no return. Put your normal life on hold to get this job done because nothing else in your life matters more than your and your children's survival and that of our beautiful Earth, with all of its animal, plant and marine life that sustains us.

Let's give thanks to our "electrosensitive" canaries in the coal mine who, despite their incapacitation and manifold challenges, found the courage and energy to tell their stories, and to Shannon for compiling them into this book, which is — in my opinion — the most important book you will ever read.

–Claire Edwards, July, 2022

*Claire Edwards worked for the United Nations as an editor for 18 years. The installation of pubic WiFi access points, far more powerful than home WiFi routers, at her workplace in Vienna in 2015 led to her involvement in the stop 5G movement. She edited, co-wrote and administered the translation into 30 languages of the International Appeal to Stop 5G on Earth and in Space, which was published in September 2018 and reached a worldwide audience. The Appeal was subsequently attacked by the mainstream media across the world and hijacked by agents of the depopulationist Club of Rome in December 2019. Claire has campaigned to stop all wireless technology and the destruction of life on Earth by electromagnetic radiation for the last five years. Her work can be found at her website: forlifeonearth.life.*

# Preface

As a child growing up in the 1970s and '80s in rural suburbia, in the United States, I had longed for the chance to travel to faraway lands–to have adventures, see different landscapes and architecture, experience different cultures, learn new languages, and sample foreign cuisine. When, at the age of 11, I learned in school about traditional Japanese homes outfitted with sliding-paper doors and futon beds laid out upon floors, I sketched these types of houses, fantasizing about designing my own Japanese-style home one day. I also wanted to ride high up in the sky on cable cars through the Alps, SCUBA dive in the warm crystal clear waters of the Caribbean, swim with fur seals in New Zealand, dive with dolphins in Hawaii, walk the winding cobble-stoned streets of old European villages while taking in the old-world architecture, view famous art pieces at the Louvre in Paris, have a picnic on one of the many bridges straddling the river Seine, ride a gondola through the canals of Venice, haggle in the bustling markets of India, and most especially, go on safari in Africa.

By my freshman year at university, I had abandoned these fantastical dreams, understanding that I was not from a rich family, and not even from one that traveled, and due to ultimately realizing how costly foreign travel and adventure really was. But a chance encounter with a fellow dormitory resident at our school cafeteria forever changed my outlook. This ebony-skinned man was some years my senior and came from the Ivory Coast in Africa, and he was a man who had traveled the world over. When I lamented that I was destined to never travel, this wise African just sat back, looked me square in the eyes and said, "You are young and you have two arms and two legs haven't you? Of course, you can travel!" But I protested that I had no money–to which he confidently responded, "You can work!" He then related the various odd jobs he had taken abroad in order to pay his way and earn his keep in exotic lands.

It struck me that he was right. I was the only one limiting myself. I could find a way to work abroad if I truly wished to travel. The one marketable skill I possessed at 18 years of age was caring for children and babies—the natural offshoot of having been raised in a large family and tasked as the eldest daughter with taking care of my own younger siblings. In addition, I had had regular babysitting jobs throughout middle and high school, so that it was easy for me to procure references if needed. And I had been studying French for a number of years already, so my language skills would enable me to apply for work with a family in French-speaking countries. Sophomore year, much to the disapproval of my parents who feared for the safety of their eldest daughter traveling solo to far-off places, found me applying to nanny (or *au pair,* as it was called in Europe) agencies for jobs in France for the following summer of 1992.

With a position secured in southern France for a three-month summer job, I then had to find a way to finance the overseas passage on my own. I worked a tedious, soul-crushing job as a temporary filing assistant at a doctor's office six to seven days weekly for six weeks in order to pay my airfare and secure a Euro-Rail train pass for visiting friends in Germany and Sweden before the commencement of my *au pair* job.

That summer was more than gratifying and my thirst for adventure, rather than being quenched, was merely whetted for further trips. Every year thereafter I found a way to travel, and in total by 2014 I had traveled to 21 countries on my own and had lived in three of those fulltime for several months, and on occasion, years, at different junctures over those years. Plus, I was able to check off a number of items on my aforementioned travel-adventure wish list. I learned French, Turkish, and some Swedish and Spanish and worked a variety of odd jobs, including several volunteer positions at farms, for wildlife conversation projects and at wildlife rescue centers, as well as working as a photographer and travel writer.

Ultimately my living abroad for the long haul did not pan-out, as I found myself chronically having to return "home" to the "States" because of difficulties with visas and trouble earning enough money to stay ahead of my ever-mounting, travel-induced debts. The romance of life traveling by a shoestring was waning and the adventures finally wearing. Retreat, coupled with some level of stability and laid-down roots, was needed. So, while I settled in the Washington, DC area to continue pursuing my career as a photojournalist back in 2003, I still made room for annual and semi-annual travel to exotic places in order to fill the lingering need for adventure, cultural education, and change of scenery.

By the age of 35 I had moved more times than my years but by 2015 I had been living in the same apartment for 7 years—the longest I had lived in any one place in my life. Still, I was content enough, loved my little home and had no designs for another move at that point.

It is one thing to *choose* travel and nomadism as a lifestyle but quite another to have it forced upon oneself. In 2014 my life changed in ways I could never have predicted for reasons fully outside of my own control. Somehow, I had become hyper-sensitized to electromagnetic fields, particularly in the radiofrequency range and also hyper-sensitized to any synthetically manufactured chemical, including any and all perfumes and fragrances. By 2015 I was forced to flee my home and city to seek a refuge thousands of miles away. Since that time, I have been made, along with my partner, to move several more times in search of dwindling places undisturbed by manmade electromagnetic, and chemical-agriculture, pollution.

I was faced with the paradox of no longer being able to tolerate most forms of travel (especially airline travel due to inflight WiFi and radar) but having to find a way to relocate anyway, even if it meant overseas emigration. At the same time, I grappled with the psychological impact of realizing that my forced exodus may prove the last time I would be able to travel at all. And this last trip would be made out of dire necessity—traveling for pleasure and cultural education would likely become a luxury of my past, possibly forever.

What I also had not anticipated (nor I doubt had anyone else) was a future world in which international travel was to become extremely restricted due to random and indeterminate border closures, and when available at all, ultimately only open to those willing and able to undergo such torments as invasive nasal-swab disease testing, quarantines (possibly in EMF- and chemically saturated environments, such as hotels or hospitals), suffocation behind facemasks, and/or injection with experimental chemically laden, dangerous, genetically altering, potentially lethal, drugs.

While, at times, as I struggled with my new and disabling health condition, I had doubted my ability to ever travel again, still I had held onto the hope of one day recovering to a point at which it might again prove possible for me. I also believed that critical-mass awareness of the dangers of current exposure levels of EMFs/EMRs (electromagnetic radiation) and chemicals would before long precipitate drastic changes in society, just as was seen with cigarette smoke in public places, lead in gasoline and plumbing, asbestos in buildings, x-ray machines for shoe fittings, and DDT spraying programs, etcetera, so that maybe even one day use of WiFi and even chemical disinfectants in public would also be prohibited.

But the new world health dictates begun in 2020 effectively extinguished all remaining hope of my regaining the ability to travel, as the added restrictions were things impossible for me to withstand—nor did I agree that *anyone* should be made to endure such biologically and psychologically invasive procedures, especially as none of them were actually based on solid scientific evidence to back the propaganda being irresponsibly spewed by politicians and the mainstream press to excuse the new measures.

I certainly was not the only one left behind due to these drastic and sweeping changes and I also found back in 2014 that I was not the only one having to flee the Big Tech takeover happening in our world, and not the only one forced to forever run from encroaching WiFi and cellphone networks in order to find radio-quiet zones in which to escape the full-body torture that exposure to these ever-expanding and planetary-encompassing signals caused me.

There are millions more such "WiFi refugees" whose lives have been fully upended, who have been forced to continually move from place to place, and lose livelihoods, family, friends, and finances in the process. These refugees include people from all walks of life, ranging in age, gender, nationality and race, but their stories do not make headline news. Their pleas for help are drowned out by "more important" news of the day—"news" largely manufactured to support the huge financial and control interests of those who own the media.

And so, these stories have to be told by those of us not held in the pocket of these controlling entities. Rather, these stories must be told by we the sufferers, or by those who care about us. But it is especially difficult for we sufferers to tell our own stories as we so often find ourselves caught up in the act of day-to-day survival with hardly any extra time or resources to devote to writing or recording our stories by other methods. And some of us are so unwell that using a computer at all is impossible. However, in spite of these obstacles we must keep telling our stories in any form we can, for as long as we can, until our voices are heard.

In this book I attempt to tell my story alongside a sampling of other WiFi refugee stories in the hopes that someday the telling will make a difference. I began writing this in early 2019 but had to take a long hiatus due to reasons that will be revealed by the book's end. Most of the book was written those three years ago so I have added this introduction, an addendum, and a part four, as a way to update the stories to bring them as current as possible to present-day—the beginning of 2022.

Thank you for reading. Please share this book with as many others as possible to help make our voices louder—to help make them count.

PART ONE

# CANARY ORIGINS

## *Chapter One*

# Our Precious Earth

*HOW CAN YOU BUY or sell the sky, the warmth of the land? The idea is strange to us. If we do not own the freshness of the air and the sparkle of the water, how can you buy them?*

*Every part of this earth is sacred to my people. Every shining pine needle, every sandy shore, every mist in the dark woods, every clearing and humming insect is holy in the memory and experience of my people. The sap, which courses through the trees, carries the memories of the red man.*

*...We are part of the earth and it is part of us. The perfumed flowers are our sisters; the deer, the horse, the great eagle, these are our brothers. The rocky crests, the juices in the meadows, the body heat of the pony, and man all belong to the same family.*

*So, when the Great Chief in Washington sends word that he wishes to buy our land, he asks much of us... For this land is sacred to us. This shining water that moves in the streams and rivers is not just water, but the blood of our ancestors. If we sell you land, you must remember that it is sacred, and you must teach your children that it is sacred and that each ghostly reflection in the clear water of the lakes tells of events and memories in the life of my people. The water's murmur is the voice of my father's father.*

*The rivers are our brothers, they quench our thirst. The rivers carry our canoes, and feed our children... We know that the white man does not understand our ways. One portion of land is the same to him as the next, for he is a stranger who comes in the night and takes from the land whatever he needs. The earth is not his brother, but his enemy, and when he has conquered it, he moves on.*

*He leaves his father's grave, and his children's birthright, are forgotten. He treats his mother, the earth, and his brother, the sky, as things to be bought, plundered, sold like sheep or bright beads. His appetite will devour the earth and leave behind only desert.*

*...The sight of your cities pains the eyes of the red man. But perhaps it is because the red man is a savage, and does not understand. There is no quiet place in white man's cities. No place to hear the unfurling of leaves in spring, or the rustle of an insect's wings... The clatter only seems to insult the ears. And what is there to life if a man cannot hear the lonely cry of the whippoorwill or the arguments of the frogs around a pond at night? ...The Indian prefers the soft sound of the wind darting over the face of a pond, and the smell of the wind itself, cleaned by a midday rain, or scented with the pinion pine.*

*The air is precious to the red man, for all things share the same breath; the beast, the tree, the man, they all share the same breath. The white man does not seem to notice the air he breathes. Like a man dying for many days, he is numb to the stench. But if we sell you our land, you must remember that the air is precious to us, that the air shares its spirit with all the life it supports.*

*...I have seen a thousand rotting buffaloes on the prairie, left by the white man who shot them from a passing train. I am a savage and I do not understand how the smoking iron horse can be more important than the buffalo that we kill only to stay alive. What is man without the beasts? If all the beasts were gone, man would die from a great loneliness of spirit. For*

*whatever happens to the beasts, soon happens to man. All things are connected...*

*Whatever befalls the earth befalls the sons of the earth. If men spit upon the ground, they spit upon themselves. This we know: The earth does not belong to man; man belongs to the earth. This we know. All things are connected like the blood, which unites one family. All things are connected. Whatever befalls the earth befalls the sons of the earth. Man did not weave the web of life: he is merely a strand in it. Whatever he does to the web, he does to himself.*

*Even the white man, whose God walks and talks with him as friend to friend, cannot be exempt from the common destiny. We may be brothers after all. We shall see. One thing we know, which the white man may one day discover our God is the same God. You may think now that you own Him as you wish to own our land; but you cannot. He is the God of man, and His compassion is equal for the red man and the white. This earth is precious to Him, and to harm the earth is to heap contempt on its Creator.*

*The whites too shall pass; perhaps sooner than all other tribes. Contaminate your bed, and you will one night suffocate in your own waste... we do not understand when the buffalo are all slaughtered, the wild horses are tamed, the secret corners of the forest heavy with scent of many men, and the view of the ripe hills blotted by talking wires.*

*Where is the thicket? Gone. Where is the eagle? Gone. The end of living, and the beginning of survival.*

—Chief Seattle, 1854, in response to an offer from the US President to "purchase" land and offer up a reservation for the natives—or face the annihilation of his people.

The talking wires blotting the hillside spoken of by Chief Seattle represented the steady and rapid encroachment of new forms of manmade electromagnetic fields at the time. Such was the density of these "talking wires" that by only a few decades later in 1890, "dense forests of telephone poles were sprouting eighty to ninety feet tall, bearing up to thirty cross-branches each. Each tree in these electric groves supported up to three hundred wires, obscuring the sun and darkening the avenues below." And "wires carrying two thousand volts were trailing across residential rooftops in the West End of London."[50]

The wires first carried voices and then power to homes all over the world. By the early 20th century, the whole world was *abuzz* with a new electromagnetic field—one non-native to the environment; one invented by man. The fields, once confined to laboratory experiments and tested for medical therapies in the previous centuries had suddenly burst through the confines of these sporadic spaces, like a demon birthed into the world—penetrating deep under the oceans in the form of transatlantic and transpacific telegraph cables, and up into the atmosphere itself as radio waves. Humans now had to adapt to a new invisible bedfellow, one that certainly did not respect the wishes of Chief Seattle to care for the Earth as if it were a cherished relative.

Electrification spread at a never-before-seen pace, and alongside it, new diseases of civilization were birthed, ones which would stay with us and grow to epidemic levels ever since—the most tenacious of which would be *influenza*, or "the flu"; something that prior to electrification and the advent of the alternating current was considered virtually non-existent.

With each new addition of electrical fields for powering homes, electric trains and city lamps, and later radar for communications and navigating machines of war at the turn of the twentieth century, to satellite and cellphone communications systems proliferation at the turn at the 21st century, also came physical symptoms in those most directly impacted by the new fields. One such new disease took shape in the form of hypersensitivity to these fields. The persons being affected could feel the changes in their bodies immediately in the presence of the EMFs and could often easily identify the source of their suffering, much to the chagrin of industrial tycoons wishing to reap the largest financial gains possible from the new inventions without drawing attention to the darker side of the new technologies.

Even long before the proliferation of these new fields, those who experimented with electricity in the preceding centuries found themselves adversely affected by their exposures to them. Famous people, including Benjamin Franklin, well known for his key-toting kite experiments with harnessing lightning bolts, wrote of their mysterious afflictions. Franklin wrote his doctor in 1757, complaining of, "a giddiness and a swimming" and "a humming noise" in his head and of "little faint twinkling lights"[51] in his vision.

Nicola Tesla, famous for a large variety of inventions related to harnessing and delivering electrical currents at the end of the 19th century, also suffered mysterious ailments in the form of hypersensitivity—doubtless caused by his chronic exposures to manmade electrical fields during his prolific experimentations with the same. Researcher John O'Neill described Tesla's illness in his book *Prodigal Genius* with the following account:

*"The peculiar malady that now affected him was never diagnosed by the doctors that attended him. It was, however, an experience that almost cost him his life. To doctors, he appeared to be at death's door. The strange manifestations he exhibited attracted the attentions of a renowned physician who declared that medical science could do nothing to aid him. One of the symptoms of the illness was an acute sensitivity of all the sense organs. His senses had always been extremely keen, but this sensitivity was now so tremendously exaggerated that the effects were a form of torture. The ticking of a watch three rooms away sounded like the beat of hammers on an anvil. The vibration of ordinary city traffic, when transmitted through a chair or bench, pounded through his body. It was necessary to place the legs of his bed on rubber pads to eliminate the vibrations. Ordinary speech sounded like thunderous pandemonium. The slightest touch had the mental effect of a tremendous blow. A beam of sunlight shining on him produced the effect of an internal explosion. In the dark he could sense an object at a dozen feet by a peculiar creeping sensation in his forehead. His whole body was constantly wracked by twitches and tremors. His pulse, he said, would vary from a few feeble throbs per minute to more than one hundred and fifty. Throughout this mysterious illness he was fighting with a powerful desire to recover his normal condition."*

Today the natural life-giving frequency of the Earth, known as the Schumann resonance, named after German physicist Winfried Schumann who first measured the field in the 1950s at 7.83 Hz, is drowned out by erratically pulsing artificial electromagnetic fields measuring in magnitudes billions[52] of times greater than this original "background radiation" field—specifically with a 1,000,000,000,000-fold increase in urban areas and a 100,000,000,000,000,000-fold increase in classrooms representing a million-fold increase in the last thirty years.

***Measuring Schumann resonance in or around a city has become absolutely impossible. Electromagnetic pollution from cell phones has forced us to make our measurements at sea.***

—Dr. Wolfgang Ludwig, physicist

All life has always been supported by this Earth pulse—the Schumann resonance, or heartbeat of the Earth—one with which we were always entrained, as we breathed with the Earth like a mother cradling her newborn child finds a common rhythm with each synchronized breath.

The 7.83 Hz frequency was discovered to match exactly that of the Alpha brain waves (responsible for relaxed, meditative states) first measured by Hans Berger in the 1920s. Since natural frequencies are theoretically infinite, ranging from a billionth of the size of an atom to the length of the entire universe—it cannot be coincidental that our brain waves match that of the Earth's resonance. Removed from this synchronization with our first mother—that of the Earth, we lose our bearings as we stumble, unable to find a steady beat amongst these new foreign, erratic frequencies.

Professor Wever, revered scientist from the Max Plank Institute conducted experiments starting in the early 1960s on the effects of removal from the Earth's frequency on groups of students. These students agreed to live in a bunker deep underground for a number of weeks at a time. It did not take long for the students removed from the Schumann resonance (which does not penetrate underground but exists only upon the Earth's surface, a magnetic field resultant directly from lightning strikes), to feel sick. They complained of headaches, irritability, fatigue and nausea.

Wever covertly introduced artificially created Schumann-resonance frequencies into the bunker and tracked changes to the students' wellbeing when exposed to the field and when separated from it. Over a 30-year-course of experimentation, Wever was able to demonstrate consistent improvement in the health of his subjects upon the introduction of the 7.83 Hz frequency and equally consistent decline with the removal of the frequency. He concluded that, "electromagnetic fields of extremely low frequency have been shown at a high statistical level to influence human circadian rhythms. This means that circadian rhythms can be used as very sensitive indicators to test the influence of ELF electromagnetic fields on human beings in general."[53]

*"If you think of it, the Earth is a giant di-pole magnet, with a north and a south pole. Everything that has evolved in nature has evolved in harmony with background electromagnetic fields. When I say everything, I really mean everything. It's not just us—it's other species as well. We are entrained with the Earth's natural fields.*

*"Entrainment is a phenomenon that happens. Think of it in the way a mother and child sleep together and their breathing rates synchronize. That's classic entrainment. We are entrained with the Earth's natural fields. It is the introduction of these new exposures that has disrupted that entrainment to the point where... Robert O. Becker (they called him the founder of modern biophysics, he was my mentor in the subject), Robert Becker used to call it 'breathing with the Earth.' With the introduction of these frequencies, we can no longer breathe with the Earth. It has the ability to affect all of our circadian rhythms—our sleeping waking cycles, our hormonal levels, seasonal changes that happen. Not just in us, but in other species. Birds and butterflies and insects use the Earth's natural fields for migratory activities, for mating activities. So, the effects that we are having at the Earth's surface with these exposures is a serious environment[al] [concern]."*

—Blake Levitt, medical and science writer, *WiFi Refugees,* RT Documentary, 2017

Many migratory insects and birds have particles of magnetite in their bodies and have been shown to be extremely sensitive to magnetic fields as they use the Earth's frequency for navigation. Not just magnetite, but cryptochrome, a light-sensitive protein, discovered in the 1990s and found in plants, animals, insects and humans, plays an important role in regulating circadian rhythms and in navigation.

Experiments over 40 years ago conducted by Robin Baker[54] showed that humans lost their innate sense of direction if a magnet was placed next to their heads.

And experiments conducted by Dr. Kuhn and his team on beehives in more recent years showed that cordless microwave-transmitting DECT phones placed in the beehives[i] caused bees to abandon hives completely.

In the year 2000, Thorsten Ritz, professor of physics and astronomy at the university of California conducted a study on robins that proved that navigation employed use of cryptochrome cells behind the eye that allowed the robins to literally see the Earth's magnetic field. Exposure to artificial radiofrequencies disrupted the robin's compass, and thus migration, at very low intensities, well below those deemed safe by the International Commission on Non-Ionizing Radiation Protection.

*"All the navigation studies, the cockroaches, the zebras, the finches, the chickens, the robins—they all have their magnetic compass disturbed by radiofrequency fields well below the ICNRP exposure limits. There's certainly solid scientific reason for supposing that magnetic fields—artificial magnetic fields will disturb the habitat of a number of species."*

—Professor Denis Henshaw-Bristol University, *Resonance; Beings of Frequency,* 2013 documentary film

Over the 25 years prior to 2012, a number of species, which rely upon the Earth's magnetic field for navigation, have gone into decline. 190 different species of birds face imminent extinction. One in 8 of the world's bird species is under threat. Four species of American bee has decreased in numbers by up to 98%. Farmland bird populations are falling by as much as 79%. Half of Britain's bee species have declined by up to 70%. 80% of Australian wading birds have declined. 62% of Asia's migratory water birds are declining or extinct. Half of all known

---

[i] With the phones turned on. Other phones placed in other control hives and turned off did not affect the number of bees returning to the hive, proving it was not the presence of the object itself causing bees to abandon hives, but the frequencies emitted by the phones. (See, *Resonance; Beings of Frequency,* 2013 documentary film for more information.)

British butterflies are under threat of extinction. 45% of all Europe's common birds have declined in numbers. 50% decline in all European grassland butterflies. 10% of the world's butterflies face extinction. 36 species of Australian shore birds have decreased in numbers by 75%. And 109 species of Arctic migratory birds have declined dramatically in numbers.[55]

While the loudest voices in the mainstream media explain away these dramatic declines by placing blame on the ever-elusive "climate change" excuse or consumer fossil-fuel burning or on pesticide usage—while all of the above may play some role in these mass extinctions and declines, the most dramatic change to our environment during this time has been due to the extremely rapid proliferation of microwave-emitting antennas. And it is this elephant in the room that is chronically ignored—at our own peril.

But other voices from the alternative press and other researchers like Jeanice Barcelo, are making the, what should be, obvious connection between the changes to our electromagnetic environment and the adverse effect on migratory birds and insects as well as human populations. As Jeanice points out...

*"...cell towers and 5G 'small cells' are being positioned perilously close to bird nests, causing some birds to abandon their nests while others may become aggressive, sterile and/or die. Hundreds of birds were recently seen falling dead out of the trees in the Netherlands when 5G transmitters were turned on nearby. Dozens more were seen dropping out of the sky in Australia, screaming and bleeding from their eyes and beaks and wailing in agonizing pain as they experienced a slow, torturous death. Hundreds of thousands of birds dropped dead in southwest America during 2020, many at White Sands Missile Range in New Mexico where an AT&T 5G transmitter is currently in operation...hundreds of thousands of birds are also colliding with cell phone towers and wind turbines. Tens of thousands of birds annually are catching fire as they fly over solar power plants where their bodies suddenly ignite into flames. When all of the above is combined with the fact that birds are now likely to be going hungry due to a lack of insects to eat [killed by microwave and millimeter waves], it is clear that our 'clean' and 'green' technologies are, in fact, lethal, and threatening the survival of many living things."*[56]

And the sap in the trees spoken of by Chief Seattle, once so revered by his people, proved to be adversely affected by artificial radio waves as well, as Bengali scientist Sir Jagadis Chunder Bose, who experimented in the late 1800s on electromagnetic fields (EMFs) and the growth of plants, found that low levels of radiofrequency energy actually slowed the movement of sap within trees.[57]

Bose contended that the "isolated vegetal nerve was indistinguishable from animal nerve."[58] Through his experiments with plants and animals Bose was able to note that all life forms were electrical in nature, growing in rhythmic pulses, and as such, respond to the slightest changes in electromagnetic fields within the environment.

Less than half a century later, John Nash Ott, famous for his time-lapse photography documenting plant growth, studied the effects of manmade electromagnetic fields on plant growth and behavior patterns in children. Ott found that both plants and children were extremely sensitive to frequencies from cathode-ray-powered television sets. Placing heavy black paper in front of TV screens to shield from the light but not block EMFs, he discovered that beans placed in front of the covered sets experienced distorted and retarded root growth. After also conducting similar experiments with rats exposed to TV radiation, Ott concluded that, "the radiation has a physiological effect both on plants and animals, which appears to be chemically mediated."[59]

Suspecting that behavioral problems in children could be linked to hours spent in front of EMF-emitting TV screens, Ott was prompted to work with a local school principal in Florida to encourage parents in getting children to spend more time outdoors than in front of televisions while also ensuring children sat farther back from the sets when they did watch TV. After making these changes, teachers noted markedly diminished behavioral problems in the children who had dramatically reduced their television exposures.

Initially met with resistance to his findings from the TV industry, with the director of research at RCA's Bio-Analytical Laboratory stating that it was, "utterly impossible for any TV set today to give off harmful rays," but by the late 1960s Ott was credited with "getting us started on the road toward control of radiation from electronic products"[60] when the US Congress passed the Radiation Control Act.

While TV sets eventually ditched their X-ray-radiating cathode tubes in the noble effort to "control radiation from electronic products," the replacement with fluorescent bulbs and LED-powered LCD displays and now microwave-radiating *smart* technology, has served to simply replace one type of harmful radiation with another, rather than fulfilling this, what has proved to be empty, promise to reduce screen radiation exposures made by industry leaders and politicians.

Following in Bose and Ott's footsteps, many young science students in the last decade have conducted experiments on plants exposed to WiFi[61] and demonstrated the negative impact of WiFi frequencies on plant growth and health, proving that biological life is still being impacted by EMRs (electromagnetic radiation) from the latest advances in technology.

As Chief Seattle so eloquently observed, what we do to these other life forms—the flowers, trees, rivers and animals—we do to ourselves, as all things are connected. If plants and tree sap respond to even small changes in electromagnetic fields, we too must certainly be affected by these same changes.

The same unrelenting force that drove Chief Seattle's tribe and so many others from pristine wild lands is the same that displaces communities and cultures today—the insatiable appetite of industry knows no bounds in its quest to obtain, own, and control all of the Earth's precious resources, including all life dependent upon these same resources. The drive for wealth and power has continued on its destructive path ever since.

Today, 40% of raw, wild land in the United States is owned by the government—meaning, government-corporate interests—stolen from the people under the false war cry of "sustainability" as excuse. This attitude assumes that the people cannot be trusted to manage their own lands—an attitude that reaches into every aspect of private lives, leaving us feeling unworthy to live upon this Earth at all, and also incapable of taking charge of our own lives. It is as disempowering as it is disabling—robbing us of our very soul fire and birthrights, while driving us to despair.

Tech industries, the most powerful and wealthy of today's corporate entities, are credited with "saving the day" as they don a false heroism while pretending that technology can solve the problems created by technology and industry.

Developers and proponents of so-called "smart" and "sustainable" tech are the actual villains responsible for the increased and relentless devastation to wild lands, tribal people, cultural integrity of societies and the health of the planet as a whole.

This false posturing of tech industries represents the essence of "green washing"[ii] in action—as conflict-mining of precious minerals, in particular *coltan* (a necessary mineral component for all things "smart") in the Congo, has created endless war with resultant death and suffering of the Congolese and native tribes on par with and surpassing that of WWII fame—all while the energy necessary to power "smart" technologies and the environmental degradation caused by artificial frequencies inherent to them is never discussed when posing these absurd "solutions" to saving our environment from destruction.

But environmental burden is not the only cost—stark human behavioral changes including tech addiction and dependency resultant from adapting these new technologies, has also had profound social impacts. Family units continue to grow apart as the increasing demands of technology and portable screens drive a deepening wedge between relations. "Community," once a physical and palpable reality, is now reduced to an online, virtual abstraction. The ever-churning wheels of industry have for the past nearly two centuries forced massive exoduses, migrations and related wars—as power struggles for resources and world domination in our now global market between vying industrial interests while the military-industrial complex guides the hand of world leaders into unending conflicts from behind the curtain of the world stage.

These power struggles effectively impose upon the world mass and ceaseless migrations, so that, forced off the land, crowded into cities, chasing job after job in a market always shifting like sands beneath our feet, the resultant upset ensures a continuing climate of instability permeating the whole world, leading to the lasting enslavement and impoverishment of its people.

---

[ii] "A review by the University of Sussex Business School finds the notion that 5G is green technology is not currently backed up by a strong, publicly available, fully transparent evidence base." See more here about how the wireless industry is ignoring the energy cost of manufacturing and installing networks and manufacturing phones: https://ehtrust.org/new-report-5g-is-not-so-green-and-could-increase-energy-use/

Uprooted families split and thrown to the four corners of the Earth, scattered like ashes to the wind, have now become part and parcel of the "new normal" as families and friends move thousands of miles apart, with no one-place to truly call home and no one-tribe within which to find belonging. With humankind anchorless—drifting upon a sea of uncertainty and upon waves of mind-numbing, soul-destroying frequencies—is it any wonder that anxiety and depression run rampant in our societies and suicide has reached epidemic proportions, even among very young children?

Electronically driven technologies are now changing so rapidly that generational gaps grow into deep chasms while children can no longer plan-out futures, as each passing year takes on an unrecognizable visage, giving lifestyles and cultures from a decade previous the appearance of being from a far-off, ancient time. In this state of deep anxiety and apprehension we are easily controlled and manipulated.

As we grasp for meaning and purpose while clinging to authority dictates, we are increasingly wary of trusting our own terribly injured instincts and unsure of our own shaky voices—now drowned out by the incessant, industry-bought media propaganda factory, spewing out insidious unimportant lies and drivel that pervade every minute of our waking lives.

*Fear* chronically grips our lives—"fear of missing out" now even has its own acronym-rooted slang—FOMO. And fear of social rejection, of being seen and also unseen, and fear of being alone and of being together, drives our decision-making while our social interactions all now take place in a virtual realm, communicated through wireless networks—held in "the cloud" where the "all-seeing-eye" of AI (artificial intelligence) watches over *all,* with "science" replacing our ancient gods. The frequencies used for these new ways of communicating engulf and alter us in profound and still unappreciated ways.

Extinction of wildlife and its supporting wild lands continues unabated as the "talking wires" abandon their unfashionable, bulky, and outdated visible cables in favor of millions of antennas of varying size stationed atop hundred-foot-high towers, and atop high-rise buildings with their reach now threatening every street lamppost, mailbox and residential rooftop with their unsightly silhouettes. These obtrusive towers now blot out the hillside in place of wired predecessors and bury their new cables underground while spewing poisonous radiation for many miles in all directions—and while even masquerading as trees and cacti.

And tens of thousands of new "next generation" 5G-millimeter-wave-beaming SpaceX *Starlink* satellites are poised to invade our stratosphere to obliterate the heavens, obscuring the very stars in the night sky, known even to the most ancient civilizations, from view—and set to irradiate the entire Earth in a vast blanket of alien frequencies, bound to forever, irreversibly alter all life forms—until a planet governed by "intelligent" machines will only be suited to support other machines or non-biological "life," so that, like the native tribes before us, and the whip-poorwill and buffalo, we too, as humans, will be driven to extinction.

Chief Seattle's lament is more relevant today than ever before with his worst fears for the Earth and humanity rapidly coming to pass. The understanding the chief shared with his people of the interconnectedness of all life needs to be relearned so that we truly appreciate that what we do to the web we do to ourselves and that by replacing the natural "web of life" with an artificial "world wide web" in the form of our Inter*net*—a net poised to entrap us all forever as it takes on a new form to become the "Inter-*net* of Things" in which all life forms are reduced to mere things, including humans—caught up in this invisible wireless net, we risk perishing like so many fish pulled out from the sea, gasping for a final breath as we vainly attempt escape.

The time is now for us to arise, take stock of our lives, cultures and world—re-evaluate, re-prioritize our values and once again decide to understand and appreciate the preciousness of the Earth and refuse to sell it alongside all of humanity to the highest bidder.

*Chapter Two*

# Canaries in Coalmines

*"HERE HE SAW THE bird fall from its perch, and, incautiously taking a breath of the air himself, his knees gave way to a small extent, but he managed to scramble out into the better air, taking the bird with him. The bird recovered within three or four minutes, and again got on to its perch. It was most extraordinary to see the rapid effect, which the carbonic monoxide had on the bird, and he was quite satisfied, after the experience with a bird in this way, that a bird was a comparative safe guide, and much to be preferred to using mice, as the fall of the bird from the perch could easily be seen."*

—*Yorkshire Telegraph and Star* (Sheffield, Yorkshire, England) of Friday 21st December 1906

In the early 1900s coalminers commonly employed canaries as a primitive type of carbon-monoxide alarm system—as these birds, having an especially high requirement for oxygen, even more so than other birds, would signal to the miners when oxygen fell to dangerous levels inside the mines. The bird, carried in a small cage by the miner, would firstly stop singing when a change in the oxygen levels was initially detected, and then, if bad enough, would fall completely from its perch from insufficient oxygen and raised levels of carbon-monoxide. Thus warned, the miners would quickly retreat, saving themselves, along with the bird, from possible death.

Ever since those times, the symbol of the canary bird has been used metaphorically to describe sensitive humans who, through their heightened sensitivity, have served to warn others less sensitive to impending dangers due to, sometimes subtle, changes in the environment.

In more ancient times when people lived in smaller communities and tribes, these sensitive few, or "canaries," were revered. It was these valued group members who served to warn the rest as to possible threats, whether they be in the form of poisoned food, contaminated water, compromised air, or approaching storms.

But in the today's world, governed by industrial greed, canaries are an inconvenience and a threat to corporate profits. They are maligned rather than appreciated, and intentionally portrayed as mentally ill, as a way to keep the public ignorant of threats posed by dangerous consumer products.

The result of canary warnings going unheeded is that the "birds" themselves, lacking the help of an observant miner to carry them out of danger, must find a way to break through the bars of their cages, imposed upon them by self-destructive societies, and find a way to safety outside of the collapsing mine—while the rest stay trapped in the mine, unaware that they are slowly suffocating.

Another commonly used analogy to communicate this same idea is that of the slowly boiled frog. Frogs placed into lukewarm pots of water will not jump out to escape if the temperature of the water in the pot is only turned up slowly by degrees. By the time the frog realizes its fate, it is too late—it has already been cooked.

## *Life Inside a Microwave Oven*

With the massive proliferation of microwave (and now millimeter-wave) radiation in our environment, it is like we are living in a giant, open-convection microwave oven, while we turn and turn around inside it–the heat cooking us from the inside-out is scarcely noticed, but it is most assuredly at least partially responsible for the ever-growing epidemic levels of cancers, diabetes, cataracts, arthritis, MS, fibromyalgia, autism, dementia, chronic fatigue and rapid aging.

Dr. Gerald Goldberg chronicled this literal cooking effect on humans in his very informative 2007 book, *Would You Put Your Head in a Microwave Oven?*. According to Dr. Goldberg the "continuous low-level exposures" to which we are all now subjected as satellite communication technologies have "transformed our atmosphere into a sea of microwave radiation," are cumulative, and the health consequences potentially catastrophic–set to create an epidemic of illness affecting 60-70 percent of the population–an epidemic, which left unchecked, that is poised to "cripple most industrialized nations" in the coming 5 to 7 years (starting in 2012).[62]

Goldberg says what is unique about microwave radiation is its, "ability to deeply penetrate into tissues and to cook or heat them up." And what distinguishes microwave radiation from other forms "is the pattern of injury that it produces, which is recognizable from the dosimetric and scientific studies."

Dosimetry studies, Dr. Goldberg explains, "reveal what parts of the body most easily absorb the radiation and become damaged. Dosimetry studies establish the pattern or signature of injury that would result from microwave exposure." In short, he says, "An individual will cook in a predictable pattern. This will show up in a pattern of disease that will progress and can worsen over time if not corrected." [63]

Goldberg includes a dosimetry model in his book that was established by the United States Department of Defense from a study conducted by the Radio Frequency Radiation Branch of the US Air Force Research Laboratory at Brooks Air Force Base in Texas.[64] The model shows the areas of highest radiation absorption/sensitivity in red, and these areas include, "the nervous system: the brain, spinal cord and major sense organs (eyes and inner ear)," alongside cartilage. Significantly, cartilage is the foundational structure for all joints in the body, and the ear lobes, nose, bronchi, and larynx.

The skeleton is the next area of the body most affected by microwaves, meaning that bone marrow is particularly susceptible to exposures, which has led to a dramatic rise in leukemia. Next affected are, "teeth, lymphatics, colon, arteries, esophagus, stomach, peripheral nerves of the body," and then, "the bronchiole tissues of the lung, gonadal tissue (ovary or testes), prostate, uterus, pancreas and spleen."

Goldberg points to two essential aspects of the heating effect from microwaves:

> *1- Regardless of the dosage one receives, the radiation will travel deeply into the body to affect tissues.*
>
> *2- Regardless of the type of exposure, the effects of the radiation are cumulative. That is, if you received a large exposure over a short period of time or if you received a low dose exposure over a longer period of time, the results are the same. The total exposure is cumulative; in essence there is no safe dose.*[65]

According to Goldberg, "fats and solid tissues are affected the most." This may explain why women seem to be generally more affected than men, as they have a higher ratio of fats in their bodies, and why breasts are more vulnerable as well.

This heating also, warns Goldberg, increases oxidative damage, uses up critical antioxidants and leads to shortened lifespans. It also decreases cellular metabolism, can shut off blood flow by causing blood vessels to "go into spasm," and causes "burn damage deep within tissues," leading to the buildup of cysts. The microwave radiation disrupts cell signaling, decreasing electrical functioning of cells.

These chronic microwave exposures are causing epidemics of chronic fatigue among our populations as the microwaves directly affect energy metabolism and shut down ATP production. Once this production, essential to supplying energy to cells for a variety of cellular functions, ceases, "literally one's brain begins to fry. The brain begins to shut down. An individual may notice a sense of brain fatigue, loss of alertness, impaired memory or concentration... The net result of microwave exposure is that the body heats up, while energy metabolism shuts down."[66]

This shutting down leads to a cascade of adaptive mechanisms resulting in the shutting down of the thyroid gland, mitochondria, and respiratory enzymes while cortisol levels rise to create anti-inflammatory effects. However, this compromise, if left to continue over long periods will lead to adrenal collapse and impair the ability for one's body to regulate one's own temperature, leading to extreme sensitivity to both cold and heat and a "profound sense of weakness."[67]

The rise of cancers, cystic diseases, cognitive impairments, eye disorders, including cataracts, and blood disorders, can all be traced directly to the rise in exposures to microwaves.

## *Slowly Boiled Frogs*

The people of the Earth (alongside all life forms and *actual* frogs) are now like frogs in a pot set to slow boil. Had the levels of microwave radiation to which we are subjected today been unleashed upon us all at once, we might have detected the change in our environment and readily jumped out of the pot—rejecting the introduction of microwave frequencies into our homes, schools, places of work and worship, and now parks, forests and other wilderness areas as well.

All generations of technology employing microwave and millimeter radiofrequencies released the second, third, fourth and now fifth (soon to be followed by sixth) generation, increasing the power levels, ubiquity and frequencies alongside each, roughly ten years apart, even though the technologies for each had been developed decades prior to release.

The motivation for releasing these wireless-communicating technologies in stages is likely threefold—one; as a way to secure the greatest profits by introducing "new" technologies and products along the way, planning out the obsolescence of each long in advance in order to force the consumer into the impoverishing cycle of endless "upgrades,"—two; as a way to forever delay adequate safety testing, as all such tests are conducted for an approximate ten-year period, so that at the end of such studies the results can be dismissed under the label of "irrelevant" to the new generation of technologies being used at the end of the study—so that the wireless industry can get away with continually releasing dangerous untested products onto the public with no oversight—three; to minimize public protest to the changes in the environment inherent with each "upgrade" to the grid—meaning, in essence, to "slow-boil the frog."

Those of us whom have become "hyper-sensitized" to each of these new generations of imposed frequencies often understand the sensitivity to be a mixed blessing. On the one hand it enables us, like the canary in the coalmine, to literally feel when we are endangered and by so doing give us a better chance at escape. But on the other hand, as the whole world becomes the suffocating mine, or the pot in which we are being slowly cooked—the heat of which to those sensitive enough to feel it becomes unbearable and makes life a living hell—an inferno heated by microwaves.

And when the world insists upon labeling us as having a *disorder* rather than appreciating the gift we have been given, by listening to our warnings and noticing when we stop singing and fall off of our perches and by helping us (alongside them) out of the mine or pot, then the "gift" ends up feeling more like a curse.

And the "sensitive" label is perhaps inappropriate, as it reduces the experience to something akin to a potentially psychological reaction rather than a real physiological one. It also ignores the fact that the subset of the population given this label at one time did *not* suffer acute illness in the presence of wireless radiation fields, but only acquired the condition *after* injury due to exposures to these same fields. Before ES (electro-sensitive) or EHS (electro-hypersensitive) and sometimes EMS (electro-magnetic-sensitivity) were socially and clinically given as labels, "microwave injured" and "microwave sickness" were used to describe the same set of symptoms experienced by the now "electrically sensitive."[68]

## EHS—Environmental Intolerance

*"Living beings depend on electricity. We are all electromagnetic beings—our heart and our brains would not work without electromagnetic fields. That's how we send impulses back and forth. That's what keeps our heart beating. So, we are electrical phenomena. And the fact that a change in the amount of exposure to exogenesis, to foreign electricity, would have no effect on our natural electrical properties makes no sense at all. It's true for plants. It's true for birds."*

—Dr. Devra Davis, 5G Summit Crisis hosted by Josh Del Sol, 2019

*"The aspects of an organism that allow it to detect and respond to electromagnetic fields, you could say... that 'electro-sensitive,' if you dissect it and break it down, 'electrically-sensitive,' you could say that that is a response to life itself. In that, for example, if we are looking at an organism to determine the presence of life, we are really looking for the presence of electricity. So, with the electrocardiogram, for example—ECG, and the electroencephalogram—the EEG, what we are looking at there is the presence of electricity from those targeted organs—the brain and the heart. When we see the presence of that—that's a sign of life. So, in a way, given that on a cellular level, electromagnetic fields are part of life itself, we could say that* ***all*** *life is electrically sensitive."*

—Dr. Erica Mallery-Blythe, *Resonance—Beings of Frequency*, 2013 documentary film

We are all electrically sensitive to some degree, however most of us have been taught to ignore the symptoms. This holds true for Dr. Erica Mallery-Blythe, a British ER medical trauma doctor, who herself, now carrying the EHS label after having been injured by overexposure to microwave radiation in the late 90s, has dedicated her life's work to helping educate the public on the plight of canaries like herself, and about dangers of EMFs in general, and EHS specifically.

Erica was a very early adaptor of wireless technology, having been given a cellphone as a young child in the 1980s. Like many others, she had always noticed getting a headache after prolonged use of the phone, but had ignored these early warning signs, taking no heed of her own inner canary.

It was not until she purchased her first WiFi-enabled laptop that she noticed other more distressing bodily pains during and after use of her new computer. The pain increased until she was no longer able to use either her computer or cellphone at all. Later, crushing insomnia, headaches, heart arrhythmias, dizziness and tremors plagued her. Concerns about brain tumors prompted her submission to an MRI scan, but the experience of the scan (something that uses extremely powerful magnetic fields for imaging) was the final nail in the coffin that forced her to retreat from society altogether in seeking recovery from her debilitating symptoms, for which, camping in isolated wilderness locations brought about incredible and immediate relief.

Telling her story to Arthur Firstenberg, as related in his book, *The Invisible Rainbow*, Dr. Mallery-Blythe described her experience moving with her husband from their home and camping from tents and out of their car as, "living like war refugees." In further explaining what that year and a half had been like she said, "You can't do the basic things you need to live. You almost feel like you're going to wake up, like it's some kind of bizarre dream... What makes me sad is to see all the people who can't escape and get to a pure environment, because if you can't get to a pure environment, it will destroy you." She and her husband eventually found a suitable cabin in South Carolina where they lived without any electricity until she recovered. Until it happened to her, Erica could not believe that as a medical doctor she had never heard of EHS.

Dr. Mallery-Blythe did ultimately recover enough to rejoin society under specific conditions. She, of course, does not use any wireless tech and limits her time in such fields as best she can. She has been a real champion of the cause, and is founder of PHIRE, Physicians' Health Initiative for Radiation and Environment—"an independent association of medical doctors and associated specialists assembled for the purposes of improving education regarding health effects of non-ionising radiation."[69]

In 2014, she wrote the following enlightening guide to EHS, called, **EHS—A Summary**. It is copied here in its entirety below.

> *We are currently witnessing the largest change to the Earth's electromagnetic environment that has ever taken place in human history. This change has taken place in the very short period of a handful of decades and continues to escalate at an exponential rate. Given that household electricity, which was the first anthropogenic (man-made) electromagnetic field (EMF), only became prolific after the turn of the century, artificial EMF has barely seen one generation from cradle to grave. The use of higher frequency microwave devices such as mobile telephony, WiFi and smart meters, have suddenly become commonplace despite almost no safety testing and decades of evidence of potentially lethal effects. This has sparked a political and scientific debate that is gathering momentum on a daily basis, raising concern about the continued use of such devices. One may assume when witnessing the vast implementation of, for example WiFi in the home, school, workplace or public domain, that ex-*

*perts have provided sufficient evidence of safety to overwhelm scientific concern. This is not the case.*

*The World Health Organisation (WHO)/International Agency for Research on Cancer (IARC) Classified RF as a Group 2 B 'Possible Human Carcinogen" (2011). Despite this, there has been no attempt in the UK at disseminating this important information to the public. Conversely, it was not even mentioned in the AGNIR government commissioned report a year later in 2012. The only safety guidelines currently used in the UK are those constructed in 1998 regarding 'thermal (heating) effects' of non-ionising radiation. These are not protective of health given the vastly documented non-thermal effect taking place orders of magnitude below these levels. They are obsolete. Other countries have responded to this information and have safety limits more biologically sensible thousands of times below ours. Mechanistic data is available to explain these effects and every bodily system is affected (as one would expect from a radiation induced illness).*

*The very broad range of RF emitting devices on the market were never pre-market safety tested and many now contain fine print warnings from the manufacturers which warn that one must keep the devices a minimum distance from the body which in some cases is incompatible with use. The public are generally not aware of these warnings or the increased vulnerability of certain groups such as children, foetuses, elderly, pregnant women, infirm and those with EHS.*

*The full paper gives an overview of facts that should be considered during the policy change that is clearly necessary, and this subsection concerns Electromagnetic Hypersensitivity (EHS) only.*

**1. Definition:** *Electromagnetic hypersensitivity is characterized by an awareness and/or adverse symptomatology in response to even extremely weak (orders of magnitude below current safety levels) electromagnetic fields of multiple types (in terms of frequency/intensity and waveforms). Relevant diagnostic coding that may be used by UK medical doctors include "idiopathic/environmental intolerance (IEI)", code Z58.4 (Exposure to radiation) under the International Classification of Diseases (ICD-*

*10), T66 (microwave syndrome)="other unspecified effects of external causes: radiation sickness."*

*[Note: All life is electrosensitive to some degree and thresholds for conscience perception will vary depending on age, gender and individual physiology.]*

**2. Common symptoms include:** *Headaches, dizziness, sleep disturbance, sensory up-regulation, palpitations, unusual pain in multiple sites, visual disturbance, auditory disturbance (esp. tinnitus), membrane sensitivity, muscle twitching, dermatological complaints, hyperactivity/fatigue (depends on adrenal status/state of EHS), restless leg syndrome, memory/concentration disturbance and anxiety (psychiatric symptoms such as anxiety and depression are likely to be secondary to the physiological effects rather than a primary cause). Interestingly, with good avoidance, symptoms tend to disappear in the reverse order that they accumulated.*

**3. Characterized by multiple sensory up-regulation:** *Upregulation of all senses is commonly noted in persons with EHS, i.e. Photophobia and/or Scotopic sensitivity syndrome (visual sensitivity), Hyperacusis (hearing sensitivity), Hyperosmia (heightened sense of smell), Hypergeusia (heightened taste sensitivity), Hyperesthesia/Photosensitivity (heightened skin sensitivity) and Multiple Chemical Sensitivity (MCS) is associated.*

**4. Exposure induced:** *EHS is a cumulative, exposure-triggered condition, and exposures are rising rapidly. Devices which emit RF and know to cause symptoms in those with EHS include: mobile phones, DECT cordless landlines, Wi-Fi/Bluetooth enabled laptop, desktop computers and laptops, Wi-Fi routers, Smart meters, fluorescent lighting, baby monitors, security systems, RFID systems and wireless gaming consoles. ELF (Extremely Low Frequency) fields (household electrical) will also cause symptoms in some individuals.*

**5. Characterized by increasing trigger susceptibility and irreversibility:** *If EHS is unmanaged and there is general deterioration, there will be reaction to an increasingly broad range of frequencies at increasingly low intensities, i.e. the number of devices complained of triggering symptoms will increase and symp-*

*tomatic distances will decrease. Tendency towards MCS will also increase and irreversibility will become more likely.*

**6. Highly prevalent:** *Estimates for the number of people with EHS vary widely, but several countries report around 4-10%. In the UK this corresponds to approximately 2.5 to 6.3 million (which is more than the number of UK wheelchair users). This is likely to be a gross underestimation (see point 7 below) given that figures are based on the number of people who have made the connection between their symptoms and EMF exposure. The number of people who have mild EHS symptoms, but have not linked them yet to exposure would be far higher. Given the ubiquity of exposure now in all environments, it can be very difficult for people to notice the association.*

**7. Rapidly rising:** *Extrapolated figures suggest that 50% of the population may be affected by 2017.*

**8. May affect everyone:** *Interestingly, the signs and symptoms associated with RF exposures from e.g. mobile phone base stations, Wi-Fi, mobile phones, radio/TV broadcasting transmitters, smart meters, MRI scanners, and other RF sources reveal that the general population (not know to be EHS) experience the same constellation of symptoms as are noted in EHS. This is a dose-response relationship. Thus, it is possible that EHS could manifest in all members of the population with enough exposure. (Please note some of the above studies are demonstrating EHS symptoms in children.)*

**9. Proven physiological condition:** *EHS has been demonstrated in a published, peer-reviewed, double blind research study, as an 'environmentally inducible bona-fide neurological syndrome', and other provocation tests corroborate this evidence. In addition, multiple papers have demonstrated physiological variations in those with EHS and genetic variations. Furthermore, mechanism are evolving that may explain the symptomatology of EHS. Therefore, increasingly, professional bodies are recognizing this as a physiological condition.*

**10. Recognised by World Health Organisation (WHO):** *The WHO states that 'symptoms are certainly real' and 'in some cases can be disabling'. Some studies suggest that certain physiological responses of IEI individuals tend to be outside the normal range. In particular, the findings of hyper reactivity in the central nervous system and misbalance in the autonomic nervous system need to be followed up in clinical investigations and the results for the individuals taken as input for possible treatment.'*

**11. Nocebo effect invalid:** *Whilst the nocebo effect (physical symptoms induced by fear) has been suggested, thorough investigation of individual histories renders this concept invalid in the majority of cases. Additionally, psychological therapies are much less effective at reducing symptoms than avoidance of electromagnetic fields (see also point 14 below) and risk perception alone has not been felt adequate to explain the characteristics witnessed. Furthermore, evidence of EHS type symptomatology in studies involving small children, fetuses and animals (where media-cultivated perceptions are impossible), also invalidates this theory.*

**12. Recognized as a functional impairment:**

** Under the disability act in Sweden, USA and Canada.*

** Legal cases are now being won for long term disability pensions/compensation (Australia, France, Spain, UK and United States).*

** Hospital facilities with low EMF have been constructed.*

** Academic bodies urge immediate protection for those with EHS.*

**The UN and the European Parliament have made clear the requirement for equal opportunities for those with EHS.*

**13. Notable persons with EHS:** *Well known, credible individuals such as Dr Gro Harlem Brundtland, Former Director-General of WHO and the first female Prime Minister of Norway and Matti Niemelä, former Nokia Chief Technical Officer have been public regarding their Electromagnetic Hypersensitivity.*

**14. Medical guidelines for management exist:** *Medical guidelines have been drawn up for doctors to diagnose and manage the condition physiologically with advice to urgently reduce exposure, and this advice is echoed by many other organi-*

*zations. Additionally, research has shown avoidance can be the only reliable form of management to improve symptoms. Currently the most reliable way to diagnose EHS is via history, i.e. it is a clinical diagnosis, but there are other tests currently being used in the private sector and in the research forum believed by their users to be diagnose or aid diagnosis.*

**15. Children have EHS:** *Many children are currently affected, but undiagnosed. Children are likely to be more vulnerable to developing EHS since their exposure is higher (as explained above), and outcomes may be worse given their developing systems and greater time for latent effects. Children with EHS must be supported at school under the conditions stated in the 'Supporting pupils at school with medical conditions' Department for Education document (April 2014).*

**16. Vulnerable groups and white zones:** *In addition to those with EHS and children, other vulnerable groups include the elderly, pregnant women, foetuses and those with co-morbidity (concurrent) illnesses. In order to protect vulnerable groups there has been increased call for designated, legally protected white zones (no or low EMF areas).*

**17. Socioeconomic impact of EHS and human rights:** *It has been demonstrated that EHS is already affecting a very large number of people in the UK (see point 6) and given that a proportion of these people will be unable to work due to their condition, revenue is being lost. Additionally, an extra burden is created on NHS resources due to inappropriate diagnosis and management of common symptoms, including in those who may be unaware they have EHS. In more severe cases, individuals are forced to live in extreme isolation, poverty and poor health. These individuals cannot access basic, life sustaining public amenities, such as grocery stores, petrol stations and health care facilities. There is there a clear breach of their human rights. We are aware that some individuals are living in automobiles and tents which can also prove threatening to health and life, especially in extremes of temperature. All EHS persons require 'comprehensive health evaluation':*

*Because of the huge socioeconomic impact anticipated for EHS worldwide, the World Health Organization has devoted considerable attention to EHS, acknowledging this condition and recommending that people self-reporting sensitivities receive a comprehensive health evaluation.*[70]

## Modern-day Microwaved "Canaries"

At 3.1 percent in the United States, 2.6 in Denmark, 3.2 in Sweden, 5.6 in Germany, 6-8 in Switzerland, 10 percent in Israel, and the highest percentage at 13.4 in Taiwan of the populations in these countries is officially estimated to be electro-hypersensitive. But unofficially according to Professor Gibson, a new study puts the number of EHS- compromised populations much higher, closer to 26% in the US, 19% in Sweden, 27% in Denmark, and 32% in Germany.[71]

But even 3% in a population of 300 million in the US, adds up to 10 million people.[72] And it only requires a number close to 2.6% (the lowest number represented in the previous estimations) for any illness to be considered an epidemic. Yet these global epidemic proportions representing millions upon millions of people with EHS are hardly ever mentioned in the mainstream media or by our governments and leaders, so that most of the public are fully ignorant of this situation and have no awareness that all of their new wireless toys (mislabeled as "necessities") are causing countless people to lose their jobs, homes, families, and to flee densely populated areas where EMF exposures tend to be strongest—in essence, to be displaced and forced to live like refugees.

There is increasing evidence that "microwave syndrome" or "electro-hypersensitivity" (EHS) is a real disease that is caused by exposures to EMFs, especially to those in the microwave and millimeter-wave range. The reported incidence of the syndrome is increasing alongside increasing exposures to EMFs from electricity, WiFi, mobile phones and towers, SMART (Self Monitoring And Reporting Technology) meters and many other wireless devices. Why some individuals are more sensitive than others is unclear. While most individuals who report having EHS do not have a specific history of an acute exposure, excessive exposure to EMFs, even for a brief period of time, can induce the syndrome. [73]

"Microwave Syndrome" or "Microwave Sickness" was ultimately renamed "Electro-hypersensitivity" (also called EMS or "Electro-Magnetic

Sensitivity"), and shares the same set of symptoms categorized under Microwave Sickness as described below:

*"Insomnia, anxiety, vision problems, swollen lymph, headaches, extreme thirst, night sweats, fatigue, memory and concentration problems, muscle pain, weakened immunity, allergies, heart problems, and intestinal disturbances are all symptoms found in a disease process the Russians described in the 70's as Microwave Sickness."*[74]

Dr. Sharon Goldberg, an integrative internal medicine physician, is educated about the effects of EMFs on health, being one of the few physicians to have also completed the advanced course in electromagnetic radiation with the Building Biology Institute. She uses her knowledge in her practice to help identify causation of otherwise mysterious comorbid presentation of illness—which means essentially a very long list of diseases at once. And this phenomenon has increased in her practice exponentially since 2014 and has been presenting in younger and younger people.

Goldberg herself experienced pain from using SMART tech. In 2014 she upgraded her Blackberry phone for an iPhone 5, but noticed that after taking a twenty-minute call on speakerphone while holding the phone in her hand, by the end of the call her finger was burning. She likened the sensation to neuropathic pain a diabetic patient would report in their toes. This experience prompted her research into the harmful effects of microwave radiation. She feels that everyone is affected by this radiation and that EHS is an incorrect label as it implies that the ability to make a connection between EMF exposures and harm is somehow a flaw in the human design. She told Josh del Sol in her 2019 interview for the 5G Crisis Summit:

*"When we think of electro-sensitivity, or the terminology that's used 'electromagnetic hypersensitivity,' which is really a scientifically incorrect term because hypersensitivity implies excessive sensitivity—that someone is more sensitive than they should be—that it's sort of an unreasonable response. But really what the science shows is that all humans are affected by microwave exposure. Certain people are electro-sensitive. What that means in my mind is that they've made the connection between their exposures and their problems.*

*"But what you have is everyone else who is affected by microwave radiation, but they haven't made the connection. So, as far as recogniz-*

*ing Microwave Syndrome in the clinic or hospital, I can tell you just some of the really classic presentations that I've seen that really should be a red flag.*

*"The most important one would be sort of youngish, college-age students, or young adults who are presenting with signs of dementia, with cognitive impairment, short-term memory loss... And it's the cognitive problem coupled with what we call orthostatic hypotension; that they get dizzy when they stand up—if their blood pressure is low, and that they're not sustaining their blood pressure.*

*"So generally, when you see those two together if you ruled out other causes... this is vey common. You see this very commonly in young adults who have had cellphones from a young age. And they'll often tell you, 'Oh my dad gave me a cellphone when I was nine or when I was ten.'*

*"So they've had that exposure for a long time. This is something that I see. And a second presentation is particularly behavioral changes in younger children. And the ones that are [presenting] extreme mood variability... particularly coupled with blood pressure [changes]. But really it can present as anything."*

Dr. Martin Pall, Professor Emeritus of biochemistry at Washington State University, has studied the effects of wireless radiation on biology for well over a decade, basing his research on nine different effects of this radiation on human health including neurological, DNA, cell death, endocrine effects, cancers, cardiac effects and early onset Alzheimer's and other types of dementia. Dr. Pall has published numerous scientific articles on his findings. Over the course of his research, he has noted a pattern of health complaints emerging as exposures to wireless radiation increase in the general population. Complaints related especially to chronic insomnia, depression, anxiety, concentration problems and sensory issues related to eyes and ears, he noted are all becoming epidemic in our societies.

The underlying mechanism involved in the presentation of these symptoms, Pall says, can be found in the effects of microwave radiation on the brain—something that has been studied extensively in animals whose neurons that normally have a thousand synapses each can be reduced to zero. From one thousand synapses to zero—this is how powerful the impact of wireless radiation on the brain can be. Studies on human populations have shown the effects to be cumulative, worsening over time.

*"So, when you have these things that are already widely occurring in our populations, and where you have things like, for instance people living within three hundred meters of a cell phone tower, which is probably forty percent of the population, having substantial impacts on these neurological and neuropsychiatric effects. We have major, major impacts from many of these exposures that we have. And all of this is covered up by the industry propaganda. It's really extraordinarily disturbing what is going on here."* [75]

Pall predicts that based on extrapolated data from available studies involving current exposure levels, not taking into account expanding networks or 5G, within seven years (from 2019) our collective brain function will crash, and that this crash will push our societies into absolute chaos. Add to the current 4G-powered 2019-level of radiation exposures, a new layer of 5G and radar-powered vehicles and the *Internet of Things* dominating our landscape, he feels that after such an introduction on a wide-scale that this timeframe of brain-function collapse could be reduced to a matter of months, launching us into a veritable "zombie apocalypse."

Add to this scenario the devastating drop in fertility levels worldwide as a result of these same exposures—the future for humanity looks quite grim. Pall feels the only solution is to halt these expansions—*yesterday*—and to roll back the technology and start protecting humanity and our world from these many sources of dangerous radiation.

The scientist and author, Arthur Firstenberg, who himself was injured by x-ray and non-ionizing radiation, astutely describes electro-sensitivity in his seminal 2017 book, *The Invisible Rainbow*, with the following:

*"Those of us whose injuries are so severe, so devastating that we can no longer ignore them, and who are lucky enough to figure out what has happened to us and why... for lack of a more acceptable term we call our injury 'electrical sensitivity,' or worse, 'electromagnetic hypersensitivity' (EHS), a travesty of a name for a disease that affects the whole world and everyone in it, a name as absurd as 'cyanide sensitivity' would be if anyone were foolish enough to apply such a name to those poisoned. The problem is that we are all being electrocuted to a greater or lesser extent, and because society has been in denial about that for*

*more than two hundred years, we invent terms that hide the truth instead of speaking in plain language and admitting what is happening."*[76]

Firstenberg profoundly sums up the dilemma with, "How many will it take before people no longer feel too alone to say, 'Your cell phone is killing me,' instead of 'I'm electrically sensitive?'"[77]

*Chapter Three*

# My Story

IT IS A CLEAR LATE autumn night in October 2014. The moon, merely a sliver in the sky, is waning to near empty just like my energy reserves. I am having another meltdown, unable to cope after multiple sleep-deprived nights in a row, and my boyfriend Sam promptly agrees to take me to the river—the one place where I am certain I will be able to get a good night's rest.

We hurriedly pack the tent, air mattress, pillows, and extra blankets upon feeling the growing chill in the air. It is going to be a cold night for camping, but facing the cold seems a better prospect than another sleepless night at home.

We abandon the apartment and my elderly cat yet again. I feel guilty leaving "Stinky" at home alone. But I need sleep, desperately—something I can no longer obtain in our third-story walkup apartment. I have to seek refuge at the banks of the Potomac River, at a campground about forty minutes' drive from our neighborhood. This has become a habit, this fleeing to the woods and water—the only way of coping, or recovering, and thus recharging myself enough to face another day of the relentless EMF onslaught in the city, and more specifically from my downstairs neighbors' WiFi router placed directly on top of a shelf just under my bedroom—neighbors who refuse to turn the WiFi off at night, even if it means much needed sleep for me, simply because it is too much of an inconvenience to push a button.[iii]

It is quite dark when we arrive, the parking lot empty—a good sign that we will have the campgrounds to ourselves, further assurance of a restful night's sleep without the risk of noise disturbance from other campers.

Finding this spot was nothing short of a miracle, something reasonably close, within the city limits, yet tranquil, largely unoccupied, and with almost no cellphone service. WiFi refugees like myself have gotten used to seeking out so-called "dead zones"—the only places left truly full of life. Ironically, what these days others fear most; losing cellphone reception, is what we covet, like pilgrims seeking dwindling sanctuary—a place to commune with the Earth where it is still vibrant and life giving, able to restore us of that which the artificial manmade world has stolen.

We fill the awkward, queen-sized air mattress (the only one I own) via the cigarette-lighter-charge-enabled air pump, barely fitting enough of the mattress into the car to complete the task. We then carry the newly pumped mattress together in the dark over the sharp gravel, over the narrow footbridge above the locks, and over stumps and rocks to our

---

[iii] This was at a time when all Internet modems still came with a WiFi button on/off switch. Sadly today, to my knowledge all new modems now do not allow customers to disable WiFi by this method, forcing customers to contact Internet providers for technical support if they want to disable the wireless features.

usual spot beneath the canopy of trees—now losing their leaves, blanketing the ground around us in crinkly, crunchy carpeting.

Once settled in, we waste no time getting to the reason for our visit—*sleep*—delicious, precious sleep. My body immediately and soundlessly sinks into the subtle, harmonious vibrations of earth beneath the air mattress, and I gratefully drift off along with the trickling sounds of flowing river a stone's throw away, glimmering slightly under that last sliver of moon, and my breath tunes in with the few remaining crickets of the season and owls hooting softy in treetops until I fall into a splendid, deep reverie.

Soon it will grow too cold for our regular exodus to the river and a solution needs to be found quickly. So, Sam nobly agrees to build a faraday cage around my bed in order to block the electromagnetic fields interfering with not just my sleep, but also with my entire ability to function, live pain-free, and un-tormented.

A *faraday cage* must enclose an area fully in metals without gaps in order to work effectively. For our purposes, namely to block out radiofrequencies from my sleeping space, aluminum screening over a wooden frame works well as the holes in the screen are small enough to block out microwaves, while also permitting adequate airflow into the cage.

And so, the screen is wrapped around the frame and around the door to the enclosure with any gaps larger than the size of the screen holes taped closed in aluminum foil. The cage works—measuring no radiofrequencies within (but dangerous levels according to my EMF meter immediately without) and provides us both with much sounder sleep—although still not as perfect as sleep obtained while camping, since the *low*-frequency fields are *not* blocked (and possibly even amplified since metals also work as EMF conductors while also serving to block out high-frequency fields by reflecting them). However, the cage, by blocking the high frequencies, at least puts an end to the night terrors, sweats and constant waking. One night we had accidentally left the faraday-cage enclosure's door slightly ajar (permitting radiofrequencies to enter), and up until discovering this mistake, we wondered why our sleep had again become so disturbed.

About six months before, in April 2014, I had returned from a trip to Bali after having attended a dance retreat workshop there. While in Bali I had discovered I could sleep (after years of chronically disrupted sleep often involving horrendous inexplicable night sweats and other sleep disturbances) even if I could not breathe (the daily trash fires in that part of the world triggered an asthmatic response and bronchial infection during my stay). I found I awoke refreshed each morning (more so than I had done in the prior two to three years back home), in spite of waking in coughing fits soon after nightly trash fires were ignited. And so I wondered what could be different enough in the environment there in Bali, compared to D.C., to so profoundly affect my sleep and sense of being well rested.

The only thing I could find that differed markedly was the lack of wireless radiation in my immediate sleeping area. The only WiFi access was about a five-minute walk from my primitive hut and the cellphone coverage was spotty as well, but the primary difference seemed to be the WiFi exposure. (At home I used an Ethernet-cabled Internet access but my downstairs and next-door neighbors all used WiFi so that at any given time there were at least five clear WiFi signals in my own small apartment, but spotty cell reception.) So, I stashed away this little observation to call upon later when I would need it in ways I could not comprehend at the time.

Prior to and during this trip I had used cellphones and WiFi without much question or restriction. However, I had always noticed an uncomfortable heating sensation just behind my ear while using a cellphone and so tended to limit my calls to not more than a few minutes whenever I could, since I noticed that the longer I stayed on the call, the more discomfort and odd sensations I experienced. I had had some previous knowledge of cellphone radiation causing potential harm, however as most everyone else likely had done, I had also assumed that the newer phones were being designed to be safer with better SAR ratings. In fact, that had been the trend in the early days of mobile-phone designs after initial concerns were loudly voiced and lawsuits filed related to brain tumors suffered by early adopters of the technology. For example, phone antennas were placed externally so that they could be aimed away from the brain and SAR[iv] (Specific Absorption Rate) ratings seemed to be improving.

---

[iv] The current SAR ratings are based on a test conducted in 1993 with a 2G cellphone (no longer supported by today's 3G-5G networks) on a 200-pound, six-foot-two mannequin called SAM (Standard Anthropomorphic Male). SAM's massive, eleven-pound, uniformly gel-filled plastic head was meant to represent the average cellphone user's brain (with no appropriate model used to represent the rest of the cellphone-using demographic;

All of this changed with the release of "smart" phones, whose new sleek designs hid not just one, but multiple, antennas inside the phone offering not only *no* ability to point the antennas away from the head, but with the increase in the number of antennas came an increase in radiation exposures as well. The new phones proved to be very powerful as they were equipped to handle a staggering number of new functions including Internet-browser navigating and video streaming.

We were all entranced with the new smartphones, myself included. Because of my professional photography background, I was particularly captivated by the added camera feature and quality of photographs available, and especially captivated by the "hipstomatic" photo app on the new iPhones my friends had. With this feature, what once would have taken me several hours in a darkroom or with Photoshop was created instantly with the touch of a screen. It was truly magical and the last thing on my mind in coveting my friends' new phones was harmful radiation exposures. I only held out purchasing a "smart" phone since I was unable to justify the upfront cost and the monthly expense at the time. But in 2012, when the competitive Android market opened up and offered cheaper smartphones at the same monthly rate I was already paying for my basic cellphone—I was in.

## *Tipping Point*

Traveling back from Bali in 2014 for thirty consecutive hours spent in airplanes and airports resulted in my total physical collapse upon reaching home. In retrospect I had probably just reached my tipping point/threshold exposure levels for EMR, which likely resulted in my adrenals crashing, landing me comatose in bed, and unable to get up for several days, during which time Sam admits to having been extremely distressed for me, but I myself remember little of those days.

---

namely smaller men, women, and children). The cellphone used for the study was fitted with an external antenna pointed away from the mannequin's head (unlike today's models equipped only with multiple internal antennae), and was held *one third of an inch* away from the plastic head and tested for a duration of only *six minutes*.

What I do remember is how my world changed after I "came to" and began to get up and get back to work. No longer could I tolerate using my smartphone, not just for calls but even holding the phone in my hand for texting and Internet searches caused shooting electrical pains to "zap" my hand and travel up the length of my arm. I noted these electric-shock sensations lasted for up to an hour after even a few minutes' use of the device.

For a time, I did not mention this phenomenon to anyone. Instead, I did what most sensitive people who grow up amongst insensitive or desensitized people do, and tried to ignore it. I had grown up as a person with heightened sensitivities, more easily disturbed by external stimulus—loud noises, strong smells, and spicy foods. As such I often came across as complaining, "finicky" or "difficult," to my caregivers but also to other children as well. I reacted adversely to many foods, chemicals, soaps, and even cut grasses to the bewilderment of my parents and doctors when I broke out into welts or hives after some exposures.

Often my reactions, particularly when of an emotional nature as I struggled to cope with over-stimulus, were treated as behavioral problems to be corrected rather than a unique physiological trait to be understood and accommodated. Thus, I learned as best I could to fit in and hide my sensitivities. Like other sensitive people, I even, at times, resorted to attempts at dulling my sensitivities with (legal) drugs, alcohol and other harmful, numbing substances. But mostly (especially since I found I was also overly sensitive to feeling the residual poisons left from drugs and alcohol in my body so that it caused too much discomfort) I tried to get healthier, believing myself to be somehow flawed, and so that if I were to discover the "flaw" I might find a way to be "normal" and no longer pretend at fitting in, but actually do so.

But I could only ever pretend to a point. Once the pain was great enough (from wireless EMR/electromagnetic radiation) I found I could no longer ignore it. I had begun to develop chronic, migraine headaches, which previously I had only once before experienced in my life during an acute illness in high-school days. Next, my sleep disturbances increased to the point of severely impacting my ability to function and maintain my sanity during the day. The sleep problems got me thinking about WiFi because of my experience of restful nights of sleep in Bali free of WiFi exposures.

So, I convinced my partner Sam to try renting cabins without electricity in Shenandoah National Park from PATC (the Potomac Appalachian Trail Club) for the weekend to test my theory—although he did not need a lot of convincing since he generally welcomed any excuse to get out of the city and into the wilderness.

I was quickly amazed at the palpable and immediate relief my whole being experienced upon arrival at these cabins. I felt happy and light-hearted, as if this incredible, invisible weight had been lifted from me. I became playful and whimsical—I felt like a kid again. Not only were we off of the electrical grid, we were far from cell towers, WiFi, and cell-phones—totally free from manmade EMR (electromagnetic radiation) exposures of all kinds. As predicted, my sleep improved immensely. Our weekend cabin rentals were happy, heady times, some of my fondest memories to this day. I never felt I was ill or that there was something wrong or flawed about me when coming to these, now sadly very rare, pristine, natural spaces.

But these were national parks, we could not move into the cabins permanently. We could not live cutoff permanently from the world either. We needed our jobs. We needed our incomes for our survival. And so the honeymoon in the wilderness always had to end and I had to go back to the torment of life inside what now felt like an invisible, manmade EMF prison.

I struggled for months with denial. Even with the contrast of our wilder-ness jaunts proving to me that I felt relief when away from artificial EMFs, I still did not want to believe that I was being so adversely affected by these fields, to which most everyone else seemed oblivious.

Then several unplanned-for double-blind tests brought the reality of my condition home to me. One such "test" happened after I had begun limiting my cellphone use and turning cellphones off in my living space, so that Sam also powered-down or kept his phone on non-transmitting "airplane mode" when staying over. On one visit Sam had forgotten that he had left his powered-on phone in his car's console so that it was after we had been in his car for not more than a few minutes when I com-plained of my left arm hurting in the same way it always seemed to hurt when I held cellphones in my hand. I wondered aloud why I would be feeling that particular electrocution-type pain in my arm while we were driving through a wooded park without cell towers or even other traffic. My arm, as it happened, had been resting directly on the console that housed Sam's cellphone, as was discovered when Sam suddenly remembered he had left it there and had left it on. Opening the console proved this fact and we both sat staring at one another dumb-founded and with little lingering doubt as to my truly being "electro-sensitive."

At other times I simply felt cell towers before they revealed themselves from their hiding places behind tree-lined highways. Proximity to cell towers of a few hundred feet or so seemed to cause a very different and even more disturbing reaction in me than close proximity to cellphones. The tower exposure gripped my head as if in a vice and refused to let go until I had gotten far enough out of its sinister-feeling grasp. I also often experienced intense pain in one or both of my eyes and even more oddly in some of my teeth—again, with an awful squeezing sensation in those areas.

## *Acquiring a Sixth Sense and Taking Drastic Measures*

I even became aware of how my body reacted to radiation from others' use of cellphones within ten to twenty feet from me. I was actually able to feel, in the form of a sort of jolt through my body, a text or call coming in on their phones just before the alerting beep or ring audibly announced the messages' arrival. My seeming clairvoyance shocked friends when I was able to announce an incoming message before their phones did and when I was able to detect when they had entered my apartment with a phone still turned on and transmitting.

In June 2014, two months after my return from my fateful Bali trip, I did the "unthinkable" and gave up not just a cellphone, but my "smart" phone—the kind of phone, from which with, many people, when polled, admit they would rather give up their right arm before parting. For me it was far more serious than losing an appendage versus losing my phone—it was give up my phone, or my life. But, "my life is on my phone," is the protest most would give if faced with this kind of decision, and I was no exception. How and when did my "life" end up reduced to so much data on a hand-held computerized microwave-emitting device? I wondered.

My seemingly drastic step to rid myself of my phone entirely liberated me from its insidious grip and I began reclaiming the life it had stolen from me. As if waking from a deep, years-long slumber, I began again to engage in the world around me. But when I did, I saw that it was occupied with cellphone zombies as if some zombie apocalypse or *Body Snatcher* invasion had taken place, of which I had been entirely unaware, and of which I had actually been, up until that point, a part. Now freed from my device, I seemed to be the only one on the street looking up and aware of my surroundings.

My friends all found it socially acceptable to interrupt our in-person conversations with glances down at their phones, touch-screen typing and clicking while distractedly chatting with me, only half listening, only partially present, always one foot in the virtual world, straddling an invisible precipice into which they may someday fall, never to return.

My connection to the Earth and physical natural world became my saving grace and integral to my recovery from our modern, toxic, manmade, artificial environment. I felt fully alive and well in the woods without buildings or cellphone reception. I discovered a rare "dead" zone in a park mere blocks from my home and visited it daily, walking with bare feet in contact with the damp earth, grounding myself, recharging my body via the contact with the Earth's natural magnetic field; the one with which we were designed to interact and recharge our bio-electrical cells.

I also found that smooth river stones warmed and laid-on my body while inclined on my bed, brought the earth to me in my third-floor apartment when I could not go to it. The stones eased my ever-increasing, WiFi-induced migraines so that it became a common sight for Sam to find me with stones on my head and torso while lying in my faraday cage—a scene that if happened upon by a stranger, would have easily led him or her to assume I had totally lost my mind. But in reality, I was regaining and reclaiming it—the one that had been zapped by EMF radiation for so long.

These remedies helped and got me through the daily torment that had become my new life after sensitization to EMF, as an "electro-sensitive"—a label to which I was still adjusting. Naturally, my symptoms and experience prompted a lot of research on the topic. And I found that ES/EMS/EHS/Microwave sickness, or whatever label you wanted to give it, something I had never heard of before that time, affected from two to over thirteen percent of a given country's population, with the US at 3.1 %, still a staggeringly high number, and if any disease had demonstrated these kinds of numbers, it would be considered "epidemic" proportions and serious measures would be implemented to help the affected population and stop the outbreak.

However, this percentage of the population—these canaries, had been largely silent, or silenced. And this official number was not indicative of other unofficial studies (studies based on symptoms, and not self-diagnosis, polling those who identified as being electro-sensitive), which pointed to numbers closer to twenty percent! Twenty is the percentage also given to the amount of highly sensitive people in the world according to Elaine Aron, author of *The Highly Sensitive Person* and also the number historically given for the portion of the population with Porphyria (a generally inherited condition, which causes those afflicted to be extra reactive and sensitive to any toxin, including EMF exposures, but can also be brought on by injurious levels of a given toxin, be it a heavy metal, chemical or excessive EMF exposure), and close to those sensitive to changes in barometric pressures (30 percent), who can literally feel impending storms.

So even without ever having previously heard of EHS, I had become personally affected, and I discovered I was far from alone in my experience. I met dozens of others online primarily via a Yahoo group, email-based forum, called *EMF Refugee*—an apt name, as most of us, of the several hundred members, apart from seeking ways to cope with the chronic pain and finding ways to further reduce our exposures to EMFs and still somehow function in society, were in fact, and are still, looking for safe havens, for low-EMF places in which we can live—in short, we are seeking refuges. The primary question being asked by forum members is "where can we go"?

This is the question my partner and I began to ask when we saw that our efforts to make my life bearable in order to continue to live in our city were not effective enough and might well never be. While I could turn off and get rid of my own phone, I could not turn off and get rid of the vast number of phones belonging to others in my immediate surroundings, nor turn off their WiFi, nor disable and take down the supporting network of cell towers, which powered the wireless world we live in.

When we checked wireless-telecom coverage maps we discovered a staggering statistic—there were no less than one-hundred cell towers (over twenty-feet tall) and no less than six-hundred relay antennas within a four-mile radius of our home. Surely there were at least *better* places to live with fewer exposures than being surrounded by so many radiation-spewing antennas?

## *Big Tech Replaces Big Tobacco; Cell Phones as the New Cigarette*

I began to feel the ostracization I had lived with for many years of my life in younger days when cigarette smoke filled the rooms of almost every public space on the planet and I was one of the minority, intolerant of cigarette smoke to the point of my intolerance being socially isolating as I found I could no longer attend events, go to restaurants, concerts, parties, bars, etc., without suffering the consequences, namely; sinus pain/infection, headaches, chest pain, feeling hung over for days after a night out (without having partaken of any hangover-producing substances) and occasional asthma attacks.

Yet when I had tried to get relief for what I called my "allergy" by seeing medical doctors, hoping for a magical drug or nasal spray to save me from my painful symptoms, I was met with disbelief. I was even told by one MD that there was "no such thing as an allergy to cigarettes." This was in the late 1990s, really not that long ago. Our collective memory seems to be very short when it especially comes to public opinion about product safety. We forget that everyone *knew* that second-hand cigarette smoke (and before that even first-hand) could *not* cause harm to anyone. And not long after, we *knew* that cigarette smoke most definitely was and continues to be very harmful to health and even extremely carcinogenic.

What most of the public does not understand, is that this knowledge only became accessible to the average person thanks to the dedication of a minority of aware people—ones directly affected by these now-accepted-as-truly-harmful effects of a particular product or toxin. These people dedicated their lives to holding Big Tobacco responsible for the often-devastating effects of their products; effects of which the industry was well aware, but put a lot of money behind keeping the public ignorant and themselves not liable. These brave people dedicated themselves to this extent, because their own lives had been turned upside down, just as mine was now being, due to wireless radiation, often because they had lost loved ones to horrific cancers and other diseases or they themselves were dying from cigarette-smoke-induced illnesses.

Once smoking bans finally went into effect in my home city (circa 2008), I was finally free to engage socially again, with all of those places that had been inaccessible to me at last no longer exclusionary. I could finally feel like a normal person and socialize again in public places; see live music, go out dancing, meet with friends for dinner—go anyplace I wished.

That blissful social participation was to last less than a decade before these same venues once again became off limits to canaries like me. Instead of smoke, rooms began to fill with invisible wireless electromagnetic radiation and likewise, once again became inaccessible to me.

In the case of cigarette smoke, even with the incredible denial and the spell the world was under, at least it was a *visible* hazard and so could, on some level be understood as a possible bother to some sensitive people, by even the most denial-ridden individual.

Not so in the case of this new threat. "Out of sight and out of mind" now had a very literal application. And I found myself in a position unable to convince the desensitized majority that their phones and WiFi may be harming them and were most definitely harming me. And arguments that a multi-trillion-dollar industry, one of the most powerful industries to date in human history, may actually, like Big Tobacco and many others before it, be lying to us about the safety of their products, tended to fall on deaf (probably microwave-damaged) ears.

## *Ms. Popcorn Takes on Town Hall*

I found myself, in attempting to explain my own plight, while asking my friends to power-down their phones when meeting with me, met with varying degrees of disbelief. It is hard for me to forget the face of one acquaintance when he stared at me with a telling "you must be totally insane" wide-eyed expression and responded with, "But radiofrequencies are all around us!" He then went on to explain how much wireless communication benefitted him in his particular line of work and how he could not envision now living without it.

It had benefitted me as well. I had relied upon it as well. I was not a technophobe or Luddite. This was *not* the reason for my protest—simply hating technology for the sake of it. It was about my *life* and saving it. It was about trying to live free from the daily torture that the wireless radiation was causing me. It was about trying to find a place where radiofrequencies were *not* "all around me" 24/7. I had discovered I could feel not just symptom-free but truly well and full of vibrant health when all manmade EMF/EMR sources were removed from my environment. I wanted to not merely survive and suffer, I want to thrive—I wanted to claim this, my and everyone's birthright—to live well and feel good and to live life to the fullest potential.

I fought to be able to stay in my town and keep my business—the one I had started from the ground up a few short years before, which was finally thriving. I wanted to continue with my budding music and dance-performance career as well, and to stay near my friends and family. I even spoke-out in protest at a town council meeting about how my legally accepted "functional impairment" was not being accommodated, in particular by my local food co-op, of which I had been a member for twelve years and an employee four of those twelve, that had recently installed a public WiFi hotspot for the "convenience" of their members and shoppers.

I also had met Theodora Scarato around this time, who helped me prepare my speech for the town council meeting. Theodora got involved with EMF issues because she understands the dangers and having children of her own, wants them to have a healthy future. She has been fighting on the frontlines[v] to spread awareness about the harmful biological impacts of EMF exposures and to take down cell towers at schools and in communities while also passing local "right to know" ordinances that help bring the fine-print hazard warnings on microwave-emitting products into the public view.

Together with Theodora and another EMF activist, Kate Kheele, on Halloween night 2014, we approached Maryland's Governor O'Malley (who we had heard was scheduled to make a public appearance at New Deal Café in Greenbelt, Maryland), to educate him about the EMF issue. In particular we wanted to set him straight after having heard him publicly proclaim that WiFi was a "human right." Perhaps access to the Internet could be argued as being a human right in today's world (and

---

[v] In 2021, Scarato was, as current Executive Director of Environmental Health, part of the group of activists who filed a lawsuit against the FCC for terminating studies for wireless product safety, and won. https://ehtrust.org/in-historic-decision-federal-court-finds-fcc-failed-to-explain-why-it-ignored-scientific-evidence-showing-harm-from-wireless-radiation/

that may have been what he meant), but certainly not access through health-destroying wireless networks.

As it happened, I had chosen to dress as a WiFi Refugee that Halloween. As fashionably as I could, I wrapped myself in shiny Mylar space blankets and wrote "WiFi Refugee" on my silver Mylar cape in bold, black marker. Even with this clear signage, I was still mistaken for going as a bag of Jiffy popcorn (assumingly as the old-fashioned stove-top aluminum-lined variety rather than the *microwave*-popcorn—which would have been rather ironic had it been the latter). It was in this rather silly costume that I was allotted my few minutes of time to confront the governor at a café table along with Theodora and Kate, and the noise from my crinkly costume interfered with Sam's audio as he stood aside to film the encounter (rendering it essentially useless).

The governor, whether or not amused by the comical presentation, did not show it, and met our passionate speech with the friendly patience of a skilled politician. He did not win reelection, so we do not know if our five-minute lecture had any impact and if it would have made a difference.

Soon after our short meeting I was forced to leave due to the café's WiFi (a café where I had previously performed regularly, singing and playing guitar at weekly Open Mic nights)—but not before taking a bow on stage with a few other winners of the costume contest.

Before the WiFi was installed at my food co-op, I could shop in peace, take my time, and fraternize with other co-op members. But after the WiFi was turned on, I could no longer function while in the store. I had to learn to bring a shopping list, even if only getting a few items or I would forget it as soon as I entered those disturbing fields. Not only would I become cognitively impaired, I would experience severe chest pains and a racing heart if I walked too near the front desk where the router was situated. I had to get in and get out as if my life depended upon my escape—not exactly the kind of shopping experience storeowners should want for their customers.

Of course, my co-op was just following a trend, trying to compete with other shops offering "free WiFi" (read: "free cancer, free birth defects for your unborn baby, free tumors, free insomnia, free hypertension, free diabetes," etc.) to their patrons. Hence, they were not the last on a growing list of stores and other public spaces, which became unbearable for me to visit.

I was suddenly thrown back to the days of being the injured second-hand smoker, but now it was exponentially worse. Wireless radiation was much more ubiquitous than cigarette smoke had ever been and the impacts far more debilitating, with its users comprising of a wider demographic, with products marketed to and for children and even toddlers; with WiFi-enabled diapers to alert parents to when a change was due—there were no limits to the industry's reach, which extended, it would seem, even to an innocent baby's bottom.

And unlike the case with cigarette smoke, I was not able to simply leave smoke-filled rooms to find relief. Even if I could choose not to use wireless technology and friends were kind enough to turn off cellphones in my presence, still I could not turn off the entire wireless grid—the cellular network required for cellphones to function.

When my efforts to educate local officials, shop owners and my friends failed, when attempts to have my United Nations-recognized functional impairment accommodated in my own neighborhood and home failed, and when my symptoms worsened, there was nothing left but to consider a drastic upheaval and relocate.

## *Escaping the Mine; Where Can a Canary Go?*

At first it seemed that we might simply be able to move somewhere on the outskirts of the city, but everywhere we looked, even over an hour's drive away, we found cell towers littering the landscape, spaced merely a mile or less apart, with seemingly no end to their imposing structures in sight. It had become increasingly difficult to find places outside the city without these networks and even public parks were adding wireless infrastructure for the alleged benefit of their visitors, leaving few options left for those wishing to escape harm.

Our search then took on a new perspective, one that encompassed not just the entire country, but also potentially—the entire world. At one moment we had even considered a move to the other side of the Earth when my newfound friend Bruce Evans in Australia offered to have us stay at his low-EMF farm Down Under. But immigrating to Australia proved no easy task and the idea was soon abandoned as too unrealistic given our financial situation and other limiting factors.

When my partner and I looked up cellphone-network coverage maps in order to seek out radio-free "white zones", we were confronted with an image that very much resembled a plague outbreak map and fittingly the areas were even represented in red—like so much blood spilled across our nation and world, it was a disquieting and disheartening illustration, and one that would prove rather prophetic in a few years to come.

But persistence prevailed and won out in the end when we at last discovered a plague-free area in southern Arizona. Yes, we found a few other locations as well and we had heard of Green Bank, West Virginia, (the national radio-quiet zone to which some EHS people have fled) which was at the time much closer to home for us, and thus a tempting consideration. But it was also much colder, more isolated, more expensive, and there were other factors to consider, like the lack of any organic foods market or substantial shops of any kind closer than a two-hour drive from the town. And we had also heard that Green Bank was not quite the paradise for EHS sufferers it was touted to be in the mainstream press, which was always happy to point to the fact that the problem for EHS people was solved—there was a place set aside just for them, much like a leper colony, a place to hide us away from the world, so that those reading these articles could rest easy and say, "well there is a place for them, problem solved," and never give the issue another second's thought.

But in reality, Green Bank locals were said to not be particularly welcoming of these newcomers; the modern-day lepers. Additionally, we found out there is, in fact, only a small radius from the radio telescope that is strictly radio-quiet, with few actual properties available for rent or purchase within that radius. Where once WiFi was banned from the entire town as well as all cellular networks, now industry and social pressures have prevailed to the point of allowing some exceptions to the WiFi ban and to permitting a cellular network to be installed albeit at a lower power density and shorter range of coverage (at likely much healthier levels of exposure than inundate the average town, a fact that begs the question that if a healthier cellular phone network alternative is possible why it is it not installed everywhere?).

Fortunately for us, I had met Ruth and Gary online while seeking out refuges. True EHS pioneers, they were living nomadically out of retrofitted mobile domiciles "on the land" as they called it, in an area spanning parts of three western states, one of which was Arizona. Ruth assured me there were safe places still to be found in Arizona and that the geographical topography lent itself to blocking wireless infrastructure and power lines, so that a person in my position actually had some options left for finding safe spaces in which to live.

It was a timely discovery as it was not only my failing health serving as impetus for our exodus, but I had received formal written notice from my landlord a couple of weeks before to vacate the property by the end of 60 days. The excuse my landlords gave for the eviction was the need to let my apartment to a relative. As my lease had been switched-over to month-to-month after the initial two-year agreement, my hands were tied, helpless to protest. This notice came after I had been a model tenant for seven years, always paying the rent on time and keeping the apartment clean and undamaged.

Maybe they did really need the apartment for a relative, but both Sam and I knew the real reason for the eviction. I had, upon discovering my reactivity to wireless radiation, tried without success to convince the downstairs neighbors to relocate their WiFi modem to another part of their apartment (out from under my bedroom) and to turn it off at night. I had also asked the neighbors to join me in opting-out of the imposed SMART utility meters. At first the opt-out had been offered free of charge[vi] to the electric subscriber, so I had been able to convince the basement apartment dwellers to make the change, as they too became concerned about radiation exposures after studying up on the smart-meter issue.

But the third SMART meter positioned along with the other meters directly below my balcony and not far from my bedroom, was linked to our landlords' account as it tied to electric use for the commonly shared basement laundry facilities, and thus could not be charged to the neighbors at the middle level, whose electrical outlets were linked to the laundry room ones. So, I contacted the landlords hoping they would agree to sign the free opt-out form. But my landlady, herself an envi-

---

[vi] Only a few months later PEPCO, the electric co., imposed an opt-out fee of $75 initial fee and then $14 monthly per opt-out. The new fees caused the basement neighbor to agree to the reinstallation of that particular SMART meter. In spite of my having opted-out I was still exposed to radiation from the two other meters. I refused to pay the new extortion rates and made certain PEPCO knew this both in writing and over the phone. Still, they charged me for the duration of my stay in the apartment, but I never paid and they never cut off my service.

ronmentalist by profession, told me that she had "looked into" the matter of SMART meters and wireless radiation and did not think there was "anything to it." She told me this in the face of my explaining to her the pain it was causing me and why I needed her help. So, even while it was a free opt-out she refused to take it. And when I also asked if she could do anything to help me in getting the downstairs neighbor to accommodate my disability, she also refused, saying that she was not in a position to tell tenants what they could and could not do in regards to use of wireless devices.

After years of being a largely uncomplaining tenant, my landlords had found I had become a nuisance and also may have thought I had become mentally unhinged. Rather than trying to help me secure my home after all my years of loyalty, they tossed me out, adding further urgency to our relocation need. While demoralizing, their callousness may have been a blessing in disguise as it increased our motivation to "get out of Dodge." Had they helped at all, a further delay in our exodus might have pushed my injured body past a point of no return.

Thus, the news about Arizona was timely, but we had to arrange a trip out there to find housing, pack and clear out in less than two months. The task at hand was a daunting one, given I was not sure how well I could tolerate a journey by airplane—the only viable option in the short timeframe allotted, but we somehow managed it all and found a new home during a week-long exploratory visit, during which we were able to stay at my old college friend's adobe home in Tucson where she agreed to turn off WiFi during our stay.

Each day during our visit, we scoured Craigslist and real-estate rental ads and we drove our rental car further and further outside of the Tucson city limits in each direction, as we found, although boasting fewer cell towers than DC, Tucson and its outskirts to be, like most cities, too high in radiation levels to prove tolerable for long-term inhabitation. Even with the mountainous landscape, in and outside of the city, poised to offer up effective shielding from cell-tower radiation, we discovered that cell towers out West were most commonly placed on top of mountain ranges in order to gain the most effective reach—allowing signals to travel up to 20 miles or more, thus nullifying the potential to use the landscape for protection even in rural sparsely populated areas far away from towns.

However, we finally, at the last minute, the day before our return flight was scheduled, found a house 60 miles to the northwest of Tucson, where at first glance only cactus and cows seemed to share the stunning yet haunting desert terrain. Further finding out that the electric utility company offered free opt-outs from SMART meters and that landline phone and Internet service were still available and cellphone service spotty at best, we felt that at last we had found a refuge—and here would be the place where I could heal from my EMF-induced injuries.

After having secured a housing rental out West, we packed up our belongings, ridding ourselves of about half of them first, and made preparations for our three-thousand-mile exodus. As if sensing the imminent move, my beloved cat, Stinky, finally succumbed to his own effects of EMF exposures, suffering a second round of blood clots in two weeks, and departed this world. He had also endured detached retinas within months after our SMART meters were installed on our apartment building, and suffered subsequent vision and hearing loss alongside decreased mobility not long after. Yes, he was in his senior years but had always had robust health, which declined significantly after the installation of the SMART meters in 2012—the same year I bought my first and last smartphone—the same year my sleeping patterns and cognitive function became affected and I suffered mysterious joint pain and tight muscles, which no chiropractor or massage therapist was able to alleviate. And the same year my plants on my outdoor garden deck began to fail, wither and die, never to be revived again. The deck, on which these plants had thrived for many years previously, was situated directly over the bank of new (stealthily installed by our utility company) SMART radio-transmitting electric meters.

## *Going West*

When the time came to move, on our ten-day-long car journey to Arizona, driving across country, we stayed over at Airbnbs or camped, since it was the only way I could ask hosts if WiFi could be turned off at night for me—something we could not do at a regular hotel. While stopping over in Missouri the host had complied with my request to turn off the WiFi, but I had forgotten to mention cellphones and had not thought about the potential problem of my host's teenage daughter using her phone in the next room while I was trying to sleep.

Feeling okay in our rented bedroom prior to the teen entering her room for the evening, I soon realized that she must be on her phone as my body immediately alerted me to the fact with distinctively uncomfortable sensations and head pain—symptoms I had learned to link to radiofrequency exposures. Pulling out my RF (radiofrequency) meter to measure the room again, the subsequent readings confirmed the heightened level of radiation—a marked change from my earlier readings. So, I fetched the host and asked if she could please request that her daughter switch-off her phone for the night as I could feel it and could not sleep. She again complied but admitted to us the next morning that she had thought maybe I was a little crazy and that her daughter especially had thought so, but she asked her daughter to do it anyway. I could immediately tell when the girl had turned her phone off and was thereafter able to sleep, actually quite well that night. We had been lucky that there had been no other strong RF interference in the area—no near cell towers or SMART meters.

In the morning after we had gotten up for breakfast before our departure, our host, still in her bathrobe, approached us at a near run, and practically bowled me over with her ecstatic hug, crying out, "thank you so much, I haven't slept so well in years!" She went on to explain that for the previous two years she had been experiencing sleep disturbances and insomnia and she finally, with the WiFi and cellphones turned off, was able to get a good night's sleep. She was overcome with emotion and tears in her eyes at the relief she felt and gratitude for our having educated her. No longer did she think I was crazy—that was certain.

Our first house rental, found on Craigslist during our February 2015 exploratory visit, sadly did not work out for us. While EMF readings seemed low at first, after moving in we noticed activity from nearby neighbors' WiFi.

The landlords had simply told us when inquiring as to neighbors that we would only have cows for neighbors—and on our visit that seemed to be the case. What they neglected to mention was the presence of a family of squatters just over the fence from our backyard.

Upon inspecting the property ourselves before renting, we had assumed the adjacent lot to be empty, as the derelict trailer and mountains of garbage did not give the appearance of inhabitability but rather of abandonment instead.

WiFi signals from the drug-addicted neighbors proved the least of our troubles after the midnight trash fires began in earnest. The fact that burning trash in that county was illegal did not bother our neighbors in the slightest and courteously delivered pleas of our own to stop the burning were met with distain and contempt. We were told to "go the *bleep* back to Maryland" in drunken shouts over the fence anytime we stepped outside in the night hoping to enjoy some uninterrupted stargazing.

## *Fleeing the Mine—Again*

The black smoke from burning piles of old mattresses, other furniture, alongside random bits of plastics and metals in an enormous fire pit at the height of the dry and hot summer sent me taking refuge in our bedroom with all windows tightly sealed and having to wear a bulky respirator as I tried to sleep through the fume-laden nights in 90-100 degree heat (the "swamp cooler", our only source of air conditioning, did not work properly with the windows fully closed, and hardly worked even when cracked open) and not fall victim to severe coughing fits. Visits from the police were unsuccessful in stopping the trash fires, as due to a loophole in the laws, squatters were immune to prosecution as only landowners could be fined. In this case the landowner was deceased and his friends took up residence after his death, long enough to legally claim squatter's rights, but as the case had not yet gone to trial the squatters were still not legal owners but allowed to stay on as they were nonetheless.

Finding ourselves helpless to stop the dramatic assaults to my health, we had to seek other housing and break the lease six months early. After all, we had not upended our lives and traveled thousands of miles just to further threaten my fragile state of health. The whole point of the move had been to find a safe haven for healing.

Fortunately, we found another suitable, wild-desert outpost less than thirty miles to the east on our dusty, dirt road, also 60 miles north of Tucson when traveling to the city from another route. In this new and much more welcoming community we found a rental house on a ranch also in a cellphone "dead" zone and hidden from the high-tension power lines by the landscape. However, this new home was listed on the real estate market and so there existed the potential for it to be snatched out from under us after the move. But the owners who lived just across the road thought it unlikely as the house had sat on the market a number of years already, so they estimated we could stay comfortably for some years to come.

In the end we were only able to stay three months when, soon after getting settled in, as chance would have it, the owners secured a buyer and we were left to search for another situation within a matter of weeks. With winter fast approaching and still not having found alternate housing, we ordered a heavy-duty canvas tent, preparing ourselves for the grim prospect of homelessness. When the tent arrived, we discovered the vinyl flooring to off-gas a chemical smell intolerable to me, but we held onto the package before reselling, just in case of desperation.

After checking into several properties along a 20-mile stretch of ranching community and having just had our proposal to rent bare land with utility and water hookups from an absent landowner rejected, our morale scarcely held through that harsh and frightening time.

By this point we had adopted two adorable kittens as well and so had a whole "family" to consider along with all the things we had so painstakingly moved across country. Even had our offer to rent and care-take the bare parcel in the *bosque* (a grove of mesquite trees) facing what proved to be an exceptionally cold winter out in the elements, even with a tent and woodstove, was a daunting challenge, one for which we hardly felt able, especially given my shaky state of health at the time.

But at the eleventh hour, against all odds, we found a house to rent and are currently (in 2019) grateful to have a roof over our heads. It is another that had been on the market a number of years and the owners still wished to sell but ultimately agreed to a month-to-month rental lease for us.

Access to the sprawling house is extremely challenging but has proved both a blessing and curse to us—a blessing because without a year-round drivable road directly to the house, the property has proved impossible to sell at the asking price, and a curse because it has meant carrying our groceries and other supplies in on foot, over and through the seasonal river, from the road half a mile away, year round, in all kinds of harsh extreme cold and hot weather, and sometimes through waist-deep muddy and cold water.

When we first were able to move in, that late November day in 2015, the road to the house, having just been redone after the summer monsoon rains had washed it out, was passable with a 4-wheel-drive truck. And so we were able to move our things in with the loan of a neighbor's vehicle up to the task—unlike our own Nissan sedan.

The property is well over one hundred acres and gives us ample space from neighbors' WiFi or trash burning and we are still in one of the few "dead zones," free from RF (radiofrequency) signals. The house is too large for us and has proved a lot of work to clean and keep up. Having sat empty a number of years we have had to tend to a lot of repairs, mold remediation, trouble with the wells (there are two separate wells serving different parts of the house) functioning, and keeping the desert brush from growing into the house is a constant job.

And we have also had to contend with "dirty electricity"[vii] in the house, to the point where it has not been easy for me to sleep comfortably indoors, causing us to resort to all kinds of creative outdoor "bedrooms" for me over the years. While not perfect, this housing situation has been the best we have found and overall living here has helped me regain my health by degrees.

We have now, in 2019, been in Arizona for over four years and my health, in our radio-quiet valley, has gradually recovered—my body having finally been relieved of the constant EMF onslaughts I faced "back home." My cognitive function has returned. I sleep peacefully and soundly most nights, and I am able to once again use a computer, drive a car, stand near the refrigerator, and cook on our electric range—all things I was unable to do for at least a year after our move to the desert. These may be small achievements for the average un-afflicted citizen and things which most of my life I took for granted, but ones that have now become a huge triumph for me.

But this valley, our refuge, which accommodates a few other sensitive canaries like me, is being threatened by our utility company with a "smart grid," with our electric company, for some inexplicable reason, no longer content to remotely read our meters via our power lines, but must now do so wirelessly even though the cost totals in the twenty millions (funds the company had to borrow), while the company claims the very

---

[vii] See Appendix A, Solution No. 9 for more information on "dirty electricity" and how to remediate it.

expensive “upgrade” will save customers money through micromanaging energy usage.

One wonders at their plans to recoup expenses (increased rates for “peak-hour” usage, data sales to third parties and government subsidies come to mind), however when asked they refuse to answer. The excuse being given for the sudden change is that the current meters in use are no longer being manufactured, although we have found this to be untrue. One hundred out of one hundred and twenty residents of our community have petitioned against the new meters and infrastructure, but our pleas have been ignored and our attempts at legal action and accommodation for those of us functionally impaired with EHS, have thus far failed.

So now we are faced with another potential forced exodus, squeezed out by the triumphant ever-expanding, ever-hungry wireless industry, we have to ask the question, all the canaries and their partners and families in my position must ask: “Where do we go?”

Left to right and top to bottom: Sam walking through the desert near our rental property. Sam crossing our river on his way to town. The author Shannon making a sand angel in the riverbed (the same one Sam is crossing in the previous picture. The river flows seasonally, mainly in summer during the rainy season. The levels sometimes reach up to 8 feet in depth.) One of the many narrow "slot canyons" in our area. Sam in a tepee, one of our Airbnb rentals on the journey from Maryland to Arizona during our move in April 2015. The author taking a dip in the Salt River, 200 miles to the north.

Left to right and top to bottom: Various stages of the Wiki-Up sleep hut, one of my outdoor bedrooms. First two showing the skeleton—bent willow and tamarisk branches, tied together in the middle. It was strong enough to support Sam's weight sitting on top. Next, we tied quilts I made stuffed with alpaca wool to the inside as insulation. Inside shot after getting moved in. Bottom left—we had to eventually cover it with greenhouse plastic to keep the rain out. Last—the Wiki-Up with finished canvas cover, one that did not prove waterproof even with the beeswax coating.

Left to right and top to bottom: Tent camping set up, homemade canvas pyramid tent with tarps to help keep rain out. Wood fame of the final sleep hut. Empty interior with our cat Mina as model. Final outside before applying the last piece of plastic corrugate.

*Chapter Four*

# EMF Refugees

ON THE EMF REFUGEE forum, we EHS members swap stories and pass along any information we have found helpful in relieving our painful symptoms. Discussions range from finding safe, low-EMF housing, reviewing EMF-shielding products and EMF home-filtration systems, finding out which cars, computers, and other electronics emit the lowest levels of EMFs, sharing the most effective natural diets, medicines and remedies for symptoms' relief, to political topics related to wireless-industry plans.

Forgotten by the world, we are on our own—there is no organization or government entity to help us out, and no public knowledge of what we deal with or how hard our lives are on a daily basis, so we have to try and help ourselves. With all of us groping in the dark, the EHS veterans help the newly afflicted, and the newbies find some solace in the comfort of support offered by other canaries; proving that one of the most useful aspects of the group serves to help primarily in terms of moral support—to help us each not feel so alone knowing others like us are out there.

And sadly, there happens to be *plenty* of others out there. Even though none of us had ever heard of EHS before becoming symptomatic ourselves, we all soon discover that the estimated two to twenty percent numbers are probably accurate. But we are, for whatever reason, spread out thinly over the planet, forcing us to connect virtually—not exactly the ideal communication method for those made ill by any or too much time spent on computers, but it is the compromise we all make to obtain much needed support.

There are a number of theories floating around on our forum (as well as outside of it) as to why we are the ones able to feel EMFs and not others. Just what is it that sets us apart? The ideas range from genetic predisposition to individual histories of heavy-metal exposures (likely from multiple vaccine dosages and dental mercury-amalgam fillings), Lyme disease or Epstein Barr infection, some type of electro-shock/electrocution experienced while very young, over-exposures to toxic chemicals, parasite or dental infections, or simply having reached a threshold tolerance level of EMR exposures. But this last point begs the question of why we personally have reached this threshold and not others subjected to the same exposures. Would others reach this mysterious threshold at some later date and become hyper-sensitized, and would we reach a tipping point where instead of EHS afflicting a minority population it would affect the whole world?

The same question could be asked as to why some people are more adversely impacted by cigarette smoke than others, or any other known toxic carcinogen. As Frank Clegg, former president of Microsoft Canada and current president of Citizens for Safe Technology, observed, "Electro-sensitivity is not like an allergy you are born with; it is an illness that builds up over increased time and radiation exposure. Just as we cannot yet explain why some individuals will die from second-hand smoke and others can live a long-life smoking 2 packs a day, we cannot explain why some individuals react to wireless radiation."[78]

## *Interview with Dr. Olle Johansson*

In 2016, as part of my quest to find more answers to my own predicament, I contacted Professor Olle Johansson, famous for his work on the health effects of EMF exposure and on electro-hypersensitivity, for a phone interview, to which he kindly agreed.

Dr. Johansson had been one of those instrumental in having EHS recognized as a "functional impairment" (the new PC jargon for disabled or handicapped) in Sweden and also by the United Nations, so that, technically speaking, any UN member by default must recognize EHS and accommodate it. So far, most countries, including the US, recognize it in name only, and not in practice. However, Sweden is one country that does take measures to aid its EHS population, but there are still limits to this aide, hence, the more severely afflicted must often retreat to wilderness places for survival. Dr. Johansson explained further:

*In Sweden, EHS, since May 2000, it's fully recognized as a functional impairment and that would mean if you would not be able to travel to a physician or medical doctor, that would be a strong violation of this special UN convention from 2007—but really on the books as far back as 2000, which states that having any functional impairment or disability, you are entitled, and I quote 'to live an equal life in a society based on equality.' It's so very simple and to achieve that if you have a functional impairment you need to have accessibility measures. If you are in a wheelchair maybe you need a wheelchair ramp. If you are EHS, maybe you need to visit a hospital that is electro-sanitized. And in Sweden you can find it in several towns...*

*In 2000, one of the ministers of the cabinet at the time, he answered some questions of a woman I knew very well, such as, 'do you regard EHS as a functional impairment?' 'Yes,' he said. 'Are there any exceptions with regard to EHS compared to other disabilities?' And he said, 'There are no exceptions.'*

*So even if we don't understand anything, we still have to give you an equal life in a society based on equality. And furthermore, the big win is, and that has been understood both in Sweden and elsewhere in the world and inside the UN, that when a country makes their country accessible to various functionally impaired people, the big win, is actually for everyone.*

*For example, if we make every country fully accessible for someone with EHS through electro-sanitation, maybe no one would get cancer in the future from EMFs. Maybe no one will have learning problems due to the effects of microwaves, etc. So, the big winner would be everyone.*

And Olle, after discovering that EMFs did, in fact, affect human health adversely, based on his own studies dating as far back as the early 1980s, which started with the impact of computer-screen exposure on skin (now known as "screen dermatitis"), eventually shifted his work to championing EHS sufferers.

Tirelessly laboring to help those impaired with this condition, he has dedicated the last three decades of his life to this end, ultimately sacrificing his career when his research rocked the industry-funded boat just a little too much. In 2017, after forty-four years at the Karolinska Institute as professor of neuroscience, Dr. Johansson was forced into early retirement with the pathetic excuse that his research on adverse health effects of artificial electromagnetic fields, as well as on the functional impairment of electro-hypersensitivity, was of "no importance." This decree came down from the likes of Vice President Karin Dahlman-Wright, who in 2018, nearly two years after her part in Johansson's dismissal, was herself relieved of her position due to allegations of scientific misconduct. (It is also worth mentioning that the Karolinska Institute, the same institute that annually awards the Nobel Prize in Medicine, receives a large portion of its funding from Big Pharma, an industry specifically dedicated to the development and research of biotechnologies of which wireless radiation technologies doubtless plays a big role.)

And if all that were not bad enough, Olle has endured death threats and even one attempt on his life—a professional sabotage job on his motorcycle that came frighteningly close to succeeding. He has since continued to plow on ahead undeterred, even after being defunded, dismissed, left without a penny, forced to declare bankruptcy, losing his home, and even being abandoned by family—he continues his important research independently, relying upon the generosity of strangers via online donations.[79]

In speaking with Olle, a kinder-hearted, more humble person I cannot say I have ever met, and in thanking him for his dedication to helping others like me, he dismissed my praise by saying that we electro-sensitives were the true heroes—struggling daily for our survival and rights in our societies, which have abandoned us.

I posed to him all of the theories I had run across thus far as to why we are singled-out from the population at large to suffer as we do. Admitting to having heard some of the same theories, he thought they were all very good questions but confessed that the funding was just not available to scientifically find out for certain if any of those reasons applied. However, based on what research he had done and of which he was aware, the findings seemed to point to the problem being with the environment and not ourselves and that it was, in fact, the environment and not we "electro-sensitives," that truly needed fixing. He explained with the following:

> *From a biomedical point of view it's natural to have what we call a 'biomedical variation'... some people react more vividly*

*to something and some don't react at all and when experiments have been done on the lymphocyte, which is the key cell in our immune system, it turned out that the changes that were observed in the lymphocytes from electro-hypersensitive persons—the very same changes actually occurred in lymphocyte from normal healthy volunteers, but they didn't have any subjective sensations.*

*And up to now there is no kind of given treatment. That also points to an interesting fact, mainly that you [with EHS] are not ill. I've been playing around with the thought recently that you are very normal and other persons that do not feel what you feel, they are* **hypo***-sensitive and they are the ones that needs some formal treatment. And again, coming back, I repeat myself, these very interesting studies with the lymphocytes, most likely everyone is affected and most of them they don't understand and feel it, because they are* **hypo***-sensitive. But you—you are normal.*

*And you have a normal avoidance reaction to a toxic environment. And the only way to deal with a toxic environment, is not, and I want to stress this, not to treat you, but to treat the environment.*

*So not to be ironic in any way, but in a way you can look upon yourself being electro-hypersensitive, that maybe in some aspects you are lucky, meaning that you will not unnecessarily expose yourself [to EMR] and maybe not get long-term effects, such as, changes in your neurological status or cancer types that have been very much discussed since the year 2016, because you will avoid exposure. So being this kind of biological warning signal as you are, or canary bird in the coalmine, you will, of course, avoid what the rest of the population 24/7 allows themselves and their kids and their pets to be exposed to.*

*The rest of us should be very happy that there are these yellow canaries in this coalmine, because they tell us something very important and it's actually quite fascinating because people more and more realize that we need to look upon this in quite another way. Of course, don't get me wrong, if you ask for some aspirin for a headache you should have it. But we must take care, according to the UN, of the toxic environment. We cannot allow you or anyone else to continue to be exposed.*

Dafna Tachover, an attorney licensed in New York and Israel, at age 43 had her life turned upside by the sudden onset of EHS. Losing her job and marriage as a result, she fled to a remote cabin in upstate New York where she finds herself a prisoner of her own home, unable to travel, taking it one day at a time, fighting for survival, she has this to contribute to this topic:

*"Humans are electric beings and there is no mechanism in the human body that protects it from the radiation. Therefore, to claim that this radiation is not affecting us is ignorant and absurd. EHS is not a disease—it is an environmentally induced condition to which no one is immune. I want to believe that the day in which the extent of this disaster will be exposed is not far. Ignoring the facts and reality do not change them and ignoring a problem is guaranteed to worsen its scale."*[80]

But it is easier and more comforting, to think that we, the canaries, are the flawed ones. Because if we are, then it means if we discover what is wrong with us, we might be able to fix it. But if the environment is the problem, and globally so, with the world unwilling to acknowledge and address the problem, then we are surely doomed.

In 2015, I participated in a DNA study spearheaded by Dr. Beatrice Golomb at San Diego State University, being conducted in order to determine if there was a genetic marker for EHS. Dr. Golomb had very personal reasons for conducting this research, as someone very close to her was afflicted—this is all she would admit when probed. My test results showed that there was actually something amiss with my mitochondria, which seemed to indicate a kind of detoxification impairment. I asked Dr. Golomb's assistant if this was a consistent finding among those with EHS who had volunteered for the study, but she was unable to say at the time, as no conclusions had yet been drawn.

Originally when we had asked Dr. Golomb for an interview for a potential film documentary on EHS my partner Sam and I were attempting at the time, she had cheerfully accepted, more than happy to help. Upon following-up to schedule the interview a few weeks later, she declined, explaining that she had done other interviews with local press, which had proved instrumental in defaming her, misrepresenting her research—

ultimately serving to threaten her position at the university to the point that she felt pressured to abandon her research. Nothing we said would convince her to change her mind. She had clearly been shaken by her experience.

The findings of my particular test still begged the question, which came first, the DNA damage (a genetically inherited mutation) or was my DNA later damaged by EMR exposures, as I have learned it most certainly is capable of inflicting? It was the classic chicken-and-egg dilemma and having no pre-EMR-exposure DNA test with which to compare, I was left with no certain answer.

## *Setting the Record Straight*

In many ways, much worse than the physical torment EHS sufferers endure, is the social ostracization and alienation experienced when coping with the constant disbelief and lack of acceptance with which one is faced when family and friends decide that it is easier to label us as mentally unsound than to face the implications of our condition—that current levels of exposure to EMFs are harming everyone. As Dr. Johansson explains:

> *In Sweden, twenty to twenty-five years ago persons with electro-hypersensitivity were actually investigated regarding their psychological as well as psychiatric profile. You said that people around you think you are crazy and that was the same in Sweden. People talked very badly about persons with electro-hypersensitivity, and called them a lot of very bad things—like they were only post-menopausal women. Then men appeared to be electro-hypersensitive and they said no, they are low educated persons. Then lawyers and professors appeared [with EHS] ... well it was just a mess. And then someone said, 'well let's investigate the persons with electro-hypersensitivity.' To make a long story short, it turns out that they were completely normal and that they did not differ at all from the other populations—apart from one point, and this is extremely interesting, they actually could stand harassment somewhat better than the other normal population.*
>
> *It's like if you have a dog, the first time you kick the dog it will get very annoyed and alarmed, but if you continue to kick it, it will get used to it and that's what is shown in this particular study at the*

> *Uppsala university that the electro-hypersensitive persons sort of counted on some slander, so they could withstand it better than the other population. But that was the only point where they differed, but otherwise normal psychologically and psychiatrically.*
>
> *...And the funny thing is I tell you the ones who don't trust you and don't believe you, they would if they were affected in that sense. It's so very obvious. I have had persons who were laughing at me and bullying me in public and then after some years they became electro-hypersensitive themselves and then they didn't bully me any longer.*
>
> *Other people look upon you as being slightly 'odd,' as the Englishman would say. For me I'm a neuroscientist and I know a little bit about the brain and the spinal column, how people behave and about psychology and psychiatry. And so, it's struck me, one thing over the years... You know I've met so many EHS over the years and their relatives and they definitely are less crazy than the general population. So, if anything I'd say that EHS people are very normal, not crazy. I've met maybe one or two out of many thousands. Many I have not met in person but I've talked over the phone or email and such, and one or two of them have been odd, yes.*
>
> *But in the normal population you would have found hundreds or even thousands that would be odd, so electro-hypersensitive people are very, very normal. And furthermore, I often remind people that the most famous electro-hypersensitive person on the planet right now is the former head of the WHO, and the former prime minister of Norway.*

The person Dr. Johansson is referring to is Gro Harlem Brundtland, who, while Director-General of the World Health Organization was known for reacting adversely to cellular phones, which gave her headaches, to the point of banning them from her office at the WHO. While giving an interview to a reporter from a major Norwegian newspaper in 2002, she began to complain of a headache and asked who had left his phone on, to soon discover that the press photographer had merely turned his phone to vibrate instead of powered-down. She then explained her reactions in these words:

*"I made several tests: People have been in my office with their mobile hidden in their pocket or bag. Without knowing if it was on or off I have always reacted when the phone has been on, never when it's off."*[81]

The following year after so publicly, unabashedly speaking of her EHS condition, Ms. Brundtland was released from her position at the WHO, for undisclosed reasons.

It should be clear to the reader by now that anyone can become EHS, it affects people from every demographic imaginable. As Olle further explains:

*"Nothing protects you from this functional impairment, not political stance, not your income, not sex, skin color, age, where you live or what you do for a living. Anyone can be affected. These people suffer radiation damage from gadgets that have been very rapidly introduced without ever having been formally tested for potential toxic environmental exposures or any other types of health hazards."*[82]

*Chapter Five*

# Birth of EHS

SIR JAGDIS CHUNDER BOSE, the aforementioned 19th century Cambridge-educated physicist and botanist, while extensively studying the effects of radiofrequency radiation on plants, found that "moderate" radio energy (working with 30 MHz) retarded plant growth.

*"Bose's conclusions, drawn in 1927, were striking and prophetic. 'The perceptive range of the plant,' he wrote, 'is inconceivably greater than ours; it not only perceives, but also responds to the different rays of the vast aetherial spectrum. Perhaps it is as well that our senses are limited in their range. For life would otherwise be intolerable under the constant irritation of these ceaseless waves of space-signaling to which brick walls are quite transparent. Hermetically-sealed metal chambers would then have afforded us the only protection.'"*[83]

"Hermetically-sealed metal chambers," aka *faraday cages*, describes precisely in what so many electro-sensitive people have had to resort to living, and in what I had to personally resort to sleeping for nearly a year while still living in the extremely EMF-polluted environment of my home city (Washington, DC), in order to gain some reprieve from what one could easily describe as a life "intolerable under the constant irritation," of now ubiquitous, much higher levels of artificial radiofrequency radiation exposure than the amounts Bose used in his experiments.

Effects of electricity on human biology were well known as far back as the 18th century during early experiments with manmade EMFs. And observations of some people being more affected by electric shock than others were written about by a physicist from Languedoc, one Pierre Bertholon, in 1780 with the following:

*"There are persons on whom artificial electricity made the greatest impression; a small shock, a simple spark, even the electric bath, feeble as it is, produced profound and lasting effects. I found others in whom strong electrical operations seemed not to cause any sensation at all.... Between these two extremes are many nuances that correspond to the diverse individuals of the human species."*[84]

18th century explorer, early experimenter with electricity, and author of *Kosmos,* a five-volume scientific work, Alexander Von Humbolt, amazed by the incredible variance in human reactivity to EMFs, in 1797 wrote:

*"It is observed that susceptibility to electrical irritation, and electrical conductivity, differ as much from one individual to another, as the phenomena of living matter differ from those of dead material."*[85]

And author of *The Invisible Rainbow,* Arthur Firstenberg, also wrote:

*"The term 'electrical sensitivity,' in use again today, reveals a truth but conceals a reality. The truth is that not everyone feels or conducts electricity to the same degree. In fact if most people were aware of how vast the spectrum of sensitivity really is, they would have reason to be as astonished as Humboldt was, and as I still am. But the hidden reality is that however great the apparent differences between us, electricity is still part and parcel of our selves, as necessary to life as air and water. It is as absurd to imagine that electricity doesn't affect someone because he or she is not aware of it, as to pretend that blood doesn't circulate in our veins when we are not thirsty.*

*"Today, people who are electrically sensitive complain about power lines, computers, and cell phones. The amount of electrical energy being deposited into our bodies incidentally from all this technology is far greater than the amount that was deposited deliberately by the machines available to electricians during the eighteenth and early nineteenth centuries."*[86]

## *On being a Highly Sensitive Person, aka, Canary*

In 1885, German physician Rudolf Arndt made the connection between the rise in "neurasthenia" (characterized by, but not limited to; headaches, tinnitus, chronic fatigue, digestive complaints, heart palpitations, nerve pains, and hypersensitivity to stimulus of all kinds) and EMF exposures. He described his patients who could not tolerate electricity, writing, "Even the weakest galvanic current, so weak that it scarcely deflected the needle of a galvanometer, and was not perceived in the slightest by other people, bothered them in the extreme."[87]

Arndt proposed that "a large obstacle to the proper study of neurasthenia was that people who were less sensitive to electricity did not take its effects at all seriously: instead, they placed them in the realm of superstition, 'lumped together with clairvoyance, mind-reading and mediumship.'"[88]

Not long after Arndt made this connection in 1894, the infamous Sigmund Freud renamed neurasthenia as "anxiety neurosis," thus (conveniently for JP Morgan, heavily invested in the continuation of the proliferation of his beloved electrical grid—that made judicious use of copper—a big mineral investment of Morgan's at the time) reclassifying it as a mental illness.

Convenient for industry, yet not so for humanity, both for the misunderstood, and mislabeled minority afflicted with being able to directly feel the effects of toxic exposures such as those that emanate from manmade EMFs, but also for the majority **hypo**-sensitive population, who, while unable to directly feel EMFs, are not less harmed by them.

In our capitalist profit-over-people society, labeling those who may serve as a warning to the public as to the harmful effects of commercial, consumer products, as mentally ill, is crucial to a polluting corporation's bottom line and survival.

It is possible, that being social animals, a percentage (estimated at ten to twenty percent) of humans genetically inherited traits of heightened sensitivity in order to protect the rest of the group, specifically by serving as "canaries in coalmines." This would have been an advantageous trait in a time in which human groups (be it tribes or larger societies) were self-governed rather than dictated-to by mega, multi-trillion-dollar corporations and a handful of powerful elite. In such groups, other humans would have the ability to recognize and revere the minority of sensitive people for their unique trait, rather than ignore or ridicule them.

Dr. Elaine N. Aron, author of *The Highly Sensitive Person* also speculates similarly in stating, "My hunch is that it [the heightened sensitivity trait] survives in a certain percentage of all higher animals because it is useful to have at least a few around who are always watching for subtle signs. Fifteen to twenty percent seems about the right proportion to have always on the alert for danger, new foods, the needs of the young and sick and the habits of other animals."[89] She adds that perhaps a higher percentage of **lesser**-sensitive people is also required, since by being less alert to dangers they would not consider consequences of every single action; something highly useful in exploration of new territory or fighting for resources. However, the higher numbers of lesser-sensitives (or "hypo"-sensitives) are also helpful in serving as a natural population check, simply because more of those (with diminished sensory perceptions) in living less cautiously are likely to be killed off more quickly!

In heralding the benefits of being highly sensitive, Dr. Aron asks her readers if their own sensitivity had ever served to save their own lives or the lives of others, making the case that this is reason enough for having such a trait. In her case she and her own family would "be dead" if she had not awakened "at the first flicker of fire-light in the ceiling of an old wooden house" in which they were living at the time.[90]

In my own case I can recall a few instances in which my heightened sensitivities were potentially lifesaving. In 2000, while living in Istanbul, Turkey, my then-husband and I used large tanks of portable propane for our cooking gas. This meant that we had to open the valve at the tank in addition to the valves on the stove anytime we cooked. It also meant we had to close off both valves at the stove and on the tank after cooking. One evening one of us had forgotten to switch off the valve at the propane tank. It was not until we were in bed drifting off to sleep, a couple of doors down from the kitchen, that I smelled the gas leak. My partner could not smell anything at all. But he did not disbelieve me when I alerted him and thus quickly went to check the propane tank only to discover the valve left fully open. He was more than grateful at that moment for my trait of heightened sensitivity, since he felt very strongly that had I not detected the leak we could have unwittingly breathed in gas all night and possibly died in our sleep from the poisoning.

Years later while living in my top-floor apartment of a large house in Washington, DC, I consistently smelled gas each time the central heating kicked in. When I complained of the smell to my landlady she simply came over, walked in, sniffed the air and said, "Well I don't smell anything," and left.[viii]

It took subsequent efforts on my part and the admission of one of the downstairs neighbors to having also smelled gas to motivate our landlords to actually investigate the potential source of the smell. As it happened, months after I detected the first leaks, it turned out that a pipe on the furnace had been fitted upside down, thus causing the gas leak. The consequence of which was, of course, that myself, and three others living in the house, had been exposed to harmful levels of gas for several weeks' time before the leak was repaired.

On another occasion my heightened senses saved my entire three-storied, seven-unit apartment building from being set ablaze by my laptop computer's spontaneously combusting batteries. One night in 2006, at 2:00 am I had been fast asleep in my studio apartment when I was awakened by a faint, electric, crackling sound emitting from my laptop situated several feet away. (Not surprisingly, as a sensitive person, I had also always been a "light sleeper," ever alert to the slightest foreign sound in my environment.) I next noticed a thin layer of smoke rising from the computer. The laptop had been fully powered-down but was still plugged into an electric outlet. I later learned that the overcharging of the batteries created something called "thermal runaway," leading to spontaneous combustion–a particular defect of this brand of batteries, which as a result had been later recalled.

I did not know any of that at the time. All I knew was that the smoke and noise coming from my computer was increasing to the point of black smoke quickly filling my entire apartment. I acted fast, dressing, calling 911, and grabbing the computer to remove from the building–all at once. While carrying the laptop down the three flights of stairs out of the

---

[viii] Someone who is poor sighted or blind would hardly dismiss more detailed visual descriptions of their environment from better-sighted individuals, nor would someone deaf or near deaf question if another with better hearing detected noises not audible to them. But for some unknown reason often when it comes to comparing capabilities related to olfactory senses, those with a heightened sense of smell are fully dismissed by those with diminished smelling capacity and the same could certainly be said to be true for those unable to feel the effects of EMFs compared with those who can.

building, it exploded in my hands. Fortunately, my reflexes were quick enough to launch the computer several feet out in front of me so that it landed on the carpeted stairway, immediately setting it on fire. At this point I yelled for help and a neighbor came to my aid equipped with a wet, wool blanket so that, together we put out the fire, long before the fire crew arrived.

Elaine Aron lists other very positive traits associated with highly sensitive people that come from having a greater awareness of their surroundings; such as, being "better at spotting errors and avoiding making errors, being highly conscientious, able to concentrate deeply, and especially good at tasks requiring vigilance, accuracy, speed and the detection of minor differences."[91] And also being very imaginative, we tend to be visionaries, highly intuitive artists, musicians, inventors, philosophers, writers and the like, and as such, most certainly contributing to society in very important, impactful ways.

Aron argues that the only thing "wrong" with such individuals, which can at times make their lives unbearable, is their inability to process stimulus as others do—in effect, they very easily become over stimulated by much smaller amounts of stimulus needed for the same effect in others to feel overwhelmed. So that, as a highly sensitive person living in a world where desensitization and overstimulation is encouraged rather than avoided, this trait can be extremely hard to handle, particularly when the sources of stimulus are ever-present and cannot be turned off, such as in the case of ubiquitous EMFs and chemicals. You then find "you are a minority whose rights to have less stimulation are generally ignored."[92]

## Porphyria

*Porphyria*, currently not a household name, something of which the reader has likely never heard before now, used to be well understood as an inherited trait affecting at least ten percent of the human population.

Porphyrins are light-sensitive molecules; bound to magnesium in plants, they create chlorophyll and are responsible for photosynthesis, and in animals they are bound to iron, creating heme as found in hemoglobin, enabling the delivery of oxygen throughout all tissues of the body. The condition of porphyria occurs when these porphyrins accumulate in excess (because of an actual lack of one or more types of porphyrins), interfering with the oxygen-transport system and ability to create energy from food/fuel sources.

The affinity porphyrins have for heavy metals is part of the reason the excess accumulation of them, in those with porphyria, can cause disease, specifically in the organs where they happen to accumulate. With increased heavy-metal exposures in our environments a person with porphyria will, for example, feel[ix] the effects of those exposures more immediately than the next person.

It is interesting to note a correlation here between actual canary birds and porphyrics; that acute porphyria results in decreased oxygen levels, so too with the canary, because of its need for higher levels of oxygen than humans, the bird will react to much lower levels of poisonous air than humans—the very trait which makes the canary useful as a warning "device" in coalmines.

Since the enzymes of the heme pathway are the most sensitive elements of the body to environmental toxins, in an unpolluted world porphyrin-enzyme deficiencies would not cause disease at all, but would simply make the porphyric more sensitive to the environment, without severe consequence resulting in the manifestation of illness.

*"In an unpolluted world this* [porphyria] *was a survival advantage, allowing the possessors of this trait to easily avoid places and things that might do them harm. But in a world in which toxic chemicals are inescapable, the porphyrin pathway is to some degree always stressed, and only those with high enough enzyme levels tolerate the pollution well. Sensitivity has become a curse."*[93]

Now considered extremely rare by the medical professionals, this trait had for centuries been known to cause extreme sensitivity and reactivity to environmental toxins in its carriers, establishing a potential genetic basis for what is now labeled MCS (Multiple Chemical Sensitivity) and EHS.

---

[ix] But it is important to note that there is evidence while some may be born porphyric, the condition can be induced through porphyrin-damaging exposures to toxins, so that 'canaries' may both be born and created later in life.

Not mentioning porphyria directly, it is still interesting to note that recent research from the past decade published in Pub Med (an online medical research database affiliated with the National Institutes of Health in the United States) in 2015, indicates a strong link between MCS and EHS as two aspects of a "unique pathological disorder." Quoting directly from the published abstract:

*"Much of the controversy over the causes of electro-hypersensitivity (EHS) and multiple chemical sensitivity (MCS) lies in the absence of both recognized clinical criteria and objective biomarkers for widely accepted diagnosis."*

The abstract goes on to list specific relevant data from the research and concludes with: "Our data strongly suggest that EHS and MCS can be objectively characterized and routinely diagnosed by commercially available simple tests. Both disorders appear to involve inflammation-related hyper-histaminemia, oxidative stress, autoimmune response, capsulothalamic hypoperfusion and BBB [Blood Brain Barrier] opening, and a deficit in melatonin metabolic availability; suggesting a risk of chronic neurodegenerative disease. Finally, the common co-occurrence of EHS and MCS strongly suggests a common pathological mechanism." (From the article "Reliable disease biomarkers characterizing and identifying electro-hypersensitivity and multiple chemical sensitivity as two etiopathogenic aspects of a unique pathological disorder.") [94]

However, in the 1940s, post WWII, with the petrochemical industry having just lost its major milk cow (profits from chemical weapons manufacturing for the war machine), they turned their sights to targeting domestic households and farms for an endless list of proposed new chemically based consumer products, for the most part containing the same ingredients as had been used in weapons of war, but now in smaller, less immediately lethal doses. These new chemical applications included use in laundry detergents, soaps, shampoos, food preservatives, crop pesticides and herbicides, lawn care, wall-to-wall carpeting, home insulation, perfumes and many more, soon to be saturating and endlessly poisoning our planet. The last thing the petrochemical industry needed was a population aware that at least ten percent of which would be immediately adversely harmed and able to feel the effects of the poisons—in effect, being poisoned more readily and obviously than the rest of the population.

As a result of the financial impact the petrochemical industry had on our economies, impacting the world-over with new jobs in manufacturing, advertising, and selling of vast numbers of new and "improved," "must-have" products; the medical profession, now backed into a corner, threatened to lose its positions and credibility, decided to collectively pretend that porphyria was now suddenly a rare trait and that no harm would come to the population at large from the influx of the new products and resultant million-times increase in exposures to poisonous chemicals.

In a society that demands conformity, sensitivity is viewed as a behavioral problem in need of correction. Far from being listened to when I complained of rashes from soaps and irritation to cleaning products and upset stomach from artificial foods, my parents felt the need to discipline, so that instead of my warnings being heeded, I was punished and grew up feeling I was inherently flawed.

This was all due to the fact that I was raised in an average American household where nobody really questioned the "authorities." If a product was available for sale in our local *Safe*way than it too was "safe" and we could trust it. We used all of the major "trusted" brands of cleaners and foods.

As a result of being misunderstood in this way, I learned to hide a lot of my sensitivity so as to not draw so much negative attention to myself. I also noticed how teachers got annoyed at my asking so many questions. Thus, with my parents irritated at my complaints, and my sensitivity singling me out for bullying at school, eventually I learned how to better fit in and suppress my feelings at times to avoid ridicule and social abandonment.

I remember how it felt to breathe in the cleaning powder *Comet* or any of the other standard commercial household cleaners my mom was still using decades later—how choking and nauseating it felt. I would ask my mother on my visits to at least ventilate the rooms in which she was working. I tried to tell her the products were dangerous and that she should at least be wearing a mask and gloves while working with them and heed the product warning labels related to inhalation and skin contact. But she would just get annoyed with me, the way she had done when I was a child and turned my nose up at certain foods (foods which actually caused my gut to ache and twist in spasms). Sighing, shaking her head, rolling her eyes, she tuned me out. She did not want to hear it.

I do not blame her now. Her reaction was normal. Nobody wants to hear—the canaries messages are extremely inconvenient and hard to integrate into current belief systems, which encourage trust in authorities of all kinds, especially, including industrial ones (and the scientists, health advisors, regulatory committees, and politicians they pay off), in spite of the number of times these corporate giants have lied to us, buried information of the type that could have saved our lives and prevented serious harm, as has happened in the case of asbestos, lead, DDT, Agent Orange, tobacco smoke... and the list goes on.

And so cognitive dissonance often wins out over reason, and drowns out the canaries' warnings. What use is my biological warning system if no one pays attention?

## *Canary Warnings Unheeded*

We canaries are extremely marginalized, given psychiatric labels, medicated (with poisonous pharmaceuticals, which only exacerbate our symptoms ensuring we are properly given the appearance of "insanity"), forgotten, cast aside, ignored—all to the grave detriment of the society at large.

And if given a voice at all, we are misrepresented, particularly by mainstream-media sources, which always are sure to cast doubt on the validity of the condition, hinting at psychological illness in full defiance of vast amounts of peer-reviewed legitimate science—research often verified even by institutes like our National Institutes of Health, all of which have proven the condition as legitimate. And the countries that legally recognize the condition as a functional impairment, and thus technically duty-bound to protect EHS sufferers under disabilities acts meant to provide them with safe access to public spaces, and more importantly to their own homes, constantly fall short.

In my own case trying to fight forced SMART utility meters onto my home, I contacted the Attorney General's Office in Phoenix and presented them with a qualified doctor's diagnosis of functional impairment due to EHS. While they were initially very hopeful of helping me, after contacting the fair housing department (HUD, which is under federal mandate to aid people like me under the American Disabilities Act), they had to break the news to me that HUD would not support my case and so their hands were tied.

I also know a woman, Liz Barris, in LA county who fought in the court systems for years to have her library disable WiFi in order to provide her with equal access to public spaces in her own town. The fight was looking good and would have set an important precedent for similar cases, however at the eleventh hour her lawyer was compromised. Here is the story of her lawsuit against the city of Santa Monica, as told in her own words:

> *I did not announce this publicly, but for the past nearly 3 years, I have been quietly suing the City of Santa Monica for violating ADA (Americans with Disabilities Act) by trying to force me to use WiFi in the library private meeting rooms as opposed to hard-wired Ethernet. It may seem like 'what's the point of doing that?' but if we get even one win on ADA, EHS and wireless radiation, the FLOOD GATES WILL OPEN. There will be an admission of harm from wireless radiation in a court of law and they are going to have to start accommodating people all over the place and other issues with involuntary exposure will have to start being addressed.*
>
> *Originally, I was offered to settle the case by the defendants, but their requirement was I would be placed on a gag order about the lawsuit and the court records would be sealed. I refused this 'offer' and demanded a jury trial in order to air this issue publicly and finally get something on the books regarding EHS. However, this whole time I have been pro se (representing myself) with the help of an attorney who would write the briefs up for me and make court appearances for me when I was unable (due to lack of legal knowledge). This attorney refused to actually take me on as a client, but her ad hoc help was necessary for me to get as far as I did. Unfortunately, I was hit with a huge bill from her in the many thousands of dollars (she usually charged in the hundreds, not thousands) seemingly out of nowhere, with no time to come up with the funds and was unable to pay it... this was actually just a few days before our next hearing. I was then told she would no longer help me. I was pretty sure I was going to win this case, but she told me she started to feel like I was not going to win and changed from wanting to help me to not wanting anything to do with it.*
>
> *I won't bore you with the details, but the bottom line is last week at the hearing, I was forced to settle the case and have it dismissed*

> *in exchange for not being held liable for the city's legal fees, which could well have been in the tens of thousands. So it is with a very heavy heart that I say there will now be no jury trial because of this 'forced' settlement. At the very least there was no 'bad' legal precedent set and I will sue again, but next time will be better armed and with all that I learned this time round and hopefully with a real attorney who will represent me or us.*[95]

A case in point to illustrate canary-misrepresentation in mainstream media; in a 2009 BBC World News program on EHS, titled "Refugees from Radio-waves" the segment is started by planting this seed of doubt as to the legitimacy of the condition (their own emphases in bold); "They **claim** that they get physically ill when too close to mobile phones or WiFi networks but few doctors actually believe."

The segment continues with David Chazzan reporting from the field, on the first French low-radiation refuge for electro-sensitives. In describing the scene, he states: "At a secret location hidden deep in the woods of southern France, there is a camp of forlorn fugitives, living in **strange** metal-shielded caravans, **refugees** from **radio waves**; people who say they can't live with mobile phones, WiFi networks and modern technology most of us can't live without. This is their safe haven."

The reporter then, after speaking with some of the EHS refugees, resorts to this obvious cheap shot, ridiculing those he just interviewed by saying, "They may be fleeing technology, but they don't hesitate to use it to work out how strong the electromagnetic fields are they are exposed to."

No one suffering with EHS has ever said they are *against* or need to flee *all* technology—simply that they need to flee or have a problem with **harmful** technology. Would this same commentator question someone's use of a Geiger counter to check for radiation in an area after a nuclear reactor meltdown? It is not any different, but this is just one example of how often this condition is intentionally misrepresented by the "authority" media outlets.

And if that is not bad enough, this so-called "news reporter" presents completely false industry-biased information by adding, "A growing number of people say they are physically tormented by wireless technology but **tests** have shown that they can't tell the difference when they are exposed to a **real** electric field or a fake one." The reporter does not then cite any actual evidence of any sort to back up this statement. He does not mention any tests[x] in particular at all.

Instead, he continues his clearly biased viewpoint with, "Most doctors tell them their symptoms are psychosomatic or that they are depressed, but some say they are real and the problem can't be ignored." This is his only acknowledgment that there is another side to the story and then he interviews a French Professor (Dominique Belpomme, Paris Descartes University) briefly who says, "There are up to five to ten percent of people who could suffer from this electromagnetic-field caused syndrome, so it is a very enormous problem and I am convinced that the European authorities will take measures."

Mr. Chazzan closes with, "But the authorities say there is still no proof even though these people say that escape is the only solution"—a statement likely intended to leave the viewer with the message that authorities do not back the claims made by these crazy-looking people dressed in silver-shielded clothing, living in metal-shielded mobile homes. And, of course, for most viewers, the BBC is the final authority on all matters and the fact that this news commentator presented the piece from an extremely skeptical viewpoint, and the fact that not one

---

[x] However, he could have been alluding to a study conducted at Essex University headed by Elaine Fox, at the Electromagnetics and Health Laboratory in the Department of Psychology, that was funded by MTHR (Mobile Telecommunications Health and Research), a company funded directly by the wireless industry itself, which found no "statistically significant" results for the existence of electro-sensitivity in the population. In the experiment a mini cell tower was placed inside a shielded room and turned on and off in several minute intervals. Persons claiming electro-sensitivity were placed in the room with the antenna and asked to tell when the antenna was on or off. But they did not vet the volunteers. Some were legitimately EHS, some merely curious wishing to see if they themselves were also EHS. And only 44 participants agreed to the study when they needed over 130 for the study to be considered valid. Also, reactions to EMR usually persist long after an initial exposure, much like any other allergy—as such, this type of test is designed to fail. Not surprising given the conflict of interest with who funded the study. Even so, 90 percent of the participants' responses *correctly* reported when the antenna was on or off. But 95% was required to say that EHS is a real phenomenon. One of the participants, Brian Stein, who entered hoping to gain legitimacy for his EHS, developed colon cancer as a result of the exposures to which he was submitted during the test. Directly after the test, he experienced intestinal bleeding and seven years later, the time his doctor told him it usually takes to develop colon cancer, he got the diagnosis. (*Resonance* 2013 documentary film)

shred of evidence was presented to backup any of his claims that EHS is not real or that there is proof EMR cannot cause harm, will be doubtlessly overlooked in the face of this evocative presentation, fully designed to cast doubt upon this condition. What I presented here was the news segment in its entirety and likely it will have been the first and only time the viewer/general public, at least in Great Britain, is presented with this subject, and by way of material so typical of the drivel churned out by media outlets like the BBC.

On the "other side of the pond" in the United States, the public's familiarity with EHS is for the most part, the result of watching a fictional drama television series by the name of *Better Call Saul.*

The series' main character Jimmy McGill, aka "Saul" has to contend with his eccentric brother Chuck's clearly "imaginary" ailment of electro-hypersensitivity. Obviously tormented, Chuck has had to quit his law practice, and has become a recluse confined to his own home. The highly inaccurate portrayal at best represents a very severe case of EHS, someone who cannot tolerate any form of manmade EMF; he is pictured in his home working by gas-lighted lanterns, typing on a mechanical typewriter and able to sense if his brother Saul is even carrying a battery-powered wristwatch on his person when entering the room (this last point being extremely farfetched in reality).

Because of the severity of Chuck's condition, Jimmy/Saul has to leave his cellphone and his wristwatch in the outdoor curbside mailbox when coming for visits. The portrayal of Chuck, while accurate in some small respects, is a complete misrepresentation of the condition—an intense exaggeration unable to even follow its own logic.

A case in point is that Chuck seems to be okay in his own home in spite of the fact that there is no apparent special protective feature in his home (no shielding to block outside radiation), so that it makes no sense that he would immediately feel the pain of EMR exposure as soon as he walks out of his front door, causing him to rapidly retreat as if a vampire struck by sunlight or a severe agoraphobic. And it is clear that Jimmy and everyone else in the show believe that Chuck's condition is purely psychological, to the point at which at the end of the series, Jimmy proves the case to his brother by hiring another character to plant a fully charged cellphone battery in Chuck's pocket, which Chuck in fact does not detect—a fact which is revealed to him (and others) during testimony in a courtroom proceeding, confirming Jimmy's suspicions (and at this point that of the viewers) that his brother's EHS was fully delusional. But it is not cellphone *batteries* that are detected by electro-sensitives but cellphones while on and in transmitting/receiving mode only.

This scene is considered one of the most memorable from the series by online reviewers, and as such we can guess that the damage as to not only discrediting the character *Chuck*, but actual electro-sensitives in general, has been far reaching.

And just to add "fuel to the fire," so upset is Chuck at the discovery of the battery and his resultant public shaming, he kicks a gas lantern off of his table to burn down his house, also setting himself alight so that he perishes in the fire. So, we have a dramatic end to fully cement the idea of this real condition as being psychological—one that only a mentally unhinged person like Chuck would claim to have.

The introduction of electro-hypersensitivity into pop culture via this highly inaccurate portrayal of EHS in *Better Call Saul* is typical, standard fare, representative of industry tactics to control information. The wireless industry, suspecting they cannot hide the existence of the growing EHS population forever, figures it best to present their version of what they would like the general public to believe; namely that EHS is a psychological rather than physiological condition, ergo, it is still safe to continue use of their products unabated and without concern for safety, all the while caring not a jot how this depiction affects already marginalized EHS sufferers. And it matters little whether or not the depiction is presented in a fictional way, with any inaccurate representations forgiven, chalked up to "creative license" and used as a ploy to intensify drama. The public mind has been influenced, the damage done, they will not recall where they heard about EHS or how, they will just *know* about it already when they next encounter similar information, and the negative judgments will already have been neatly engrained in their psyches.

As the whole world becomes the mine filling with poisonous, invisible EMR, like the gases in coalmines, threatening to collapse all around us, the canaries' many cries for help continue to fall on industry-influenced ears, while the bodies fall off from their proverbial perches, and pile up ignored and forgotten...

On the following pages the reader will find the stories of the modern-day canaries, it is my hope you will listen and heed their warnings.

PART TWO

# THE CANARIES

*Chapter Six*

# Meet the Canaries

*"ALTHOUGH THESE PEOPLE are sick, their illness is difficult to diagnose, and it's even harder to convince others that it actually exists. Their symptoms include cluster headaches, nausea, chronic fatigue, a burning sensation on the skin, and a metallic taste in the mouth. Sufferers claim the cause is wireless technology. There's no known cure, and the only way to alleviate the symptoms seems to be to distance themselves from electronic devices and the influence of omnipresent wireless networks. Electro-sensitive people insist that WiFi and cellphones are inflicting constant harm on humans, animals, and nature. Their testimonies are not the only evidence that non-ionizing radiation may not be as harmless as we have been led to believe.*

*"Scientific studies that have been conducted on plants, insects, and mice[xi] suggest these electromagnetic waves may be damaging living organisms. Scientists[xii] from around the world have appealed to the UN,*

---

[xi] The most recent decade-long study of cellphone radiation on mice by the National Toxicology Program, concluded in 2018 and funded by the FDA, found "clear evidence of cancer" caused by the radiation. The FDA, who ordered the study, tossed it out as "irrelevant" since it was conducted on animals and used older technology. As we are exposed to a new generation of wireless frequency each decade, any decade-long study into long-term health effects can be conveniently and summarily dismissed by industry. See: https://microwavenews.com/news-center/ntp-nyt

[xii] This is a reference to the International EMF Scientist Appeal. This appeal, first submitted on May 11, 2105, has been signed by 248 scientists from 42 nations and called upon the United Nations to urgently "protect nature and humankind" from these EMF exposures. See www.emfscientist.org

*warning of the negative, long-term effects that electromagnetic fields could be having on animal and plant life. At the moment, electro-sensitive people have no choice but to flee to the woods or distant rural areas that wireless technology hasn't yet reached. This often means leaving their families behind. Such sanctuaries are not easy to find, however, and they are becoming scarcer by the day. Sufferers warn that they are just the first to have detected the problem, which they expect to get worse and affect more and more people. Their message is not to stop progress, but to proceed with caution, making sure new technology is really safe before it is made widely available."*

–*WiFi Refugees*, RT documentary film synopsis, 2017

## AMY

In 2015, after we had settled into our new home in the desert, we were anxious to get out and meet other members of our rural community. We had heard about a weekly, casual social event held at the local community center dubbed "Coffee Hour", at which area residents were invited to gather each Wednesday morning for relaxed conversation and (you guessed it) *coffee*, accompanied by a tempting array of breakfast pastries.

After driving several miles along a bumpy, dusty road, we finally found the hidden community center thanks to the thorough directions of a friendly neighbor. Cascabel is the kind of place that does not wish to be found and potentially have its peace disrupted by an over influx of outsiders, hence the lack of any type of signage along the main road (intentionally left unpaved, serving as a further deterrent to unwanted through-traffic).

Upon entering the quaint, attractive, tin-roofed, brick building, situated next to a towering ancient saguaro cactus, we were greeted by about five or six intimidatingly handsome and strapping, blond-headed, blue-jeaned, wholesome-looking members of the Foreman family. (We would later find out that the entire numbers were not present that day, and that the large family actually totaled nine people, including seven children.)

Their blue eyes and inviting smiles all directly turned their attentions on us—the newcomers. We were immediately struck by (and nearly experienced culture shock from) the fact that none of them were holding or looking at any kind of wireless device. And, as it turned out, everyone we met that day, in addition to this family, were able to maintain eye contact for the duration of entire conversations—something neither Sam nor I had probably experienced in several years back in wireless-technology-addicted Washington, DC.

We were, of course, soon asked by Jesse, the father of this impressive-looking family, what had brought us to this far-off desert place.

Upon hearing our story, his eyes grew ever-wider—the reasons for his surprise not indicative of the usual disbelief with which we were most often met, but only because coincidentally, his wife Amy, struggled with the same condition. And it was due to this condition (in Amy's case both EHS and MCS) that the family had also moved out to the same area not two years before—all the way from their farm in Missouri.

Not long after this surprising encounter, I met Amy for the first time. And when I did, she immediately greeted me with a hearty, warm, welcoming embrace, full of excitement and relief to find a fellow canary way out in this ends-of-the-Earth location.

Amy's story was far more dramatic than mine, since her symptoms had escalated more quickly and severely after reaching threshold chemical and EMF exposures. She had come very near to death's door, but thanks to finding a safe haven to which she could relocate—namely, our pristine valley—she and her family made their exodus in time to literally pull her back from the abyss.

She has since recovered enough to handle small amounts of EMF exposures and regain some semblance of a life carved-out for herself and her family on the modest homesteader's ranch they had purchased and were kept busy working.

Except that now, fast-forward four years later in 2019 (making over five years of peaceful living in our valley for Amy), my chance discovery of our electric utility company's undisclosed plans to "smart"-grid our community threatened to drive Amy and myself, along with a recent small influx of other safe-haven-seeking canaries, from our homes, forcing us, once again, into exile with nowhere left to hide, and facing the very real possibility of homelessness.

Desperate to save her home (half of which the family themselves had invested their own hard labor into building), her farm and her life, Amy has spent part of 2018 and the entirety of this past year (2019) relentlessly dedicated to stopping the SMART-grid from encroaching on our lands. Sadly, so far, her enormous efforts seem to have been in vain and she and her family's futures are uncertain.

The following is a moving synopsis of her plight as told by her eldest daughter, Louisa, taking the form of a final appeal to our utility company, Sulphur Spring Valley Electric Cooperative (SSVEC), to reconsider their SMART-grid plans and allow us to keep our current digital (AMR) power-line-communicating meters. Amy, while still in Missouri, had reacted to their former home's "smart" meters with seizures and so, cannot risk triggering these life-threatening symptoms again.

> *My name is Louisa Foreman. I am the oldest of 7 in a family of 9. I come today from a personal standpoint. I normally like to address things from a factual*[xiii] *standpoint, however I have felt the need to come forward and speak for those who cannot speak for themselves.*
>
> *I do not currently live in Cascabel, Arizona anymore, but my mother, Amy Foreman does. Just over five years ago when my family and I lived on a fourth generational dairy farm in Missouri, she developed a life-threatening condition known as Electromagnetic Sensitivity while living in our home with WiFi, cell service and a live breaker box... This illness often accompanies many autoimmune conditions, and the symptoms are terrifying. There is currently no cure or treatment known to the medical establishment, except to avoid any exposure to radiofrequencies, wireless radiation (or even at times, electricity), at all costs. This illness has often been likened to an intense allergy to peanuts. The only thing someone can do to avoid life-threatening*

---

[xiii] My understanding of Louisa's use of "factual" here is intended to mean "unbiased", rather than to suggest that the information she presents is not factual, as she most assuredly presents all of the facts of her mother's illness due to microwave exposures and the subsequent effect on the family without exaggeration or any elements of fiction.

*symptoms to peanuts is simply avoid them, no matter the measures that have to be taken.*

*In November and December 2013, my younger siblings and I had to watch as our mom endured symptoms of EMS: ever-worsening crushing chest pain, shortness of breath, dizziness, debilitating migraines and introductory grand mal seizures. We watched as she quit her job as music and vocal instructor at the community college; as she dropped out of the symphony as a cello player; as she stepped down from playing the piano at our church.*

*She came out one evening and in between uncontrollable spasms and tears gently told all of us 7 kids (then aged 18 to 8 years old) that this would probably be her last night alive and that she loved all of us, and the one thing she hoped was that we would continue to love the Lord for the rest of our lives. We cried and sobbed and told ourselves it couldn't be true.*

*Then a woman in Tucson who was familiar with this strange set of symptoms called us saying she had a place for us, and after several months of searching, we came upon Cascabel, AZ—a unique place so remote it didn't have cell service, and the internet was still hardwired. We set up an old-style corded phone and did a check of our meter and our local utility [SSVEC's] standard for those meters and found that all the meters in Cascabel were AMR meters and posed no risk to my mom.*

*My dad sold all his dairy cows that he had spent years on in Missouri, milking twice a day, and we committed to a new life of healing in Cascabel for our mom and our hearts were lifted as we embraced this miracle of a new chance at a life for our family.*

*Mom began to reach out to scores of others living with this condition of EMS, who were terrified and filled with fear about these paralyzing symptoms. Hospitals were not an option, since for some, even the presence of electricity set off a flare-up of these symptoms. Some couldn't leave their living situations and eventually experienced worse and worse grand mal seizures until they finally died. Over the years living here, several folks with this condition visited Cascabel and some moved here and were able to live in a tent or a trailer with no electricity (if their condition was severe) and nearly everyone who visited said they had never experienced anything like Cascabel. The complete freedom from cell towers and wireless or radio transmitters instantly gave their bodies a chance to heal.*

*Fast forward to the present. Over the last 200-plus days, since Cascabel's original appeal to the board of directors in October of 2018 in opposition to the proposed smart grid, our family has been living in uncertainty and anxiety about our future and the life ahead for our mother and several others like her. We have been told that Cascabel is on the pilot program for the rollout of the smart-grid with the installation of wireless smart meters that no longer run on a hardwired system. And even if our own family could personally opt out, meters from neighboring properties would still cause damage to our mom's health.*

*We are all too familiar with the horrifying experiences that took place in our family during our mom's last months spent in Missouri. It was a miracle she survived and is here today. We cannot go through all of those months again, wondering if she'll end up being just another deceased individual, victim of the ruthless neurological damage of Electromagnetic Sensitivity. It is too much.*

*Our family and many others in Cascabel have done everything we can to come up with any solutions possible as alternatives to the proposed smart grid. We have offered two options to SSVEC over the last several months: 1) Purchasing enough analog meters for Cascabel and having a meter reader come out twice a year to read the meters. All expenses would be paid for by our family, if needed, even if it meant selling property on our multigenerational farm in Missouri to pay for this option. Our mom's life is worth it. 2) Working with Landis and Gyr to install wired PLX meters, of which we have been assured by representatives at Landis and Gyr that this is a simple and viable option.*

*Currently, there is no more dialogue permitted between SSVEC and Cascabel. Those suffering from EMS have no direction about the future of their life. Our family wonders how much time it could be before our mom slips back into life-threatening seizures as a result of an introduction of wireless frequencies.*

*We at Cascabel have long stood behind SSVEC as a local cooperative, and we have supported their stance against smart meters in the past. We know SSVEC is owned by those they serve and we treat that position with SSVEC with honor and respect.*

*Please consider Cascabel's alternative options and please offer an answer to us, so we can know how to prepare for the future of our lives and many others.*

In addition to having submitted reams of irrefutable evidence proving harm[xiv] caused by SMART meters to the utility company, Amy added the following sample list of EMF Refugees whom have contacted her for help in finding safe havens, to the growing pile of documents ignored by SSVEC. These snippets of stories compiled by Amy and included below with her permission, speak to the enormity of the problem so many disenfranchised EHS/EMS[xv] sufferers find themselves facing.

***EMS Patients Searching for Safe Land/Home in Cochise County***

*These people have all reached out in person, on the phone, and over email regarding land in the Cascabel area. Over a hundred more have written me, but do not know how to leave their current part of the country, including many who have ended their own lives. I am the only person on this list who lives in a safe house on property I own. Sixty-plus EMS sufferers and their family members are listed here.*

---

[xiv] Including house fires caused by overheating, spontaneously combusting, microwave-transmitting meters, to the point of earning these "smart meters" the moniker "consumed meters", maybe not so "smart" after all. And tellingly, the damage, and even loss of life, caused by these fires is not covered by insurance or utility companies. In fact, the largest insurance underwriter, Lloyds of London, will not cover any kind of harm, biological or otherwise, caused by any radiofrequency-emitting piece of equipment.

[xv] During our fight against the SMART meter program, we discovered that EMS (Electromagnetic Sensitivity) is more readily accepted as a legal term over EHS and thus we employed it for any legal documents and evidence submitted to SSVEC.

**Jennifer P.** *(with husband and 3 children, homeless; living in minivan): "If by some miracle you hear of a place near there that might be good for us please let me know . . . We did drive by mile marker 2 but that was right next to a dairy farm, so not good for me. I keep hoping that the next place I see will be the 'one' but it never happens. I'm so desperate and I can't tolerate anything."*

**Heidi F.** *(homeless, staying on another person's property): "I cannot thank you enough for talking with me, and trying to help, including calling Julie W., in order to try and find a way to put the trailer on her property [on Ocotillo Road north of Benson]."*

**Jen P.:** *(homeless; selling all possessions to buy an RV she wants to park in Cochise County area). "My plan is to park the RV in So. AZ . . .not in the city of course. . . . the parking I have located does not have APS or TEP for power, so there is no opting out of smart meters. The parking where there is opt outs: has cell towers, or aps substation EMFs. . I would like to speak more with you about what you are doing or wanting to set up. There is large demand in AZ."*

**Sarah (with husband and two children**, *living in a tent): "We still do not have a home and are still outside. It's been 10 months. I am much better than before, but cannot find a place outside (except for one pristine location, far away from everything-which won't work for the kids)."*

**Kay S**. *(for her sister, who is severely ill and must leave her home): "Many thanks for all the information. I have sent it to my sister, and hopefully she will make progress getting to Arizona."*

**Julia W. on behalf of Sarah C.,** *who cannot use the Internet (homeless; currently living in someone else's basement: shaking, headaches, vomiting from neighbors WiFi.)*

**Leslie M. for her friend's family:** *"Hi there, I have a dear friend... she and both her late teenage children are extremely ill, sores in their mouths, difficulty breathing. They . . . have nowhere to go. . . She needs housing– I think one of your units . . . for her and her children would be perfect - she'd just need land?"*

**Doug and Katie M.**–*purchased land south of Benson for severely EMS wife, now threatened with SSVEC smart meters.*

**Janet P.**–*homeless; living in toy hauler/cargo trailer in a remote area, now threatened by SSVEC smart grid.*

**Anthony S.**–*homeless; living in small metal trailer in a remote area, now threatened by SSVEC smart grid.*

**Mark D.**–*homeless, no car, no phone, I met him in person by chance, and he explained his illness and desperation. (I do not know whether he is still living or not.)*

**Katie** *H.–homeless, in truck, parking along roadways--constant headaches and vomiting.*

**Jeff L. and wife**–*homeless, living in car, no long-term safe place to park after over a year of searching southern Arizona.*

**Craig M. and wife**–*will be forced to move from his Portal, AZ home because of new cell tower. Took him 8 years to build his safe home there, after years of homelessness, living in his car. Hoping he can move to Cascabel before the tower goes up.*

**Angus from MCSF** *(for* **two** *of the members of his group--***many more** *spoke with me at a teleconference event he hosted for me to give out information about our area and safe housing solutions.)*

**Annie** *H: --"I have a home but needs a lot of fixing to be able to live in. . . . This disease keeps you away from friends and do not enable you to go places. Very lonely life. I am glad someone address this issue of shelters. We could have a village with only MCS and E.I."*

**Rachel S. and son** *(homeless; living in trailer, just had to flee the remote area where she was parked because of RF from neighbor's new WiFi. Can no longer be in Sierra Vista area because of two-way metering)*

**Rachel** *H.: possible eviction from current rental, searching for land with no cell towers or smart grids in Cochise County--this search has been ongoing for over 5 years.*

**Shannon R. and partner:** *Renting house in Cascabel, now threatened by SSVEC RF smart-grid installation. "We could maybe camp on BLM land for a bit. I just need a place to use my computer, and be able to access wired Internet and phone for work. Of course, we need water too, but we can fill up some jugs here and there. I don't know."*

**Jesse F. and wife** *(that's me)* **and 7 children:** *Living in safe home in Cascabel: now threatened by SSVEC RF smart grid installation.*

**Stephen S and Lan L,** *homeless and living in compact car*

**Eric L.**: *cannot live in his home; living outdoors in a remote area, sleeping on metal cot on the ground.*

**Dianna S:** *searching for safe land in Cascabel area for three years.*

**Maria August:** *Failed to find a safe place. Tried to find a place in Cascabel, but nothing was available. Took her own life March 12, 2019. " anyone asks, you can say I ended my own life. But it would be more accurate to say I died from Electromagnetic Field (EMF) poisoning. I am not ashamed of my actions. They were based on compassion for my own suffering . . ."*

**Susan:** *homeless; living in RV trailer in remote areas.*

**Lorraine:** *homeless; living in Airstream trailer in remote areas.*

**William R.**: *homeless; searching for safe land near Cascabel after having to leave Mescal tent-camping location because of cell tower EMF.*

**Helen and John C. for son and daughter-in-law and their family:** *"The location is appealing to me if there is any possibility that she could improve, and (most important to me) meet other people who are creative and dealing with this mysterious and depressing condition. I am reaching out to you . . What do you think? I feel horrible asking, and hope to not offend you or be too forward but am hoping you can share how you connected with [other] folks suffering from EI, TILT, EMS, MCS."*

**Linda L:** *searching for safe land near Cascabel for years.*

**Charlotte A:** *searching for safe land near Cascabel: " [Need to ask] Tim for a plat, just to be thorough. It is disappointing. Did you say someone else has land for sale . . .?"*

**Chris C:** *homeless: searching for safe land in Cascabel.*

**Kristin C. and husband and two daughters:** *homeless: living in a motel room and searching for land: "Sadly, back in May of 2012, while on an electrical job, he was severely injured and almost died. . . . We arrived on May 15th of this year and are currently in a hotel that is making us sick. . . . I'm shaking from EMFs because the hotel fills up on the weekends – no doubt this place is TOXIC. There is no way we can continue living here. But, I always get tired thinking about another move. And where to this time? We've gone from the suburbs to sojourning and are sooo ready to settle down somewhere. I don't know how [people] can deal with this illness without God. Sadly, I think that's why there are so many suicides in regard to it."*

* * * * *

## CYNTHIA

As Amy's list so glaringly highlights, the plight of EMF refugees/canaries too often results in homelessness. In the autumn of 2015, we met one such "bird in flight"— Cynthia, a 50s-something school teacher from northern California, came out to our rental ranch to see if she might be able to find a suitable home in our valley, or, at the least, a warmer stopping-over place for the winter, as the campgrounds in California had just begun to close down for the season.

Cynthia's onset of EHS the year before (in 2014), ended with seeing her living out of her car, like a nomad, traveling from place to place, camping out in forests and state parks, in her quest to find symptoms' relief. After a year of living thusly, she has found she prefers sleeping outside and has gotten so she cannot sleep indoors anywhere comfortably even if in a low-EMF environment. Her hope, when we met her, was to continue with this lifestyle until she recovered her health enough to move back to her town and resume teaching. Four years later, after our meeting, to the best of my knowledge this has still not happened.

Cynthia's symptoms had started with reactions to her schools' powerful WiFi network, with a WiFi router for her classroom placed directly overhead in the center of the room, under which she so often stood for long periods of time while using the overhead projector. When she began to feel poorly, she knew something was up, and even told her kids, pointing to the WiFi router on the ceiling, saying something to the effect of, "we are being irradiated, this is not good... they shouldn't have that in here." In her own words she describes the onset of her EHS symptoms to me and my partner:

"I was coming home from school, and I was just exhausted, couldn't hold my head up, I know (as a teacher) you are tired at the end of the year, but not in October and not like that."

Suspecting a possible reaction to the WiFi in her classroom, she then had her natural-medicine doctor test her for EMF sensitivity and received confirmation that she was, in fact, reacting to EMFs. Her doctor subsequently wrote a letter to her school asking that the WiFi router be removed from her classroom but for many months no action was taken and it was during this time when Cynthia's symptoms started to intensify and begin to have a cascading effect...

"Then I started noticing this burning. I'd get this burning," continues Cynthia, pointing to the middle of her torso, "and then like someone was pouring cold ice water across my back..." (pointing to behind her shoulders), "quickly, so my body temperature was like this and this, my body temp was way off, my thermostat was way off.

"It's not a hot flash, [it] doesn't feel like a hot flash, it's not like I was sweaty, like working out hard and sweating. It's like my cells are being changed. Sometimes, a woman who has either gone through radiation therapy or has a friend who has, will say that, [and this] is what my friend says, it feels like they were burning her."

Normally Cynthia experienced these symptoms only at school due to the WiFi exposures there, but then she took two weeks off in late January 2015 to visit her mother who was ill and in hospital, and she started to notice the same burning and fluctuating hot-and-cold sensations while at the hospital and also on a Saturday, whereas before it had only ever happened during her work week.

"So, I was like, why am I getting them [the symptoms] in the hospital? Then I looked up and went around the hospital corridors and there they are, every 15 feet there's a router, and I look up and there are antennas on the hospitals...

"So, I started adding it up and then I get home from spending two weeks with my mom and walk into my apartment, which had been safe up to that point, and I start feeling it in my apartment, and was like, 'it's everywhere!' Then I really knew I had to do something about this."

What she did about it was to call a recommended "building biologist," someone who specializes in EMF testing and EMF mitigation in homes and other buildings. Below, she describes her experience of having her home analyzed by the building biologist:

"The first thing she does when she walks in, is explains a bunch of things to me—the RFs, and the dirty electricity. She gives me a little education and then she brings out her really good meter. She turns it on and it just goes 'nnnneeee!"' (Here Cynthia demonstrates with a high-pitched screeching noise which sends our two cats, who had been curiously sniffing at her feet, high-tailing it out of the room.)

"I looked at her, and I ask," Cynthia continues with her narrative, gesticulating with wide-eyes and an open mouth, "'what's going on?' And she [the building biologist] turns to me and says, 'You probably should not be in here, you shouldn't be living here.' I think I started crying, I was like, 'Are you serious!?' And she says 'I've never had such high readings in the residential area.'

"The culprit was radiofrequencies and it was from a 900-foot-away cellphone tower on a Pacific School of Religions' rooftop. That was the closest one, but she presented me with the antenna-search information and showed over 500 others in my area. And you know on the radiation meters, the lowest reading is .0005 and .0006, so my apt was at 6.0 and 7.0 with 38,000, 26,000 and 18,000 in the main living space I was in. And so, I left and that was the last time. I never saw my apartment again."

Soon, while at school, Cynthia began to sense if her teenaged students had left their phones turned on after entering her classroom, even though she had specifically requested that they be turned off and had even placed a large sign at the classroom door to that effect. By way of demonstration, she air-draws parameters about 4x4 feet to illustrate the size of her sign and explains with:

"I was asking them [the school, to get rid of WiFi] and explaining to my kids every period. I made this big neon-green sign which read, 'Please turn your cell phones all the way off!'

"I can feel them walking past me, just like you, I can feel them. I can feel the shiver and the heat, mostly it was the shiver, cold, cold shiver at that point. I'd walk around the classroom asking 'can you turn your cell phone off?!' And I'd tell the kids at the beginning of class... I had good classes, honors classes, and they were sensitive kids, they were really respectful. Half would turn them off the first day, but a lot of them didn't have them off—they were supposed to, but didn't."

We both relate to experiencing the shock seen in others when they realize we actually can tell if they have a phone turned on. Generally, people disbelieve our ability to sense the transmitting phones, so they think no harm can be done if they leave a phone on vibrate instead of powered-down—it just seems too incredible to them that anyone could feel a phone powered-on. And when they discover someone can actually feel their cellphones' EMFs, they view the ability as being on par with something paranormal like clairvoyance or some other kind of sixth sense. And we never know if we are then revered or feared for our unusual powers.

Cynthia's school never did honor her request to remove the WiFi from her classroom to enable her to teach pain-free. Relating more about her experience she continues her story:

"I've been a late adapter [of technology]. I've questioned it. I felt like it's the relationship and how the teacher presents the curriculum and it's the design, creativity and essential questions and it's the interactions [that matter]. The tools can be helpful, but they started pushing it, I mean *pushing* it [tech] in the 80's. I was resistant. I couldn't even find out how to turn the damn computer on most of the time. Then I'd get frustrated and I just didn't want to deal with it. I don't want to deal with this, you know. I want to teach...

"I read somewhere that our bodies react to EMF differently. It depends on the frequency and the duration, etc. And there is a cascade of symptoms, which was my experience too—first it was this, then it was this. And then I made some mistakes, with certain things in my apartment. And then I was overexposed to magnetic fields, and then after that I was dizzy around magnetic fields—and before that, I hadn't had that at all.

"It's really hard. Sometimes I'm just standing there like, 'what next, what next?' I tell people I run from society, but that's not really correct. I run from people's devices. Everyone has a device. So, it seems like I'm running from devices. And I wonder how to explain this to people. It's like I have to calculate every move I make in society—every time I walk into a store; every time I go to get food; every time I stop at a gas station; every time I need to make a call. I don't know what's going on, but the payphones aren't working anywhere. I can't find a payphone that's functional."

We tell her we have found a few functional payphones in Tucson (at least they had dial tones, we did not try making a call), but had not seen one for a very long time back in Maryland/Washington DC, or anywhere on the east coast or on our cross-country drive during our move. It does indeed seem that they are disappearing along with landlines in general. Telling more about her frustrations with trying to find a payphone to use, Cynthia continues:

"All the way from northern Cali, I've been stopping and calling and either they don't put the call through or I call the operator and they say sorry, try another pay phone. Or the pay phone is broken—it's torn off, or the cord is broken in half or something. And I think, 'well who is doing this?' I almost feel like Verizon or Comcast is sending people out to destroy the payphones."

Cynthia tells us more about her exodus from her home, job and former life in Berkley:

"My Dr. friend, the naturopath, said the first priority is you have to get in healthy places, and that's what I've been doing, that's what I've been looking for...

"When this started, early on, I started to try and camp [close to home]. I was camping in the Redwoods in this beautiful park, near Point Reyes and Salinas, at the base of a beautiful redwood, and I laid down on my mattress and I started to feel 'buzzy' and I started to get hot, and I am like, 'what? I'm in nature!' And the neighbors, I saw them on their devices, and I'd even talked to the park ranger before, and asked where was the nearest cell tower. She pointed to the nearest peak and I thought, 'well that's far enough away.' But there were probably antennas scattered all over to bring the signal down here [at the park] because we could get it [the signal]. So, it wasn't safe and I had assumed because it was nature and the Redwoods that is was safe and that was a shock—I was reeling from that. I don't get it, because the stuff is traveling farther and farther too."

When asked about how she supports herself on the road, without a job or home, and also to tell us about the reactions of her family and friends to her ordeal, she responds:

"I had paid for disability insurance, as a teacher, not social security, but private, so right now that's being a bit supportive... But my family, and friends, it's been mixed... My family, my mom has been very genuinely concerned and supportive and worried because I'm out sleeping in a tent—my sister, whom I love dearly and my brother too.

"My sister, she's been mixed. She questions me sometimes (and that's been extremely hard), that I am making it up. And I'm like, 'why would I leave behind a life I loved, why would I leave a place I love?' Finally, I felt like I fit in perfectly. Why would I leave my job when I have six more years before retirement? And I've been planning this and I've invested for thirty years in this profession. I just want to say 'What the...?'

"Why would I completely disrupt my life? Absolutely, completely disrupt it in every major way. Why would I eject myself from the life I loved? Why would I do that? It doesn't make sense. I would ask my sister that when she would question me about this.

"My brother, we haven't been in a lot of contact. I wrote them both a long letter about this, documenting it, because I wanted both of them to know, because my mom was so ill and we have to start dealing with her and I want them to know what's going on and I want them to get the same information. My brother didn't respond for months. I got nothing from him. Like I said, we'd not been in as close of contact.

"But I got a letter a month ago from him and it was very sympathetic and he's a fireman, he's a captain in LA County and they've been ramping up the cellphone antennas on the rooftops...

"In northern California they are everywhere [antennas on top of fire stations], but they've been putting up more. And they have someone in their union, who noticed this was bad. And apparently if a man has been working in a fire station with the antennas on it for something like five years and they are trying to get pregnant with their wife, they can't. They have miscarriages, they can't get pregnant. So that's one piece of information I've heard about this. So, the firemen put up a big fight. They have a strong union and powerful lawyers and this one fireman leader, who is like, 'you will not, you cannot do this.' And they fought it and they won.

"So, he [brother] wrote me this letter and he said," (At this point Cynthia chokes up, her voice breaks as tears stream down her face.) "'I completely understand, I know these issues are real.' And he underlined 'real'... that was like, that felt so, you know, *amazing,* for someone you care about, for someone in your family to validate you completely, no question, '*I know these issues are real*.'

"He briefly explained what the fire department went through and he says, 'We won, they are going to take them and remove them.' They have won. And it was like, *complete* support."

I can understand the depth of Cynthia's emotions when speaking about her brother's understanding and support. Complete and total support of this condition is not an easy thing to come by, even from close friends and family members. I have heard many cases of EHS sufferers being totally abandoned by both, after learning about their impairment, either because it is too hard to be supportive or simply because of disbelief.

So often we are met with the pretense of support when others make a show of accommodating and "supporting" us, stating merely that they believe that we *feel* this way, and so will stand behind us by signing a petition or accommodate us by turning off a cellphone or WiFi router in our presence (still a rare occurrence so that we are overly appreciative of these breadcrumbs cast at our feet), while sometimes in the same breath after voicing their support, will also admit to not believing that EMR is capable of causing biological harm.

In one such example, a neighbor of ours in Cascabel went to some lengths to point out that he had turned the WiFi off for our visit (fully unnecessary to explain since I would have noticed if it had been left on) and talked about how he was a Quaker and Quakers, he told us, were taught to "stand aside" for others, even when they disagree with another's position.

And so, I was also meant to be grateful for his allowing me to speak my mind when explaining to his wife (at her request) the dilemma my partner and I were facing in regards to the potential smart-metering of our valley. The neighbor made this (his gracious allowance up to that point) abundantly clear when, after I had spoken on the topic for roughly ten minutes, he stopped me mid-sentence, pounded an angry fist on the table directly across from me, and shouted, "Enough! I have let you go on long enough!"

Our neighbor did not wish to hear anything more about the health effects of electromagnetic radiation because he himself had asked his "brilliant physicist" friend if there was "anything to" the EMF issue and his friend had answered in the negative. The neighbor then let me know he had "stood aside" for me and made it clear that this was the limit to his "support." I was not allowed to speak further on the topic that day.

However, he still respected my right to *feel* as I did and he would continue to "support" me and others in our valley who "felt" this way. Before departing, I was able to briefly insert myself by calmly stating, after the storm of his angry tirade had subsided, that regardless of what he believed, we who were suffering the effects of EMR exposures (and not imagining it) were simply asking for safe places to live and work—we just wanted to be accommodated.

Clearly there is a real difference between lip service to being supportive and actual, practical accommodations being made for those in need. People who participate in self-congratulatory "acceptance" of EHS, feeling ennobled that they have refrained from outwardly labeling us as crazy or trying to have us committed to psychiatric asylums, take part in a different brand of shunning when they merely pretend to show signs of sympathy but do nothing of consequence to actively help our plight.

Nor do they make any changes in their own use of harmful technology and this alone sends a clear enough message to canaries like myself. As long as there are consumers willing to purchase unending amounts of new wireless gadgets, the networks upon which these devices are supported will continue to proliferate unabated so that those wishing to get away from its invisible, torturous hand will be left with no recourse.

This type of covert shunning can be more psychologically damaging than the overt variety, as it leads to a lot of confusion and self-doubt about who is truly on our side. It makes us feel more invisible and our suffering seems in vain if nobody is going to listen, so again we ask, what use is it being a canary?

Cynthia did not find a suitable refuge to provide her with a home in Cascabel when we met her. Over the years, I have sporadically kept tabs on her progress via a friend of hers with whom she checks in regularly as she herself stopped using computers and cellphones completely, relying mainly on still finding landlines for communicating as she continues her search for a permanent refuge. Last we heard, she was dividing her time between camping in parks over the summers and doing a work-trade on an off-grid farm in the mountains outside of Globe, Arizona in winter, where she is able to live in very basic shelters without electricity—while still seeking a longer-term living solution.

* * * * *

## LIZ

As a striking blonde, statuesque woman, with a confident yet genuine and caring presence, Liz Barris, reminds me of an ancient Viking warrior princess or goddess. Perhaps she was one in a past life, but even if not, in this life the battles she has fought as a formidable, determined EMF crusader should earn her the title. I previously mentioned Liz's three-year-long lawsuit against the city of Santa Monica for violating the American Disabilities Act by refusing to allow hardwired Internet access in public spaces, namely in Liz's local library, and instead exposing citizens (including those with EHS) to dangerous EMF radiation in the form of WiFi—but what I did not yet mention is her incredible résumé as EMF activist and lobbyist.

Liz is the founder and director of the following organizations: the People's Initiative, a non-profit educational organization on EMF topics; the American Association For Cell Phone Safety, devoted to implementing brain cancer warning labels for cellphones; Citizens For Radiation Free Community, for lawsuits against utility companies related to EMFs emitted from wireless devices and infrastructure; and Stop Smart Grid, for lobbying efforts against SMART grids.

As far back as 2007, Liz started her intensive and extensive battles in Washington and all over the United States in order to stop SMART meters, 5G, cell towers and to hold the FCC, FDA and telecoms responsible for allowing health-destroying destructive technologies into our public spaces, workplaces and homes, and for failing to provide warning labels on their products, especially concerning the safety of our children and expectant mothers.

To list just a small sampling of her numerous battles and achievements over the years—in 2008, she personally wrote a legislative bill called The Children's Wireless Protection Act,[96] which Liz explains was, "the first of its kind to be brought [forward] in the US calling for brain cancer warning labels on cellphones, requiring hard-wired Ethernet Internet connection in all schools and a 1,500-foot setback[xvi] of all cell towers from all schools." The bill failed in Washington but was picked up by a handful of other states including Maine. In Maine the bill passed by a majority bi-partisan vote after a six-year flight, but the wireless industry stopped the bill from being enacted (and thus passed into law) by flying out their own lobbyists who were able to persuade legislators to reverse their votes when it came time to vote on the enactment of the bill. The legislator, Andrea Boland, who authorized the bill was punished by the wireless industry for daring stand up to it, and so showing itself as the bully it truly is, as Boland was "gerrymandered" out of her district—meaning she was forced out of her seat as a Democrat when her district lines were redrawn from Democrat to Republican—something with which the wireless industry specifically threatened her upon passing the bill.

Liz lobbied in Washington a second time with a citizens' lobbying group she formed as founder and director of the American Association For Cell Phone Safety[97] in 2009, to lobby for The Children's Wireless Protection Act with a new bill containing more focused language, again for brain cancer warning labels on cellphones. Her group then published a paper

---

[xvi] Liz admits that even with the 1,500-foot setback the ability for the telecom companies to simply as she wrote to me, "crank the power up (and they do) and make up for the distance." Compounding the problem currently as an update in 2022 she adds, "Also with 5G we now have transmitters in front of people's homes, especially with the OTARD ruling being passed by the FCC which now allows for neighbors to put them on their homes without any input from the community."

titled "The Legislators Guide to Warning Labels of Cell Phones and the Layman's Guide to Non-Thermal Effects from Wireless Devices and Infrastructure."[98] The paper served to help bring the first legislative bill "regarding warning labels for 'non thermal effects' from wireless radiation emitting devices in the US, to the state of Oregon."

The bill won a majority vote again with the legislators as happened in Maine, but this time after the Chairman and author of the bill was threatened by the wireless industry to be gerrymandered also, as had happened in Maine, the Chairman refused to let the bill go to vote and so it died—thus the bullying tactics used by the Wireless Industry worked—again.

These early defeats did not stop Liz Barris. She went on to file two mass tort lawsuits (2012-2015) against the two largest utilities in California—PG&E and Edison[99]—for biological harm caused by SMART meters and SMART grids. Retaliation against the lawsuits came in the form of a corrupted congressman (who was a lobbyist for PG&E prior to taking office) filing a bill in Congress to *mandate* both SMART meters and the SMART Grid for *all* of the United States. Responding to this proposed bill, Barris and others organized a national citizens' lobbying group (StopSmartGrid.org) for the purpose of blocking this bill. This time "Team Barris" won and the bill was successfully blocked[100], after a two-year concerted effort to do so.

In 2014 by way of a non-profit organization, Liz was able to fundraise for an attorney to help her local community successfully block a major cell-tower installation in the center of Topanga, California.

And in 2015, Liz hired an attorney to block another cell tower planned for the local fire station, one that was part of a federally funded cell-tower rollout, involving 150 towers[101] for LA County. Liz worked with members of the Fire Fighters Union and citizens of LA County to successfully block the 150 towers. This year-and-a-half battle was the one to which Cynthia referred in the previous story, whose own LA-county firefighting brother related to his sister in the letter she received from him after a long absence.

It was the winning of this fight that served not only to block the towers and prevent untold further harm and possible deaths to LA County citizens and firefighters, but unbeknownst to Liz, helped mend a damaged relationship between two siblings and bring comfort to a canary struggling to survive and also be understood by her family.

In addition to lobbying, Liz has organized many protests against SMART grids and in one case gained high-coverage media attention because, without her knowledge, one of the SMART-grid companies in question was owned by ex-Vice President Al Gore—ironically the same ex-VP who pretends to care so much about the environment he felt compelled to make the nauseating propaganda film, *An Inconvenient Truth* (which I feel would be more appropriately titled as *An Inconvenient Turd,* due to its contents being a load of crap).

Liz has also testified at government hearings as expert witness on health effects of EMF exposures and given countless PowerPoint presentations to communities, government agencies and schools on the subject.

Liz became painfully aware of these important issues regarding EMF exposures after she herself was personally affected when she began to get "strange sensations" in her ear in 2007. Over time, her reactivity to EMFs as an EHS canary, increased to a point where she had to quit lobbying, as the wireless radiation exposures in government buildings were too difficult for her to handle.

On December 10, 2017 she detailed her challenges in the following email newsletter:

> *I feel you are owed an explanation as to why I dropped off the face of the Earth so to speak, shortly after our trip to DC and the simultaneous acts of our lawsuit being thrown out along with my doctor telling me she thinks I have cancer, but I refused the chemo and radiation up front so she then refused to biopsy because she said that could spread it faster and if I wasn't going to do chemo, then I'd better not chance a biopsy.*
>
> *My depression was overwhelming, partly because I knew the loss of the lawsuit would mean my slow, downward demise via smart-meter pulse, but also the EHS completely prevents me from being in society and well, there are no words actually to even describe this situation. My bread-and-butter work is here and I am surviving, but in constant pain from the pulses and frequencies, the cell antennas and towers, etc.—depressed because I can't go anywhere or do anything as there is wireless radiation EVERYWHERE, coupled with my doctor's diagnosis*

*makes me sort of feel like 'the walking dead' as they say. So I must say all this changed my direction into focusing on trying to make more money so I could maybe buy some land somewhere and save myself, but that never happened so I am still in the same place.*

*Also, because of my EHS, I have had to come to terms with the fact that I can no longer even lobby anymore. The last time I did that was in Maine and I had to take breaks and go lie down on the grass outside because I got so sick from the wireless inside the buildings, but there were cell towers outside too, so it was really hard to escape the radiation. I've heard people can actually get better from EHS if they find a safe place before the EHS takes them down. I may be at that point where I will have to give up what little income I have and just leave and go to some cave somewhere and dig in for a few years until I am better—if I get better, which, of course, hope springs eternal. I honestly don't know what I'm going to do at this moment, but I just wanted to send you all an email and tell you what was up with me. I love you, you are great—never stop, never quit and we will prevail eventually. I do hope to live to see that day.*

I was first put into contact with Liz in 2015 when I was trying to fight SMART meters in my own town (in Washington, DC) and attempting to stay put and not be forced into exodus. I asked Liz during our first phone conversation why she chose to stay in LA and not "get out of the mine," as it were. Her courageous and uncompromising answer was, "because this is my home." And as such, it was her right to live in her home and town without being physically assaulted by microwave radiation. I really admired (and continue to admire) her fighting attitude, but I myself felt nowhere near strong enough to fight to the degree she was doing, and as I already related, decided I could no longer endure the torment, or bear exposures in government buildings either, as she finally could no longer do in 2017.

But this setback, including a cancer diagnosis, might have slowed Liz down and caused her to change her approach but it has not stopped her. In the same newsletter Liz announced the findings of a groundbreaking scientific study[102] headed by Dr. Gunnar Heuser, of which she has been an active participant and through her nonprofit helped to fund. The study made use of fMRI[xvii] (functional magnetic resonance imaging) scans to measure brain functionality (rather than simply showing objects like tumors in the brain as happens with regular MRI) of EHS subjects vs. non-EHS subjects. It is the first known study of its kind, intended to make visible what is invisible—changes in the brains of those with EHS.

Below is a sample of two fMRIs. The brain shown in the left image is that of Liz's (with EHS) and the one to the right is a non-EHS volunteer. This diagnostic tool, tellingly is used by US government doctors in diagnosing "Havana Syndrome",[xviii] the name given to the symptoms of microwave-sickness experienced by government employees working at various international embassies (including one in Havana, Cuba) from 2016-2019 believed to be due to intentional targeting of the embassy personnel with covert pulsed-microwave weapons.

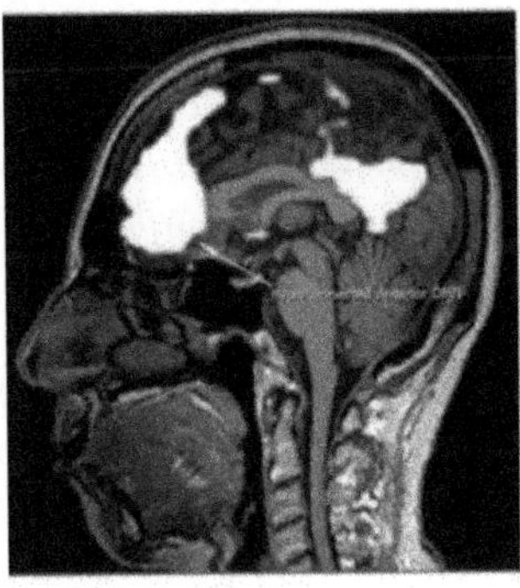

**Figure 5:** Lateral view of case no. 5.

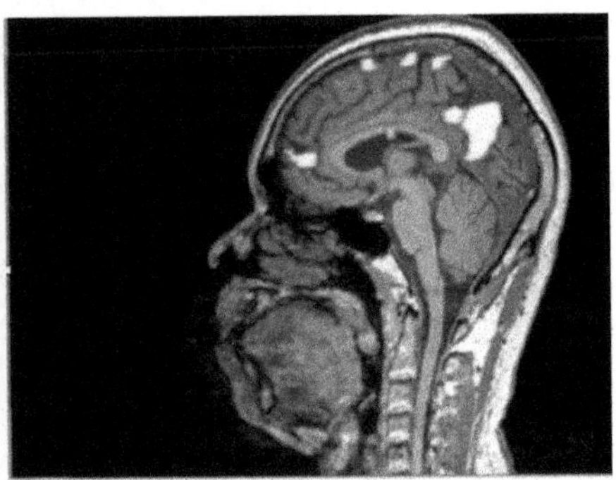

Figure 11. Lateral view of a normal fMRI. Note absence of abnormal white areas seen in our patient group.

---

xvii MRI scans of any type make use of very powerful magnetic fields, and in many cases has worsened or even triggered EHS in patients who have undergone this type of diagnostics (as happened with Dr. Erica Mallery-Blythe). So, while it may prove helpful in demonstrating EHS as a truly biologically based syndrome, I disagree that it should be used as a regular diagnostic tool as the study proposed, as this has the potential to worsen the condition.

xviii So impactful were these attacks that the State Department carried out an investigation in 2019 to find the cause of the symptoms of the Embassy workers that ranged from vertigo, dizziness, headaches, fatigue, hearing loss and memory issues—all symptoms also consistently reported by people with microwave illness or EHS. The National Academies of Sciences, Engineering and Medicine found that "directed, pulsed radiofrequency energy (as) the most plausible mechanism" to explain Havana Syndrome. (San Francisco Chronicle, December 6, 2020, "Radiation Ties to Ailments of US. Envoys," Ana Swanson and Edward Wong.)

All ten EHS study participants who submitted to the scans showed similar abnormalities, "often described as hyper connectivity of the anterior component of the default mode in the medial orbitofrontal area." [103]

This visual evidence of EHS as a biologically based physical, and not psychological, condition (caused by exposures to microwave radiation), Liz hopes will pave the way for winning future lawsuits against the perpetrators of these biologically destructive technologies (and as such, most definitely constitute assault by electronic weaponry), and help canaries everywhere, particularly by ending the debate on the existence of EHS.

* * * * *

## TIM

I met Tim on the EMF Refugee forum sometime in 2018. We spoke several times over the phone upon discovering we had both found refuge in Arizona. We never did meet in person although we made attempts to do so on a few occasions.

When we first met online, Tim was in his late 20s and had been working some years in Silicon Valley as a developer of iPhone apps up until the onset of microwave-sickness symptoms in 2018, during which time he was forced to quit his job and move to Scottsdale, Arizona in order to take refuge at his mother's home there.

For a time, Tim was able to successfully manage his symptoms in her relatively low-EMF home. That is, until January 4, 2019, when he was hit hard by radiation from a newly activated 5G antenna. On that day, a day Tim remembers all too well, he and his mother were making their way around the neighborhood, taking their daily walk together, when on a corner only a few blocks away from his mother's home, Tim experienced intense pressure in his head and horrendous pain in his abdomen. Doubling over in pain, he fell to the ground. This all happened very near to a new metal box positioned on a lamppost.

A few days later, upon passing the same strange lamppost, the nightmare repeated itself and thus Tim's suspicions were aroused that the lamppost could be serving as a stealthy container for a new 5G-antenna array. And so, he contacted AT&T, the local wireless telephone provider, and had his suspicions confirmed that they had in fact, "turned on" the new 5G network, on January 4th.

At that point, Tim knew it was time to move on—yet again. Last I heard, he was living out of a tent in another, more remote part of Arizona, while struggling to get by and hoping to find a way to make a living that would not expose him to sickening levels of EMFs. Tim has had to sleep in a tent in all seasons—a very daunting task in Arizona's hot and cold extremes, but so far, that is the only way he has found of finding relief.

Tim escaped the suffocating mine of 5G networks, but his 60-year-old mother was not so fortunate. While sympathetic to her son's plight, she did not heed the warning and stayed behind. Within 6 months of the 5G being activated, she (a non-smoker) developed lung cancer and died.

* * * * *

## EVA

I learned of Eva's story after she was put in touch with me through a mutual acquaintance. Her story is particularly heartrending as it perfectly describes how the ceaseless encroachment of manmade radiation forces so many like Eva to live truly as refugees.

Eva has had to constantly struggle in order to survive while living for two decades as a nomad in very harsh conditions on her quest to find safe housing and stability. Her world was totally upended due to microwave sickness—she lost friends, home, career, pets, etc., when she was forced to take flight. She currently lives in Oregon but, as most of us do, in a compromised situation, as it is the best she has come by in our irradiated world. Eva still hopes to find a safe and permanent housing solution and to live out the rest of her life free from the daily torment imposed on her by ubiquitous artificial electromagnetic radiation. Here is a brief account of how EHS or radiation sickness has, as she tells me, "pulled the rug out from under her life," as each day became a "scramble and struggle to survive," told in her own words below:

> *I have been injured and my life destroyed by manmade, radiofrequency radiation. I became sick from the radiation from my cellphone in 2000, after using it for two years, when it went digital. I turned, and kept off my cellphone for an unrelated reason about a year later, and my illness improved about 80%, but worsened again after continued exposure to other sources, since radiation is ubiquitous—from emitting devices, towers, and installations of all kinds.*
>
> *[My] symptoms include face pricking, teeth pain, flashing pains through the head, burning pain on top of the head, spine agitation, inner exhaustion, insomnia, popping in the nose, pain in the palms of the hands and soles of the feet, slowed heart rate, and others.*
>
> *I was a grant writer and designer, and struggled to maintain my job, as I became increasingly intolerant of the computer.*
>
> *I was forced to leave everything I knew and loved behind—my boyfriend, cats, a home and city I loved, my family, career aspirations, hobbies and everything I owned, and go on the run for my life just to find a place to sleep at night and where I would not be electrocuted.*[xix]
>
> *I was 38 years old, and I had to figure out where to go and where to turn, because I could not stay where I was. I made phone calls and did research, and my boyfriend found a place, miraculously, where I could stay and work (on a farm), that did not have cellphone towers. It worked out pretty well, but I could not stay [long term], and needed to continue looking for permanency.*
>
> *It was a nightmare of instability, uncertainty, and destitution—being harassed for sleeping in odd places, and for staying too*

---

[xix] The more correct term would be irradiation, but electrocution aptly describes the sensation of being irradiated for many canaries.

*long in the forest. I met someone with the same condition, who had volunteered with the Northern California tree sitters as a way to escape the city he was forced to leave for the same reason as I. We lived off of his Social Security income, and I contributed to the driving.*

*We were refugees—with nowhere to go. We scoured the country for four years, living and sleeping in a car or a small camper—in canyons, in church parking lots, and on sides of roads.*

*We traveled thousands of miles, going north in the summer and south in the winter, finally finding relative safety in a forest in the Northwest where we met a landowning couple that offered for us to put trailers on their land. I managed, but was not well due to power lines running nearby, and I contracted additional illnesses due to the harsh living conditions.*

*We moved several times with our trailers, but still in the same area until the land on which we were staying was sold.*

*Over twenty years, I lived in a car, a camper, a truck, a tent, and a tiny trailer—without plumbing, facilities, adequate heat, and usually without running water. A few friends remained supportive and non-judgmental during this time, which meant the world to me, so rejected I felt by society.*

*I became so susceptible to radiation at one point that I nearly died after exposure to the microwaves from a friend's short cellphone call as I waited outside--over the following days, my body became so poisoned that I had to lay on the wet ground to discharge the radiation to save my life, as the energy lodged in my cells and tissues. I had classic radiation poisoning—insomnia, nausea and brittle bones. It was two years before I could go into town again.*

*Now, I live in a house, but my residency remains precarious as it is filled with radiation and no place exists where one can escape it.*

*Caption: Clearing with trailer, Eva and farm owner.*

**Housing conditions for the electrically disabled:** above – no heat, phone, water, electricity or plumbing; below – living in a car. Engine catches fire going up a mountain hill in search of safety.

* * * * *

## MARIO

I met Mario on a social-media group for electro-sensitives in 2014 before I found the EMF Refugee Yahoo group. A French Canadian in his late-50s, Mario was living in Montreal at the time, ready to make an escape from the incessant EMF onslaught in Canada. Mario, who realizes now that he was always sensitive to EMFs (and understands that we all are, even if we do not directly notice, as we are all electrical, energy beings), became aware of his reactivity once "hyper-sensitized" only two hours after SMART meters were installed around his neighborhood, at which time he found himself collapsed on the floor in total pain.

The next day he went out to purchase Mylar sheets and placed them over the floor, walls, and everything, as shielding, but then realized he could not go outside, where he could feel the burn from all of the antennas, like a burning in his brain. He knew he would have to move away from Canada at this point. In the meantime, he went through 40 pounds of salt per week for his daily saltwater baths (5 lbs. per bath) in which he often spent hours, in order to gain relief from the EMR onslaught.

Mario spent his career as a documentary filmmaker and as such spent some years in the late nineties working for television programs on a series about scientists and their passion, which translated into their research. In so doing, he made the connection between research funding sources and what scientists were able to research and how industry-backed studies through the funding influenced their work. The TV station could not use Mario's work once he brought in this element of truth to it, so he was forced to leave that particular project.

Also, while working and living with Inuit tribes in Northern Canada for two years Mario uncovered corruption while working with the Inuit school board and so was transferred to Montreal in order to keep him from exposing this corruption in his films.

It was this innate enquiring nature and ability to question the system that prompted Mario to leave the school system at age 15, although this did not stop his continuing education at the university level studying communications and cinema, which led to his work in the documentary-film field.

And it was Mario's questioning attitude and realization of industry-government corruption that prompted his studies into the health and environmental impacts of manmade EMFs. For this reason, well before becoming hyper-sensitized, he had awareness of some of the issues related to EMF harm and personally only owned and used cellphones for very brief periods; specifically, on two occasions, once in 2000 and again in 2001—in both cases to aid him in searching for apartment rentals.

Otherwise, Mario never found a need for owning cellphones and he knew they were harmful. However, he did have a WiFi connection and used it. While reminiscing about this, Mario tells me that he first installed a WiFi modem in his apartment in 2001. He says he did not need it, that nobody needs it, and to this day he is not sure why he used it. It was not until 2013 that he began to do more research into communications technologies when he understood that WiFi was not good. So, at that point, he hardwired the connection to his computer, but never turned the WiFi off inside the modem. He did not do this because he was, as he says, lazy about it, until he experienced serious pain from it. Then he finally turned it off, but this action came too late, by 13 years.

Once hyper-sensitized in 2014 and with the realization that he needed to get out of Canada quickly, Mario's research into safer places led him to setting his sights on Uruguay where he planned to go for respite as well as study into quantum physics and electromagnetism. What really sold him on the country at the time was an impassioned speech he heard by the then-president José Mujica—a moving speech that made it clear the president did not intend to sell out his country to Big Tech and Big Agriculture interests. Mujica had thus far refused "smart" utility meters for example and the country offered a lot of beautiful off-grid organic farmland as well as sparsely populated countryside with many expatriates from Europe, the US and Canada, already making claim to portions of the countryside in pursuit of their natural-living homesteading dreams.

So, Mario packed up his life in Montreal and made the leap of faith all the way to that very far-off South American land by the end of 2014. After so doing, Mario initially experienced a degree of relief in his new surroundings and sent me pictures of his new "friends"; a wide variety of exotic colorful flowers, birds, lizards and butterflies. He seemed excited about his new prospects and chances for regaining his health and life in

Uruguay, but admitted to me that sadly the population, like everywhere else, was "buying smartphones like crazy." Still, he trusted he would find a suitable living situation before long.

In November 2014, Mario was staying 20 kilometers from Minas in a *campo* that happened to be in a valley, which fortunately blocked a lot of signals from surrounding areas. But any time he walked too far from his camp he felt himself being microwaved. Nothing compared, he told me, with what he had had to contend back in Montreal, but still not good enough, as he did not feel his nervous system quite able to cope in that area for long. However, he also reported great improvements to his health from what it had been. And his next move was to get a driver's license so that he could travel more easily around the countryside in search of the best possible healing refuge.

We fell out of touch for nearly five years after those first emails exchanged in 2014. In 2019 after reconnecting, Mario filled me in on his life. Uruguay had ultimately not worked out as hoped. He had been disappointed at how much RF pollution there turned out to be, even in more remote countryside areas and he found the population quite complacent about tech industry plans for 5G and further 4G encroachment. In spite of their president's initial promises to the contrary, it appeared that the country was quickly being bought by Big Tech, with no signs of its citizens putting a halt to plans for total irradiation of the country.

Also, Mario found life as a hermit socially challenging, with the few times he did venture into town, when absolutely necessary, he found himself the brunt of townspeople's jokes, wrapped as he was in a metallic EMF-shielded cloak, they teasing referred to him as "the apparition of the Virgin Mary," both because of his unusual attire and due to his rare visits.

His next move, a year later, took him westward to the Andean mountains of Bolivia where the rugged topography and, as he put it, "a certain technological delay," offered much more promise as a refuge for someone suffering with electro-sensitivity to the degree he was then. In Bolivia he was able to find healthy places, in which to live without having to be totally cut off from society. Yet, when he went into town people thought him to be ten years older than his actual age and because of his staggering gait (a result of microwave injury) they supposed him to be constantly drunk as well. But Mario soon found an unexpected solution in his quest to recover his health. He related the rest of his story to me in the following email:

> *Then something interesting happened. In my search for solutions to my wretched condition, I had heard that a traditional and sacred Amazon beverage, ayahuasca, was performing small miracles in health. So, I invited a Bolivian therapist, Karina Luna Pizarro del Castillo, to meet me in my isolated house in the countryside. She has been studying and working with these ancestral medicinal plants for many years and is extremely electro-sensitive too. So, she presided over an ayahuasca ceremony for my benefit that night.*
>
> *The story may seem a little fantastic, but to my great amazement, the next day I was able to see an improvement in my tolerance to high and low radiofrequencies.*
>
> *On the way back to Samaipata the next day, she asks me if I could not let go of my protective cloth. The country people talk, she said. So, I resigned myself to burying the precious metal cloth in my backpack.*
>
> *As I wanted to stay in the village, we went to a super nice hostel, built according to the golden, number one law—without WiFi and quite far from the antennas. On the other hand, my Accoustimeter detector showed an ambient level of energy sufficient to make me suffer. I still rented it for the night.*
>
> *When I woke up, I was fresh and rested. I had slept like a baby. I lived there for two months, until Bolivia activated the 700 MHz on its territory. The radiation level then doubled in my living space. I moved from place to place. I went back to the countryside—as a result, my radiation tolerance, memory, vitality and mood had greatly improved. I enjoyed the restaurant and even the disco. I was said to be ten years younger. I was now my real age, I guess. And I no longer staggered as I walked.*
>
> *I was going to Santa Cruz, which as a microwave, has nothing to envy from US cities. I was there for two or three days, and I had some difficulty concentrating, but it was tolerable. Above all, I no longer felt pain. I estimated the improvement in my condition to be about seventy percent. But this voluntary exposure to noxious*

*waves may not have been a good idea. I ended up seeing a decrease in that tolerance. But, nothing like my pre-treatment situation.*

*A few months ago, workers started pouring cement on a small vacant lot in a residential area. When asked by citizens about the purpose of the construction, they said nothing. But in Samaipata, secrets don't last long.*

*It was a tower for the new generation of antennas. We may live in the country, but people aren't stupid. Also, I was asked to hold a conference for citizens, another to inform elected officials and for a radio program.*

*We then had a 'socialization' meeting with Entel, the mayor of Samaipata and concerned citizens. The Entel representative had underestimated the intelligence and knowledge of citizens by presenting them a WHO* [World Health Organization] *position from the previous decade. He had to forfeit the arguments presented by citizens, like; the 20,000 health studies that contradict this WHO position; the international studies that assert the drop in value of properties adjacent to antennas; the impossibility for Entel to acquire an insurance policy for damage caused by electromagnetic radiation; the arguments of Dr. Jose Luis Cortes, who joined us from Spain via Skype; those of the citizens who refused Entel's anesthesia and who responded with articles of law, among others from the Bolivian Constitution, and finally that of the mayor who solemnly affirmed that according to the law, its role was to respect the concerns of citizens, which it shared in light of the information it now had.*

*The angry Entel representative then asked the secretary of the town hall to register that Entel was completing his progress efforts in Samaipata—a thinly veiled threat. The victory was short-lived. The town hall retreated. Entel abandoned its antenna site in this area, but installed others elsewhere.*

*When they test there, the RF level goes well beyond the capabilities of my Accoustimeter* [EMF meter] *for hours. I have seen swarms of bees fleeing in chaos. In the countryside where I live there is now almost as much signal as in the village. Twice during the tests, I found that small finches had abandoned the nest before term.*

*I take comfort in the fact that ayahuasca has helped me a lot physically and emotionally among other things. These experiences have taught me that this hypersensitivity is rather a gift of life. In fact, when the pernicious effects of technology are mitigated, then there are tremendous opportunities for evolution that might otherwise have been unknown.*

*Today, I conduct an ayahuasca ceremony every month. Participants range from depressive to electro-hypersensitive. In order to understand the therapeutic aspects of ayahuasca, I did some research. Scientific evidence supports my assessment of the improvement of my condition.*

*I will mention here only two studies that seem to me to be the clearest and most revealing. First, Brazilian researchers Vanja Dakic, Renata de Moraes Maciel, Hannah Drummond, Juliana M. Nascimento, Pablo Trindade, and Stevens K. Rehen, published in December 2016, which stipulates that 'harmine,' a substance present in the ayahuasca infusion stimulates the generation of human neural cells. It examined the effects of harmine alkaloid in cell cultures containing human neuron-generating cells from pluripotent stem cells. After 4 days of treatment, the neural cell culture pelvis proliferated and increased by 71.5%.*[104]

*The following article, published in July 2016, bears a title that I think is perfectly suited to the situation we are concerned about. 'The Therapeutic potentials of Ayahuasca: Possible Effects against Various Diseases of Civilization,' Ede Frecska, Petra Bokor, and Michael Winkelman review scientific knowledge about the potential role of ayahuasca in treatment of degenerative diseases caused by oxidative stress, such as Alzheimer's and Parkinson's. It also covers its properties as a myelin agent, antidepressant, adaptogen, anti-tumor, neuro-protector, neuro-regenerator and support for autoimmune or inflammatory diseases.*[105]

*There is a very interesting way to improve the condition of electro-hypersensitives in particular, but also to help the many other victims of the sicknesses of our 'civilization' in general.*

*Today, I am looking for a new place to live because they are installing* [RF emitting] *street lamps even in the little community of*

> *40 families where I live. But I can still go to places at 30 minutes from Samaipata where there is no signal. Until SpaceX covers all of the planet with its satellite weapons, I imagine.*

Following up his email with a phone conversation, Mario tells me he sees his electro-hypersensitivity as a gift, something that has really opened up his heart and eyes. In his case he feels the universe sent him this gift so he would wake up—either wake up or die.

"I was really in pain," he says. "We [with EHS] are really lucky because we feel things." Remembering back to when he could not feel EMFs he calls it "another world."

"We say we didn't feel anything because we didn't know how disturbed we were. I was another man when I got to Uruguay. I didn't know how confused I had been. After getting out of the [higher intensity of] microwaves in Canada, I began to think clearly."

Everything in the universe is electrical, Mario tells me, and he says, "You are electromagnetic, you are energy, not just blood, skin and hair. Your body is your vehicle. You are not your body. You are eternal. We are not material. We are playing a game in a material theater."

Ayahuasca ceremonies proved transformational for Mario and helped him to see his sensitivity as a gift while helping him physically heal as well as opening him up to other spiritual dimensions. It does not completely cure his EHS, he admits, as there will always be microwaves around with which to contend. But he tells me now, he is "good." He can handle trips to his small town where there are a lot of antennas. Santa Cruz (the biggest nearby city) though, is "another story."

But he is learning how we all consciously or unconsciously manifest our realities. Because everything is electrical, he explains, "You have to use the positive and the negative together." Mario says the New Age movement that focuses solely on the positive fails to understand the importance of the negative in our electrical world and so this interpretation of "the law of attraction" fails in its results. To illustrate this point, Mario says to tell someone who is focused only on the positive to go home and disconnect the negative wire to electrical wiring and electronics and see what happens.

"Polarity," explains Mario, "is part of the world, so get used to it and use it to create a better world for others.

"If you look at Tesla's experiments, zero-point energy is the other reality, the one we don't see. But we are connected to this world and we can use this energy. We do use it and we could use it in technology but bankers don't like it because you can't put a meter on it to control usage. But the quantum structure of the universe is in this dimension or plane, 3D or 4D, depending on your interpretation of plane or dimension.

"The big polarity—negative and positive makes all life work in material life, without this there is no life, you need it, it is electrical, not 'negative is bad and positive good', it works together.

"How to connect with the zero-point field? Your body knows how. You just say, 'I'm ordering myself to be at zero point even if I don't know how.' With practice you will realize in a few seconds within your body that you are in the zero point and after that you express with words your desire—what you want to change in your reality including the negative, which might be 'even if I don't know how'. That's the negative ['I don't know how'] but it can be much stronger than that and the outcome [for change] even stronger... you do it [create reality] even though you don't know you do it.

"Microwaves interfere with the process by creating mental confusion, so you can't think clearly. So, you are prevented from really creating your reality. Well, you are creating it but not controlling it. *That* is the attraction law."

Now Mario is so clear-headed and accustomed to manifesting things he wants for his life, it feels as if he does it without thinking and he has come to expect, what others might view as, the unexpected.

An example he gives is seeing two tourists one month before his birthday pass in front of his home with small "sausage-type" dogs and thinking how he would love a dog like that as companion for his other larger dog. But, he thought, those dogs tend to bark and he did not want a yappy dog. So in this example, the positive would be the idea of having a certain type of dog, and the negative the idea that that type of dog barks too much. So rather than blocking the negative thought, he invited it in as part of the process. One month later, on his birthday, a dog identical to the ones he had seen pass by his house showed up at his door. It did not seem to have an owner, and while Mario looked diligently for the owner, it was never found. So now this dog, who, very interestingly does not bark, lives contentedly with him and his other dog.

This is one of many examples Mario now experiences on a regular basis as he enjoys this creative process called *life*, not regretting for a moment that his EHS affliction brought him to a place he sees as a kind of paradise where he eats wonderful tropical fruits and nuts, meets interesting people and has fulfilling life experiences.

* * * * *

## OLGA

Olga Sheean is a potent creative writer, visionary, author, EMF educator, and relationship therapist, who has contended with the adverse health effects of EMF exposure for much of her adult life—effects that culminated in an acoustic neuroma—a benign brain tumor on the auditory nerve—that brought the issue to a "head" (*her* head, as she wryly points out), intensifying her electro-sensitivity and propelling her to explore the deeper drivers of human dysfunction, and ultimately, similar to Mario's experience, Olga has come to understand EHS for the gift that it has been as a catalyst for change, providing valuable insight into how we can create our own realities.

In 2019, I contacted Olga after reading her witty and poignant story, uncompromisingly chronicled in her incredible 2017 book, *EMF off! A call to consciousness in our misguidedly microwaved world.* I felt an immediate kinship reading about her journey, as many details paralleled my own, even ones unrelated to coping with EHS. We had both been *au pairs* in France during our younger years and had worked as photojournalists focusing on wildlife conservation issues. And we both shared a love of the ocean and wild, windswept beaches, and also relished spending quiet hours voraciously reading or playing *Scrabble* with our respective partners. Olga is also originally from Ireland, the country of origin for one branch of my own ancestry and a place I also called home for one magical year from 1996 to 1997.

I also really appreciated her dry, typically Irish sense of humor, and her positive approach to healing, coupled with a generally fearless attitude in facing unique life challenges.

Olga starts off her *EMF off!* book with the following powerful introduction:

*"Although this book focuses on addressing the adverse effects of man-made EMFs in our environment, it is designed to bring* all *of you back 'online,' activating those parts of you that have become disconnected. Many of the solutions provided here will also help with personal crises and problems, since empowering you has a positive impact on all aspects of your life. The book explains that EMFs represent just one doorway to healing, awareness and wholeness, and why we have focused on technological supremacy at the expense of our humanity, autonomy and spiritual connection. It also offers inspiration for living with the kind of conscious connectedness that generates more break-throughs, love, magic and fulfillment than our wireless devices could ever do.*

*"I hope it will ignite in you a fierce determination to become the powerful individual you are neurologically and spiritually designed to be, while reminding you of the infinite possibilities that unfold when we collaborate to create the world we want. We are all in this together."*

She adds an important footnote:

*"This book is not anti-technology; it is pro-humanity, pro-sanity and pro-survival. Many people believe that WiFi is the only way to access the Internet, having never known anything else, or having forgotten that things used to be hard-wired, not so long ago. As a writer and online therapist, I find my computer invaluable, but it is safely hard-wired to our internet modem, providing all the benefits of online access without any of the adverse effects of WiFi microwave radiation."*

Olga's own journey with EHS, as mentioned, led to a frightening and life-changing diagnosis, which two years later saw her "going under the knife" in Vancouver, Canada. She relates her frustrations at having to stay in an EMF-saturated hospital for four grueling days post-op, where she struggled to get any kind of sleep, let alone of the healing and restorative variety, while grappling with hospital staff fully ignorant to the hazardous EMF-toxic environment working contrary to their patient's recovery—an environment similar to the ones that caused her tumor in the first place. As Olga so aptly summed up in her book, "Trying to recover from brain surgery for the removal of a tumour, while being bombarded by the very same microwave radiation now known to cause these tumours, is one of the many ironies that become apparent as I peel away the layers that cover up the deeper truth."[106]

She sees the current addiction to always-on digital connectivity as a symptom of a deeper truth—that of our disconnection from not just our natural world and each other, but from our selves—from our inner wisdom and guidance. For a very long time we have given away our own knowing—our own authority, to an external source. This disconnection first began, as Olga related to Josh Del Sol in the 2019 5G Crisis Summit, "when Christianity was forced upon our ancestors. And at that time, they were forced to convert, were forced to defer to a higher author, an external authority. Over time... we lost connection with our innate wisdom, our own spiritual compass, if you like, our moral compass, our emotional integrity. And we've given away that piece of ourselves to the outside; so, our own hotline to God, our own connection to the universe and intelligence, whatever it was called back then—we were forced to give that up.

"...More and more we were giving away authority to the outside. We were losing the connection with our deeper selves, what our bodies needed, how our bodies functioned, and how nature supported us in that—the healing properties of nature, the spiritual connection that we got through nature.

"...Because of the low self-worth that has been programmed into us, that changed everything. And it put us into a state of reactivity as well, and desperately seeking approval, acceptance and connection. You can't get nourished emotionally or spiritually if we're not connected. This is partly why the digital age is so appealing to people, because they're seeking those connections. That's one of the reasons why it's also so difficult to give it up. We're desperate to fill in those pieces that have been missing for a very long time; part of the reason we are still so needy of the social media and all of that stuff, is that we need comfort. We want something to make up for what's missing inside. So that gives us something. It's immediate gratification, since it's all the time. It's always there. It keeps us with a sense of engagement of belonging, but not really, because it's just gadgets."

Olga gave up her own "gadget" (cellphone) after it caused the acoustic neuroma that required surgery and left her with permanent deafness in her right ear—a high price to have paid for her own lack of awareness as to just how harmful artificial radiation could be. She shares my own sentiments in relating to Josh how not having a phone for her has been "wonderfully liberating." "I haven't had one for many years," she confesses, "and I love it. I don't miss it at all. The simpler life is for me, the healthier and more relaxed I am."

She admits that she knows many electro-sensitive people who are "severely affected by wireless radiation" but still own a cellphone. However, they explain it is "just for emergencies." But, as Olga points out, "you still need a cell tower for those emergencies. It's not going to come on just when you need it." She says we need to be in alignment with our desires. If we want to stop the wireless industry from harming us, we need to take action by not financially supporting the industry's wireless networks by owning cellphones.

And Olga strongly feels that getting back in touch with nature and stillness, free from distractions, will bring us back into true connectivity with ourselves, our planet and one another.

Making important changes in order to regain our inner authority, getting back in touch with our true selves and living in integrity is challenging but, as Olga told Josh, "We have to ask ourselves: 'What's it going to take? What's it going to take for me personally? ...How bad do things have to get before I'm willing to change? Does somebody close to me have to get cancer? Do I have to lose a loved one? Do I myself have to get cancer? What's it going to take for me to see that it's less awful to go down that road of self-responsibility and find a better way, than to continue with the pain or the loss or the ill health that I'm feeling?' Sometimes it just has to get bad enough for us to make that switch, because our programming is so deeply entrenched."

Olga, of course, speaks from experience—she personally knows all too well how bad it had to get for her to make important life changes. Hopeful that her brain surgery would provide relief from debilitating symptoms before the removal of her tumor, she soon found that post-recovery presented its own challenges and learned that much of what she had been experiencing pre-surgery was due not only to the effects of the tumor pressing down on her eighth cranial nerve, but also from EMFs in her home environment. The persistent buzzing she experienced pre-surgery was replaced by an equally persistent and even more irritating and painful, "zapping on the nerve where the tumour used to be."[107] She found herself in a state of chronic dehydration while also having to regain balance and readjust herself to her world without the aid of hearing in her right ear.

Perhaps even more harrowing for Olga than the physical pain of trying to cope after her surgery while still contending with EMFs in her environment, was realizing how reluctant friends were to accommodate her inconvenient-seeming needs. She relates the social experiences in her book after explaining the physical torment EMFs were causing her:

"So, when (former) friends tell me that turning off the WiFi or cell phone is a hassle, or when I sense their reluctance to go out of their way to meet up in some low-EMF location, I reflect on the rather more significant and permanent hassle of having a brain tumour. I understand that they can't relate to what I'm feeling, but do they think I'm just looking for attention? Or am I just not worth the extra effort? Countless other distractions compete for their attention—meetings, Facebook conversations, morale-boosting Tweets, places to go, things to do, e-mails to answer, clients to cultivate, friends with no limitations. Having travelled the world and worked for prestigious international organizations, I'm used to being mobile, independent and self-sufficient, and I feel diminished by how much I must now apologize for: *I'm sorry I don't have the energy. I'm sorry I can't drive to your place. I'm sorry I can't go to that café. I'm sorry for the inconvenience*. In our wireless world of constant connection and instant accessibility, my sensitized system is like a digital dinosaur and I cannot compete."[108]

While having linked some of her symptoms with EMF exposures pre-surgery, Olga had not felt herself particularly "electro-sensitive"—it was post-op and subsequent nerve damage that turned her into a human EMF-meter. She explains what it is like to have been thus transformed:

> *It's another year before I realize that I'm being exposed to high levels of microwave radiation (from WiFi and a cordless phone) from the apartment above us, where our landlady lives. But it's only when we go on holiday to Mexico for some relief from the crushing strain that I realize what's been going on. As we check out the various available hotel rooms, trying to get as far away as possible from the WiFi at reception, I realize that I can physically feel which rooms are receiving a signal. We check this with a cell phone, noting which rooms get the most bars, and the hotel's technical officer joins us with his own device for checking signal strength, confirming what I've been feeling. A big piece of the puzzle suddenly falls into place.*

*After blaming my diet, my lifestyle, my constant worrying and the surgery itself, I realize that my ongoing exposure to microwave radiation has been causing almost all of my symptoms—and creating many new ones. The damaged nerve in my head has become a sensitive antenna that picks up all forms of electromagnetic radiation. I can tell from 10 meters away if someone has a cell phone turned on. I can feel the signal from a WiFi router. And I can tell when a digital electricity smart meter is transmitting one of the many high spikes of microwave radiation that it emits, day and night.*

*The damaged nerve has become my most reliable radiation meter, zapping me whenever I'm exposed to a strong electromagnetic field. Other symptoms—intense cranial compression, nervous agitation, jumpiness, tics, spasms, insomnia, headaches, stabbing pains and anxiety—confirm that I've found the true culprit, now that I understand how it affects the body. The more research I do into electromagnetic radiation, the more I discover about its insidious, invasive nature, the many biological effects of exposure and just how much scientific evidence there is (many thousands of independent scientific reports!) confirming the harm being caused. Finally, things are starting to make sense and it's a huge relief to know that I'm not losing my mind or imagining things.*

*Yet this newfound awareness is like a double-edged sword. While it's reassuring to know what's been making me so ill, the rest of the world seems oblivious. Everyone around me is charging ahead with normal life, addicted to their wireless gadgets, and living their lives online with their smart phones stuck to their bodies like limpets. The more I learn, the more I recognize the dangers and the more isolated I become.*

*The world has no patience for this. Having free WiFi everywhere is so much fun that no one wants to be bothered by something that's not affecting them. Really, Olga, what is your problem?*[109]

True to her fighting-Irish nature, Olga used her writing prowess to relentlessly petition local and international leaders for change. Her papers titled, "No Safe Place," detailing the plight of electro-sensitives, sent to the Mayor of Vancouver, and another addressed to the World Health Organization called, "World Health Organization: Setting the

Standard for a Wireless World of Harm," received a lot of attention. But in spite of her best efforts, including filing a petition to the British Columbia Human Rights Tribunal along with a "vast amount of scientific documentation confirming the harm caused by radiation from WiFi"[110]—a petition that was ultimately dismissed due to conflicts of interest between the tribunal's corporate lawyer and wireless telecom corporations—Olga ended up having to move a number of times in search of her own "safe place," including at least one daunting Transatlantic move.

Hopeful of retaining her home in Vancouver and of educating local officials and neighbors into making necessary changes to accommodate her, like myself (while I was still living in DC), she and her husband had to move away from the city to healthier, more remote environments. In those remote locations she also encountered many of the same challenges that my partner and I have had to address—dealing with the elements, biting-insect infestations, and other unexpected sources of EMF pollution. She has been there and done that, and has the t-shirt to prove it.

Having done her part in helping to raise awareness of a politically corrupt system, Olga, now focuses on promoting change at a deeper level, working with individuals in her counseling practice to dissolve human dysfunction and cultivate greater self-responsibility, wiser choices, creative evolution and personal autonomy. Focused on healing at all levels of humanity, she shares her enduring wit and wisdom through numerous articles and via her blogs, books and interviews[111].

Olga's life and work serve as a shining beacon of hope to us all. She firmly believes we can all consciously create a new reality, and at the outset of 2020 she set up a platform on her website called 2020 Vision[112], where anyone can write their own unique hope and vision for 2020. She still has faith in the endurance of the human spirit and our potential for human collaboration or "co-creation." For her own life, she has her sights set on the perfect place on a beach overlooking a vast sea of blue.

* * * * *

## RUTH and GARY

During my quest to find an EMF refuge somewhere on the planet, I happened upon the truly inspiring nomadic EHS warriors—Ruth and Gary, living out on the remaining wild lands of the American western frontier. It was because of what they generously taught me and my partner Sam about EHS and how to heal from radiation-exposure damage that influenced our own decision to move to the American southwest. After our move to Arizona, we were able to meet up with both Ruth and Gary on many separate occasions during the last four years, and hear their own incredible stories.

I cannot tell Ruth's story without first telling Gary's, because it was Gary who, in Ruth's own words "saved her life." Were it not for their chance encounter in Ouray, Colorado at her place of employment at *Wiesbaden* in 2002, Ruth may have never uncovered the source of her then life-threatening insomnia and may not be with us today—a fact of which Ruth is ever keenly aware and as such, feels forever indebted to her close friend Gary.

But Ruth is not the only person Gary has rescued. Devoting himself over the last three decades to helping other electro-sensitives reclaim their lives and find safe havens, with what he calls his Smart Shelter Network (reclaiming the word "smart" used in the *true* sense of the word, not the industry-corrupted one) and in protecting our last wild lands with his Land Steward program.

Combined, the Smart Shelter Land Steward program covers an impressive 10,000-square-mile service area from Aspen to Pagosa Springs, Colorado, to Moab, Utah. And as of 2016, there were 1,400 entries in Gary's address book—entries he accumulated over a period of twenty-two years in every way imaginable, 800 of which are identifiably EHS—people with whom he is in regular contact. Offering to them free advice and support, Gary subsists on the generous donations of some of the more financially fortunate individuals in his network.

In his 70s and devilishly handsome, standing at a full 6' 4"—Gary Lee Duncan cuts an imposing figure. Something of a revolutionary, philosopher and inventor—uncompromising and not one to mince words, his tough exterior protects a very generous and gentle heart. But it is not uncommon for those with EHS to develop a hard shell as it becomes necessary for survival in the face of the incessant barrage of insults and rejection from naysayers, and in coping with inescapable torturous EMF exposures. With nobody standing up for us, we have to learn to defend ourselves however we can.

With an equally imposing and incredibly diverse résumé, Gary, further proving this condition is not limited to any one demographic, lists his many credentials—a trained physicist; green home builder and cabinet maker; an FCC-licensed radio broadcaster—he was once news director of four public radio stations, operating two public television stations (one of which was on an aircraft carrier in Vietnam); and as a virtuoso folk guitar playing, singer songwriter he also founded and produced the Telluride Bluegrass Academy and Workshop Series. Finally, as a creative writer, he shares his quirky sense of humor and outlook on life while also educating about EMF harm via his *Desde la Terra* stories from the ADSC Standard (electro-sensitization and the Meridians) and *En Terra* (surviving and thriving in the wild as a solution) fictional publication series.

However, once Gary became ill from his toxic environment with the full symptomology of EHS and MCS, he had to leave behind several lucrative careers, the last one having been in the cabinetry industry in Beverly Hills, returning to Colorado in 1993 in order to regain his health.

Gary tells us the story of his own EHS diagnosis[xx]:

"You can look at someone like me with this knowledge base of EMFs and say, 'well someone who knew all this stuff certainly would have recognized the dangers of it.' I was trained in an electromagnetic-field degree at the university of Colorado in Boulder with two of the men who worked on the Manhattan project—invented the atomic bomb. I can calculate an electromagnetic-field intensity from a power line four miles away and that was at age twenty. That didn't work. And then when I was seventeen I got an FCC license to operate a commercial radio station, and it included massive information about 'Don't expose yourself or anyone else to radiofrequency radiation, this stuff will kill you'... Well, THAT didn't work."

---

[xx] To view our filmed interview of this segment, see: https://www.youtube.com/watch?v=zzZPY8J0a8o

Gary shakes his head, chuckles, clearly bemused with his former ignorant self—for what Gary was referring to as "not working" was the warning information with which he had been educated about EMF-exposure hazards.

"So, I ended up in the hands of Dr. Roland Holzworth of Montrose, who did know the toxicology of it, because he was German. Germans knew about electro-sensitization and chemical sensitization back in the 50s. Roland is a German-licensed Naturopathic physician and trained in EMF toxicology and he listened to me whine about my physical complaints for a year. I was just devastatingly ill. I had terrible intestinal problems and dermatitis, bizarre dermatitis effects and chronic fatigue. I had the chemicals out of my life but was still living in this EMF stuff and he would say..."

Gary puts on a comical matter-of-fact, German-accented voice and continues his story, imitating Dr. Holzworth, "'Vell, you chemically sensitive, you electro-sensitive as vell.' I thought, 'you know this guy is a little bit out there, we met at a drum circle, I think this is New Age wacko crap.'"

Gary's initial reaction to this doctor reveals how much of his own bias and built-in denial around the subject clouded his judgment at the time, admitting as much when he next refers to himself in a self-mocking tone saying, "I mean this is a physicist with an FCC license. 'Well, this is wacko crap!'" Shaking his head in disbelief, laughing at the recollection, he continues with:

"One day Roland looks at me and he says, 'You know I'm getting really sick and tired of listening to you whine all the time. I'm telling you what is wrong and you are not listening. So 9:00 o'clock tomorrow morning I'm coming to your house with my EMF meters and I'm going to show you what is making you sick.' He had me backed into a corner, there's nothing I can do. So 9:00 am he shows up. I'm living in this fourteen-foot travel trailer. I had everything you have in a conventional house. I had an electric refrigerator, an air-conditioning system, a desktop computer, and a cordless telephone. Anything you've got in a house—I had it, and all crunched into this little tiny space. I could barely sleep between all this electronic crap," Gary gestures with his hands to illustrate himself squeezed in by it all caving in around him and explains, "and it's all plugged into these octopus outlets. We already knew that I severely react to electrical fields of 2 volts per meter, which is a tiny, tiny, electrical field.

"Roland got through surveying my house with his meter. The average level of the electrical fields in the house was 3-4 thousand volts per meter. And I looked at these two figures, because I am a mathematician and I said 'well this appears to have a large discrepancy!'" Here Gary lets out a loud laugh. "Still not enough [evidence for him]. I had this weird thing that when I laid down in bed I would get this inflamed sinus drainage, it felt like somebody pouring gasoline down my nose and lighting a match, I mean PAIN... This would wake me up, I'd be on my feet in my trailer with tears coming out of my eyes it was so bad.

"When I would sit up in bed a couple of minutes, I'd be fine and when I'd lay back in bed here it would come again," Gary gestures with hands indicating pressure, an attack of his head. "I hadn't told Roland anything about that, he's running through my house with the meter and all of a sudden he stops at the bed and puts the meter in the middle of the bed and the meter goes down to zero and he moves the meter up the headboard where I slept with my pillow and the meter peaked and he moves it back down and it comes back down three or four times and I'm looking at the meter going 'holy crap'...

"You know even physicists can read electromagnetic-field meters and what had happened was there was an aluminum strip coming all the way across the trailer that I had covered with a piece of wood and going to the other side of the trailer it was passing one of my octopus outlets with about eight cords plugged into it, and it was picking up the electrical field and it was transferring it through the aluminum and when I laid down in bed it was right here [pointing to top of his head] it was about one inch from my cerebral cortex.

"And he knew that when you take an electro-sensitive and you expose them to high electrical fields it has terrible reactions in the sinus cavities. That was the source of my problem.

"And he says," again Gary continues in a mock German accent, "'Okay ve fix this.' And I said to myself, sarcastically, '*Right.*' He went outside, drove a brass rod into the ground, ran a copper wire into the house, buried the end, touched it to the aluminum strip and he says 'Okay, lay down.' And I said 'Like hell Roland, this hurts.' And he says 'Oh no, vont hurt now.'

"And he starts talking to me as I lay down and he says 'How do you feel?' and I thought 'Oh my god, this the first time I've been able to lay my head on my pillow and not have a headache and inflamed sinus. And I said, 'well, well it's gone' and he said, and his eyes get really big like this, and he says 'Vatch out!' and he lifted that ground wire up off that aluminum strip and a bolt of pain..."

Gary illustrates with a slap of his hands across each other, "shot through me like somebody had pulled the trigger and I jumped up out of bed, and THAT is when my denial about the effects of electromagnetic fields finished.

"If the electromagnetic reaction in the electro-sensitive was psychosomatic, then what we would have is a whole bunch of people who already knew about electricity and its hazards who then imagined themselves into feeling this, and I am as well trained in what should have prevented that than anyone I know.

"Denial in alcoholism and electrical sensitization is exactly the same, and it is absolutely bullet proof. It took not reason, not information, but PAIN." Gary illustrates through hitting his fist against his head and continues.. "When Roland lifted that thing off of there, a shot of pain went through me and took the denial out the bottom of my feet with it and it has never returned, as an aspect of this disease."

Finally seeing the light and no longer in denial about his condition, Gary's new discovery ultimately resulted in his getting back to nature and abandoning so-called "civilized" life, trading the many "conveniences" (ultimately *trappings*) of the modern era for a much more healthful (albeit more primitive) type of existence.

His diverse background and many acquired skills proved indispensable in helping Gary forge a way for his new life in the wilderness as a WiFi refugee. The building skills in particular proved crucial in his process of retrofitting an inherited fourteen-foot Nomad travel trailer for off-grid living. But Gary was also fortunate to have learned hunting and wilderness survival skills from a young age.

"You know I had a background in boy scouts from the time I was four years old, my mom and dad were taking me out camping and teaching me survival skills and we hunted. We didn't hunt for sport. No, we hunted to eat in the winter. We lived very naturalistic lives. Or I don't think I would have known how to survive out here on the wild, much less be comfortable with it. I'm perfectly at home in the middle of nowhere."

When asked what it was like for him to move out into the wild, unhesitant and enthusiastic Gary eagerly responds, "Oh I'd died and gone to heaven! I didn't have to put up with neighbors. I was back in the environment... I've always been an outside boy. I'm also extremely solitary and very deeply introspective and to have this day after day after day of peace and quiet and freedom from pain and the debilitation, it's basically the psychotropic effects, the depression and that stuff... I had died and gone to heaven!

"And to find out that after all of these years of trying to be responsible, making really good money, and ending up having it completely destroy my health and my life, I found out that if you just let go and learn to be a minimalist, and you listen to the people that really know—you can have a wonderful life, you can do things for other people and you can do the things that you always wanted to do.

"This is an incredibly sick culture. That's what's behind why this is happening. To pursue that—to try and ameliorate your own desires with the capitalist American middle-class expectation, is basically like trying to hit your head with a sledgehammer so that you'll have another bright idea."

### *Gary meets Ruth*

While Gary was camped out near Montrose, Colorado, in 2002, he discovered what he calls a true salvation for EHS sufferers—natural hot springs. Without knowing the exact science behind why the hot springs offer such relief, it is speculated that the absorption of minerals through the skin helps to repair damage caused by EMF exposures (which can disrupt mineral and electrolyte balances in the body) and helps to discharge and remove excess EMF charge along with accumulated toxic heavy metals and chemicals.

So as part of Gary's healing regimen, he regularly frequented the *Wiesbaden* hot springs in Ouray, which boasted an underground vapor cave carved into the side of a mountain. The caves themselves also serve as a natural shield from EMFs making the hot-spring visits for Gary sheer bliss. Ironically while Gary was soaking below ground in what he calls "heaven", Ruth, sandwiched in a corner between computers, cordless telephones, fax machines, copiers and the like, working at the front desk in the reception building aboveground from the caves, was experiencing a kind of EMF "hell" up above.

Curious about Gary and his regular visits, a bright and perky Ruth cheerfully prodded him for the reason. Gary happily acquiesced and related his EHS/MCS story and how the cave springs really helped him to heal. In this way Ruth and Gary struck up a friendship over a two-year period.

Ruth, a Colorado transplant from Oklahoma, an attractive, petite, but tough woman, grew up with an uncomplaining can-do attitude typical to "Okies." As a result, one might say that she had an overdeveloped sense of responsibility, which ultimately got her into trouble with her health when she soldiered on at her job in spite of her growing, alarming health problems.

One day in 2004, out of the blue, she told Gary that she had stopped sleeping. This confession was not all that surprising to Gary who had noticed that, "the color was gone out of her face, the sparkle was gone out of her eyes,'" and that she had, "just looked slower and slower." He says that after a while when you have environmental allergies, you get so you can identify other people suffering with the same. Gesticulating with his face and body, leaning over, and stooping, with tilted head, he mutters, when recalling how had Ruth appeared to him, "Ugh, there used to be a human being in there and now there isn't."

Ruth chimes in to reminisce about that time: "I knew that I was having trouble. Something was not right. I wasn't sleeping well. And at work... I worked at the front desk and was on the computer and phone etc., and I would hold the mouse. My boss had bought a wireless mouse and from day one I could not hold onto that mouse.

"But I didn't have a clue, I knew nothing about EMF, hadn't even heard of it. But I had become friends with Gary and he had started talking about chemical sensitivity, cause he's also highly chemically sensitive too. I didn't know anything about that, so I was slowly learning about chemical sensitivity and one day I quit sleeping–bam! I didn't sleep and couldn't sleep. Gary said, 'I think I may know what is wrong with you.' And he came down to my apartment. I lived two blocks from work so we walked over, with these [EMF] meters and went through and found horrendous magnetic fields in my apartment."

Gary, who by that time had already been living with EHS for a decade, began educating Ruth about the dangers of overexposure to EMFs and tried to convince her that EMF exposures were, in fact, the reason she was not sleeping. At first Ruth was hesitant to believe him. It just all seemed too improbable. She had been going through menopause and her employer and colleges told her it was menopause, but something intuitively told her this was not true, that it was not just menopause. Ruth explains what she decided to do next:

"So, what I started doing was getting in my car in the middle of the winter. This was a Colorado winter and Ouray is in the mountains at 8,000 feet, but I'd drive out into the valley where it was a little warmer and I'd try to sleep. And then I would go to Gary's micro [trailer] on my weekend days off and after about two or three nights I'd start sleeping. And I'd go back to town and I wouldn't sleep. And I'd leave [town] and I'd sleep. And I'd go back to town and I wouldn't. And slowly it finally started sinking in to this thick head that he [Gary] may be onto something. After six months of this I finally realized he's right. But I think, 'what am I going to do—I'm going to quit my job? I've got no support, what am I gonna do? And all this time I just keep getting worse and worse."

Gary tells us he had to move her out "onto the land" away from the EMFs of town, a total of nineteen times before Ruth was prepared to make a drastic change in her life in order to finally put an end to what was now, after six months, life-threatening insomnia. Gary paints us a vivid picture of what those early days of Ruth's visits were like:

"I would bring her out to my places. She would get on my bed. She would have some tea and a little something to eat, then you would just watch Ruth..." Gary demonstrates by leaning over and crashing his head on an imaginary pillow, "...Conk! And she would just sleep for thirty-six hours, and wake up. She would be not fully reset, but fine, and stunned that she could sleep and then say, 'well I've gotta get back to work!' And I'd said, 'Ruth, I wouldn't go back there.' 'Well, I have to.' She's being responsible, so she goes back to Wiesbaden, and the minute she finished a shift at work her sleep function stopped and the exhaustion set in. I would see her in the middle of the week because I was going back to the hot springs and she just looked like death warmed over."

Finally realizing she could not continue living as she was, Ruth, not knowing what was next or how she would manage, took a leap of faith and quit her job, giving the standard two weeks' notice. But even then, she did not know if she could make it until the end of the two weeks. Toward the end of the last week, Gary, intuiting that she may not be okay, called her at work to ask if she thought she could make it to the end of the two weeks, when Ruth confessed that not only did she not think she could make it until the end of the week, she did not think she could make it until the end of that night. Understanding the severity of her condition, Gary told her to hang on—he was coming to get her.

Gary explains, "And I said, Ruth, if you will get a replacement, and go home and pack your bags, I will get in the truck right now. I will meet you at the curb and we will put your bags in the truck. We will come back to my trailer in Moab, but you have to agree if we do that, not to go back. And she said 'I don't have any choice. I don't understand this but I have to do what you say. I can see now that I can't stay here.'"

Ruth chimes in on Gary's account to add, "I don't think I knew how sick I was, it took me two weeks to get up to where I was able to even function. And when I felt well enough, I went back and I moved out of my apartment and I left. I never went back. And I lived in my car and my tent.[xxi] I don't know if I could have done it on my own. I didn't have the energy or resources. I probably would have just stayed there and died."

While arguably saving Ruth's life, being on the land was not all a bed of roses, as Gary relates:

> *It's a difficult thing to understand that you can take someone like Ruth (and this is indeed what she ended up experiencing), and take them out onto the land where they are safe, and have them be completely devastated and exhausted. And at first, they start sleeping—a lot. And the color comes back and then the energy comes back, but they are still devastated and the recovery is going to take a long period of time. And it's almost a physical impossibility not to have to go back to town, not to have to deal with your apartment, not to have to go to public hearings to get your unemployment benefits...*

---

[xxi] Ruth lived in her car and tent along with her three cats for several months until she was able to get a micro trailer of her own.

*Every time she would do that* [go back into town], *the impacts of being around the electromagnetic fields would devastate her. It's an irony that the cleaner the electro-sensitive gets, the more susceptible they are to being re-inundated on these necessary trips back into the contaminated environment. And that went on for Ruth for years. She still lives with it. We all live with it.*

*We get stronger and stronger and better and better the more pristine our environment is, but we become more and more susceptible to reintroduction to those fields. This is typical with every allergenic reaction.*

*I'm a twenty-nine-year clean-and-sober recovering alcoholic and one of the most famous things in recovery circles with alcohol is that the longer we stay clean and sober the more susceptible we are to the devastating toxic effects of ethyl alcohol. There is a myriad of stories where somebody took one drink and was dead the following week after having been sober for thirty years. This is classic. The disease is progressive whether we practice it* [abstinence] *or not and exactly the same thing happens with electro-sensitization.*

*The idea that life in the wild lands is a panacea is not at all accurate. Ruth went through obvious isolation and an extraordinarily deep sense of fear, loneliness, isolation, and deprivation—this thing where you realize there is nowhere you can go anymore, you watch the families around you... She had her entire family disown her over something they couldn't understand, and all of the sudden she has no income, and at the same time she had the knowledge that there was no way that she could hold a job. So what are you going to do? Then, the storms start settling in and then it rains and the mud is so thick you can't leave and even if you wanted to go back to town you can't and the challenges of living in these wild environments just become absolutely overwhelming.*

*I think in retrospect the only way that any of us ever survived is that any kind of alternative was completely stripped away. And we had to confront the devastation and the fear and the technical challenges; the income problems and the health problems; the exhaustion and the depression. It drives all of us to a deep spiritual bottom, where you have to completely abandon every-*

> *thing because everything else has been taken away from you. And for those lucky enough to experience it and survive it, out of that comes renewal and a definition of an entirely different life. I think that is what's gone on with Ruth—it just tears away all of the expectations that you ever had and replaces it with something else that hopefully is going to take us and a lot of other people back out of this. Ruth's case is an excellent example of it.*

When asked if Ruth liked her old life she responds with her infectious laugh, "Yes, but I would never go back. I thought I was happy in Ouray. I hiked. I did photography. I did all of this stuff. I love this lifestyle [she has now], but it's NOT for everybody. And yes, there are things I would change. If I wasn't so sensitized, I could go to different places. Traveling would be much easier. But to go back into the insanity of town—absolutely no way would I ever go back."

She then relates to us more of what it was like when she first retreated into the wilderness for her recovery: "Because I was so sick (I was horribly sick) I spent days in bed. I went to town about once a week. It's all I could handle. I couldn't travel. I quit my photography. I really had no life. I was just existing. Gosh this [camper] truck, it's given me back my life. And somewhere along the way that truck—it's not just a truck, it has become my home, it's my little sanctuary. And I can't imagine giving it up [the lifestyle], even if I could, going back to something else. I could maybe live in an adobe house out in the wilderness—yeah, I could do that, absolutely."

Ruth has, in fact, searched for some years to find a safe property on which she could have a permanent home, but has repeatedly come up short—always there is interference from cell towers, power lines, wind farms and SMART utility grids.

When we meet her in 2015, for the previous five years Ruth had recovered enough to travel while living out of her camper truck, which, being an artist, she has painted with beautiful eye-catching mountain and desert scenes—a sure conversation-starter at rest areas, she is often able to meet strangers and educate them about EMF hazards while talking about living out of her "little sanctuary." Ruth has found a way to travel safely on her own terms. Unable to travel more than one hundred miles per day due to the EMF exposures inherent to any motorized vehicle and increasingly along roads and highways, she travels at her own pace. But she is also limited by her ability to find safe sites along the way, a growing problem with the ever-expanding encroachment of wireless infrastructure for an industry so hungry it is unable to leave even remote wilderness areas untouched and out of its clutches.

One thing Ruth does regret is not being able to visit family back in Oklahoma. Being thusly restricted she was unable to visit her family even when her own mother was terminally ill. Not fully understanding her fragile state of wellbeing, which is so easily upset by modern-day communication grids, her well-meaning son had offered to come out and pick her up and drive her back; her response to which was:

"Where do I stay? How do I even travel with you that far? It's a thousand miles. And it's like, I can't go. I *cannot* go. And I didn't. The thing with trying to go east of here, states like Kansas, Oklahoma, they have no public land. There is no public land—all privately owned. Where can we go? That was the thing when I was going to go see my mom and my kids, it was like, even if I could get there, I have nowhere to go. I can't stay in their house. Where can I go? And you don't want me going down, cause if Ruth goes down, then what are we going to do when I'm a thousand miles from anyplace safe."

When Ruth references "going down," I know exactly what she means—full adrenal collapse, not an easy thing with which to cope or from which to recover, and certainly not in a place hundreds-or-more miles from any non-EMF-polluted white zones.

Initially Ruth's three sons were disbelieving of her condition, saying they did not think there was "anything to it," but acknowledged that she *believed* it and that was as far as they would go on that topic. One son has been particularly unaccepting as Ruth explains:

"I have one son who basically doesn't talk to me. He thinks it's all in my head and I've gone off my rocker. I can't do anything about that. He's an adult. He's forty something and I don't have the energy to deal with that. He's going to have to accept it or not. But the funny thing is he's an electrical engineer. They were taught in school... they are taught to shield all of these delicate components, but [they think] it can't affect a human being. But that's what he was taught and he firmly believes that, so he thinks it's all in my head. He was taught when they made a distinction between ionizing and non-ionizing [in terms of harmful effect] and that information is so old now, it's almost out the window."

Because of the difficulties Ruth faces in traveling east of the Rockies, she invites her family to come and visit her out in Arizona. And finally in 2015, the year we first met with her, she excitedly tells us how, after ten years, she was able to see one of her sons and meet her two grandchildren for the first time.

"I have not been back to Oklahoma in ten years. But my son was going on vacation, and he's got two little boys and I was able to meet them out here. Oh, that was so neat. I loved it. I got to see my son whom I hadn't seen in ten years! I don't think they understand, you know, mom [referring to herself] does not travel. I just cannot."

I relate to her story. Some of my family and friends understand better than others, and some simply believe that I *believe* I have EHS and probe no further. I also have not been back home to Maryland in many years and the full weight of realizing I will probably never be able to return sometimes is a heavy burden to bear.

Gary tells us that Ruth is the poster child for electro-sensitization, and that even the discounting judge with the Social Security disability administration was able to identify her as such, explaining with:

"This is a woman who is completely destroyed. She's cute, attractive, petite, sharp—completely devastated by this stuff and has been completely torn apart by environmental allergies, stripped to the bare threads, left abandoned to sleeping in her car with three cats in the middle of the winter in the Colorado mountains. From that she's pulled herself up by the bootstraps; she's learned to be a carpenter, and built two houses for herself. She's learned to be a mechanic and completely restored a 1976 camper truck, which she now lives in. She's become an expert in electro-sensitization; she has all the meters and knows what they are for, and is now an advocate for this [EMF/EHS education]... Ruth is without a doubt one of the most successful recovery stories that I know of in the electro-sensitive community."

Holding a seemingly radical viewpoint, after his own experiences, Gary's stance on wireless-device usage is simply uncompromising, openly addressing what others dance around from fear of falling completely out of public favor or losing prominent positions, Gary states in no uncertain terms: "There is no such thing as the safe use of an irresponsible device. Therefore, we only associate with those who don't use or own cell phones."

He does not see the public waking up to this issue until they themselves taste what it is like to become sensitized to EMFs:

> *And the unfortunate fact of this is, because all of that* [information on EMF health effects] *has been covered up, and we haven't been taught the way we ought to teach and the science has been repressed—what's going to have to happen now is the global population will all have to become electro-sensitized in order to 'get it,' that this is what's happening, the symptom spectrum of EHS... especially memory failure, sleep disturbance, induced anger and hostility, fibromyalgia and inflammation issues are now populationally generalized and deepening.*
>
> *The documentation about the problems with electricity goes back to its origin and has been industrially and militarily suppressed, which means governmentally suppressed, because democracy is an illusion only in the public's mind in America. The 2012 Bioinitiative published 1,800 medical studies documenting exactly what we are talking about here. So the idea that no information, no evidence, no research about this exists, is ludicrously inaccurate.*
>
> *You can believe what you want to, but I think where this comes down to is that these information wars cannot be won and the result of that is really simple. Look at Ruth. Look at Shannon. Look at me. And if information worked it should have worked on me. It didn't work on me. What worked on me was pain.*
>
> *We are driven by the capitalist ethic, and the capitalist ethic is fueled by keeping you sick, so that the last thing that is going to happen is public education about this.*
>
> *What is unfortunately already in full swing is the global meltdown because now the entire populace is becoming symptomatic to electro-sensitization because of their own wireless proliferation, their overuse of cellphones, which are now monstrously addictive, WiFi, cordless telephones and car boosters.*

***Back to the Land***

Meeting with Ruth and Gary on wild lands, and seeing how simply and integrated they live with their natural environment, we are met with the undeniable stark contrast to modern, manmade, artificial living versus living deeply connected with the natural world. And it does not take long to realize which existence is truly better, healthier—fully giving the impression, being out on the land, that things are as they should be.

And we are confronted with the realization that our birthright, just as the birthright of all living creatures, entails the right to clean and freely available food, water, air and shelter. The idea of corporate and government land ownership and subjugation of land and people seem a rather foreign and destructive, oppressive, artificial system in contrast. And it is hard not to relate to the native peoples of this land when the "civilized" Europeans came and brutally forced them off of the wild lands on which the natives had subsisted for millennia in harmonious balance with the plants and animals of these environments, bringing the very foreign ideas of land ownership and industrial exploitation with them.

Owning the land, or the sky and waters, was as Chief Seattle stated in his speech presented at the outset of this book, a strange idea to the native people, but one to which modern man has sadly acquiesced.

The sky above is now chronically streaked with persistent chemically laced contrails blanketing the land with aluminum nanoparticles and radioactive barium. Clean drinking water is in such short supply that rather than being offered to all living beings freely as in pre-industrial times, it has been commoditized, sterilized and irradiated—now rendered totally devoid of life, it is sold to the people of the world in plastic bottles.

And being sold air in bottles next, is not a far-fetched concept given how destructively toxic air in cities has now become, with many so sick from resultant chronic low oxygen levels and damaged lungs, they often have to resort to just that—breathing artificial air, or rather oxygen, from bottles or canisters.

And of course, for a very long time now, modern man has relied upon commercial stores from which to purchase food for his sustenance—"food" mostly adulterated and altered beyond recognition from what the word used to represent. No longer can man gather his food from a pristine, supportive environment, but must instead "earn a living" through gaining employment within a giant corporation and effectively become a cog in a vast machine, working at jobs that further serve to imprison, subjugate and degrade him along with the rest of humanity and our wild Earth. "The end of living, and the beginning of survival," as Chief Seattle warned, would happen once man was removed from his natural environment and kept prisoner in reservations and cities, is indeed upon us and describes life now for most of the world's people. And the loneliness we feel thusly removed from the natural world and from other animals is all-pervasive in society today.

For Ruth and Gary, the ideology of deep ecology is a real, daily practice not some abstract theory, as it is for most of the rest of us. Having found true freedom and peace, they live the type of lives for which many of us strive but never attain, as we chase the golden carrot of freedom, wealth and leisure our societies promise us, always dangling just out of reach. But Gary and Ruth, living on the land as they do, have reclaimed their birthrights.

After so many years of his life sacrificed as a cog in the industrial, capitalist machine, working for "the man," Gary now starts his days as he sees fit, usually with a song. Pulling out his guitar, he greets the dawn in this way, demonstrating to us firsthand his motto of "I won't do it unless it's fun."

And some days he dances the "Tango Libre" with Ruth, and even I am graced with Gary as dance partner on my encounter with him on the wild lands. *Tango Libre* is what Ruth and Gary have dubbed their free-form liberation dance. Having found a way to celebrate their lives in the wild, they have danced this dance over many terrains in all seasons, even in the dead of winter over snow-covered landscapes and in sub-zero temperatures. When not celebrating thusly, Gary spends much of his time intentionally in deep solitude, truly communing with nature—connecting with the earth.

Ruth and Gary feel strongly that the land has healed and sustained them and in return they must help heal the land. They do this via their land stewardship program. Often in direct cooperation with forest services and state land directors and personal, they are able to work-trade in caring for the land enabling them longer stays on various areas of wild lands. Their stewardship methods involve non-chemical, manual, invasive plant control and removal, collecting garbage left by tourists and preventing and reporting misuse of lands wherever possible.

I was privileged to witness firsthand the latter responsibility carried out in action one day when the peace of our afternoon walk in the wilderness was abruptly disturbed by the horrendous noise of an ATV quad illegally driving off-road (deliberately far off from other designated ATV paths created specifically for their use), and in the process utterly trampling and destroying the fragile desert plant life surrounding us. Upon seeing the ATV driver leaping over delicate dunes, speedily approaching us, Gary stubbornly and bravely planted his towering frame in the driver's path, shouting and holding his hand out in an authoritative gesture, bringing this particular trespasser to a halt. With the engine finally quieted, Gary's booming voice brought home to this encroaching stranger the severity and seriousness of his sins, and made it abundantly clear, in no uncertain terms, that he was to leave at once and never repeat such an offense.

I could relate to Gary's instinctive, aggressively protective behavior. Once, while still living in Washington DC, during a walk in Rock Creek state park, my one and only natural reprieve from the insanity of city living, I came upon a group of teens. All clad in matching t-shirts, they had clearly been using one of the many picnic areas for an afternoon summer youth event, likely from some local church or other youth group. To my utter horror I saw that they had made a game of taking turns at throwing, from a pre-determined distance of several feet away, a hatchet into an ancient oak tree. Without pausing to consider consequences I ran over to the group, yelling at the top of my lungs to stop what they were doing at once! Justifiably startled, they immediately obeyed, submitting to my authoritative tone. I continued to publicly chastise them for their cruelty in harming this innocent tree, explaining that what they were doing in fun, besides being most definitely illegal, could very likely kill this possibly hundred-plus-years-old tree.

It is very difficult in these instances for those of us who not only understand life's interconnectedness but profoundly feel it and who truly appreciate our complete reliance upon the health of our natural surroundings, to relate to the completely careless, thoughtless and often sadistic behavior of the disconnected, desensitized masses.

Having been able to safely retreat to wilderness for Ruth in her "Ruth mobile" and Gary in his "chariot" for well over a decade, suddenly starting in 2009, these safe sites have begun to rapidly diminish after a dramatic change to the environment was noticed by Ruth and Gary on or around June 1st of that year. After June 1, 2009, many electrosensitives who had found refuge in the wilderness areas of Western Colorado suddenly began to experience splitting headaches, asthmatic conditions, fatigue, memory loss and sleep impairment. At this time Gary, Ruth and others were forced to flee into even more remote areas in order to find relief.

Gary says about the sudden change, "Although definitive and accurate information regarding what changed has never been completely forthcoming, it seems to coincide exactly with the national and regional conversion from analog to digital transmissions for commercial broadcast; radio and TV, and possibly for mobile communications, and deployed notably and which we traced, without exception, to the transit of these signals via power-line broadband/plug-and-play services over the regional AC [alternating current] power grid—turning all power lines into disastrously dangerous and destructive RF [radiofrequency] transmissions. This is in direct violation of FCC prohibitions requiring licensure and regulation of all such transmissions, which has not only never been enforced, but with very few exceptions, not even mentioned, and certainly not by a digitally lobotomized media now treating their devices as food source."

And in 2016, the year of our visit to meet Gary for the first time, safe radio-quiet sites used by Gary, Ruth and many others, are again under threat due to the proliferation of wireless infrastructures, now reaching into remote wilderness areas—thanks to the assumption by government and industry leaders (themselves fully disconnected from nature and wellbeing, not at all fit to make decisions that directly impact our natural world or public health) that American visitors to national and state parks and public lands could no longer handle any time away from their wireless devices, not even on wilderness retreats. One may ask; what is the point of camping, "communing with nature" and leaving your home at all if you cannot unplug, get free of your addictions and leave these things behind on such excursions**?**

Compounding this problem is a new threat to these sacred spaces—the sudden trend for retirees to sell their homes, trading in for RVs and living in these RVs "on the land" alongside people like Ruth and Gary, who depend upon safe wilderness sites for their survival. These retirees who are choosing this lifestyle often due to having had their pensions stolen from them by corporate robber barons, are then no longer able to afford a mortgage, and forced into early retirement, or also disabled in some way, suffering health problems that make them unable to continue to work, doubtless from their exposures to wireless technologies. Of course, they bring their wireless tech/microwave radiation with them in the form of cellphone boosters attached to their roaming homes and all of the wireless gadgets contained within them. Bringing this radiation with them as they do, works to effectively introduce the very toxic exposures Ruth and Gary and many others like them go into the wilderness areas in order to flee—so that there no longer is a place to which to escape, and this affects not only human refugees but wildlife as well.

As Gary sums up: "We thrive and have for two decades by finding and caring for the cellphone-radiation-free corridors remnant in the American wild. And the bees have done the same thing. Take it from us, like us, they're alive, happy and well, just not anywhere you can get cell reception."

And these newcomers certainly do not understand the concept of land stewardship. They do not commune with their surroundings. They do not appreciate where they are, but instead become part of the problem of encroachment, by merely viewing their beautiful surroundings as something to quickly capture through their smartphone lenses and post immediately to social media to claim their bragging rights. In effect, they may as well not be there at all, since they are hardly present. Why not just stay in RV parks and look at photos of wilderness online instead? It would be essentially the same experience for them and this way they could spare those sacred places their destructive and disrespectful presence and keep wilderness areas wild.

Ruth and Gary, having survived things most of us can hardly imagine, and just like soldiers who have gone into battle side-by-side, share a deep connection with one another as well as with the lands they inhabit. Forever bonded by their experiences, while no longer living together, they remain closest of friends. And so, while facing an unknowable future, they still meet at least semi-annually and dance the Tango Libre.

*Chapter Seven*

# More Canary Warnings

## ***Canaries on the Silver Screen***

GREEN BANK, WEST VIRGINIA, home to the National Radio Quiet Zone, being unique in the world as a radio-quiet zone situated within an inhabited area (other quiet zones are located in hostile, uninhabitable environments), has become a focal point for WiFi refugees seeking asylum. As such, there have been a number of EHS sufferers (some from overseas), to relocate to the area. And as it has probably become the single place in the world with the highest concentration of electro-sensitive canaries, it has also become a focal point for documentaries about electro-hypersensitivity.

Green Bank was also a place Sam and I considered for our move, but as I explained earlier in this book, it did not turn out to be the WiFi refuge we personally were seeking, while also being cost prohibitive, as real-estate prices, due to the scarcity of available properties, are quite high, and rentals few to none. While Green Bank has reportedly helped a number of canaries, the amount of space available in truly radio-quiet areas near the observatory, or even in the town itself, is certainly not enough to house the millions of refugees worldwide seeking asylum. As such, its prominence in media as *the* solution to this problem is very misleading, as well as disappointing, to the many who are crying out to have their very real refugee statuses taken seriously.

Many more radio-quiet zones need to be established, especially for the sole purpose of making a refuge for the WiFi refugees. As it is, while Green Bank has a (limited) radio-quiet range, the available housing in the area is both expensive and mostly outside of that radius while also presenting problems of non-radiofrequency EMF exposures in the lower-range frequencies that manifest in power lines near homes and dirty electricity on home wiring. And while other refugees certainly welcome others seeking the same, the majority of the town's population is not necessarily so welcoming or interested in the influx of outsiders in need of what little refuge the town and surrounding areas offer. Also, the radio-quiet status in the area only exists due to the presence of the Green Bank Observatory, something totally dependent upon government funding—funding that can always be taken away. If that happens the radio-quiet status would be taken away with it, making way for the influx of wireless-grid development afterward.

One of the documentaries made about electro-sensitives in recent years was done by RT (Russia Today) documentary filmmakers, who also visited Green Bank, West Virginia to interview canaries for their 2017 film, *WiFi Refugees*.[xxii] Even though the filmmakers focused the film primarily on Green Bank, and did not address the lack of available space there or in the world at large for the 2-10 percent of populations in need of refuge, the presentation of canaries was, at least, sympathetic and accurate—unlike Hollywood fictional depictions, like the example of *Chuck* in *Better Call Saul* or the BBC mini-doc portrayal previously mentioned.

The RT crew, in addition to interviewing electro-sensitives, also met with Michael Holstein, the business manager for the Green Bank Observatory for an interview for *WiFi Refugees,* who explains in the film that the observatory is used to study radio waves from space (hence the need for the radio-quiet zone status to minimize interference from artificial radio waves)—radio waves so weak that, he says, "the energy given off by a single snowflake hitting the ground is much more powerful than the radio wave that the astronomer is trying to receive." This is an important point and could serve as a good rebuttal for anyone claiming that our use of artificial radio waves for telecommunications is not a big deal since we are constantly "bombarded" by radio waves from space. But, those waves are infinitesimal in comparison to any frequencies introduced into our environment by man.

---

[xxii] By pure coincidence my partner Sam and I, when first attempting our own EHS documentary that never fully materialized, in 2015 came up with the nearly identical title, *WiFi Refugee*—a title I decided to use for my book instead, *before* I learned about the film.

The RT filmmaker asks Michael, if he thought that humanity could ever give up their cellphones or WiFi, to which Michael offers an interesting and important response:

"That's an extremely interesting concept because, and I'll go back 15 years, fifteen years ago we wouldn't even be having that discussion, because it didn't exist... And no one knew that they were missing out on anything. The extremely quick influx of wireless devices into people's lives has almost just taken over. It is a convenience, you know. That's what it is. A comfort. A convenience. And its interesting to watch people from outside of this area (Green Bank) come into this area, for even a short period of time and feel the burden of constant connectivity being relieved from them. They enjoy the fact that they are not constantly being pinged by their cellphone or their email or Facebook or whatever it might be. We don't have that issue here. I haven't spent the last 15 years being addicted to it and then having to withdraw from it. I just don't have it and generally find it rude when I go elsewhere and people are constantly on that instead of interacting socially."

***Sue Howard***

Sue Howard is one of the electro-sensitive canaries included in the *WiFi Refugees* film. She, her husband, and daughter, explain in the film how they had to create a totally shielded room for Sue in which to retreat from the ubiquitous EMFs in their New York suburb after she became affected.

Sue says, "My world became smaller and smaller and smaller until I ended up living in a box... We built a room... it looks like aluminum foil [the shielding material] but it's not... There were days I didn't know if the sun went up or the sun went down and I couldn't run electricity in there so I didn't have light... I could intermittently use a flashlight."

Her daughter Kathryn talks about visiting her mother in her "box": "When Christmas came around, we had Christmas in the 'silver room.' We made the best of it but still you don't want your mom to be in constant pain."

Sue's husband related the experience of his wife's confinement due to EHS, saying that there were, "so many things my wife could not do. [She] even missed our daughter's graduation from college. It was 3 to 4 hours away. She couldn't take the drive and she couldn't stay overnight. Living in a shielded room... living in her car during the day, somewhere its safe. It's the kind of thing you don't really tell people because one, they will think you are crazy, and two, they definitely won't understand. I mean why bother?"

Sue explains what it feels like when exposed to EMR, "When I'm around wireless, I feel prickly all over my skin, my skin gets prickly. *Prickly* sounds like a cute word—prickle, tickle. No, it's really quite painful.

"All these sensations are very painful. When I am around a very strong magnetic field, oh my god, I feel it in my head. It's like a real hard pressure... When I'm around dirty electricity, my skin burns, it's like this burning—it's like its being cooked. It really feels like a burning."

Sue demonstrates the levels in their partially shielded home in which all wireless is turned off and other electronic appliances are kept to a bare minimum. Using her radiofrequency meter, she shows levels at about 3.6 millivolts per meter. She says the average background levels in most suburban areas are around 400 millivolts and up to several thousand in some hot spots.

In 2017 when the film was made, over 7,500 WiFi kiosks were going up in New York City bringing background levels of radiation up to 20,000 millivolts per meter in those areas. Sue's daughter Kathryn also demonstrates the levels in a New York subway, where there is now cell service *and* WiFi, showing levels close to one thousand.

Sue eventually discovered Green Bank and began to divide her time between New York and West Virginia, increasingly spending more time in Green Bank than at "home," even though this means that she and her husband (who has work obligations in New York and not yet ready for retirement) only see one another every two weeks for a couple of days. In Green Bank, Sue's husband explains, Sue is able to enjoy a social life, whereas there is nothing for her at home, where she is mainly confined to four silver-lined walls.

### *Leo Halepli*

Leo, a young-looking man, possibly in his early 30s, is also interviewed in *WiFi Refugees* as an EHS resident of Green Bank, who relocated to Green Bank in order to recover from his disabling symptoms. Once a high-functioning successful businessman and Ph.D. student, his condition forced him into full retreat for the foreseeable future. As he explains for *WiFi Refugees*:

"About when I was 25, I was living in London in the UK. I was doing a Ph.D. I was doing some consulting. I was a semi-type-A kind of character—you know successful in life. Pretty much overnight, I was at a conference and I was staying with a friend and I had a fever episode and my life changed dramatically in the weeks that followed. I had to basically tell the school I wasn't coming back. I had to hand over my work assignments. I said, 'you know I'm going to take a three-month break. I'm sick, the doctor is going to help me get better.' That's what you think. We did a bunch of tests—nothing really extremely out of range. I had a thyroid issue and the doctor said, 'take this medication, six weeks later you'll be fine.' Six weeks, six months went by, nothing happened. I kept going from doctor to doctor to doctor. I had extreme fatigue. I could hardly get out of bed. I was unable to leave the house. Remember that I am a trained researcher, I couldn't write. I was basically not functioning as a social, productive human being."

### *Jennifer Wood*

Jennifer, a Green Bank refugee resident, was interviewed for both *WiFi Refugees* and another documentary (not exclusively about electro-sensitives, but about the impact of new technologies on our world in general) done the same year (2017) by Werner Herzog, called *Lo and Behold; Reveries of a Connected World.*

For *WiFi Refugees* Jennifer shows off her modest cabin with no electricity and a woodstove for heat, nestled in the woods, which she (a former architect) had built upon "skids" so she can move it around. She says she has already had to move it a couple of times.

Before becoming a WiFi Refugee, Jennifer worked as a registered architect, all over the world, and spent many years in Nepal. She was working for several years in Hawaii when she got sick. She and her husband during that time, went back to Nepal thinking they could better diagnose her illness since the US doctors did not seem able, and maybe have a better chance of recovery there. But they had cellphones in Nepal too, although Jennifer did not know that was the reason for her illness yet. She remembers reacting to her cordless phone while going through the process of trying to uncover the root of her illness, and noticing her symptoms increase but believing herself insane to think it might be the phone as culprit. She thought to herself at the time, "electricity does not make you sick," but noticed she could not be on the computer anymore or watch TV. But since there were so many things she could not do at the time; she did not realize how related the electronics/EMFs were to her illness and gradually pieced it together when what she calls her "attacks" were linked to exposures.

Describing her EHS symptoms for the documentary, Jennifer says, "Overnight it started around 10 o'clock at night, it was like a *bam*, everything felt, here in my solar plexus and my stomach, everything just got stuck and stopped. There was a strange metallic poisonous taste not only in my mouth but it was in my whole body and at the time I had no idea what that was, but it was the same sensation I get when I am around really strong WiFi, which is almost everywhere now in the world, but some places are worse than others. When I'm in a crowd of people who are using cellphones or WiFi it gets much more intense.

"So, it's the same feeling, you feel pins and needles in your head and a metallic bar in the forehead and it's a kind of nausea that you can't describe. It's not the kind of nausea I used to feel when I was well, like if I got from food or the flu or something. This was a kind of poisonous nausea that's impossible to describe. It felt like my body's detoxification mechanisms just shut down. As if this poison got inside me that could never come out."

In the documentary, *Lo and Behold*, Jennifer summed up her story as follows:

"I became really ill from radiation sickness in 1996, I lost 50 lbs., I nearly died three times. And I became reactive to all of the wireless radiation signals when all the cellphones [infrastructure] went up, they went up in massive numbers in 1996. I tried to do all kinds of treatments, I moved, I lost my career. I was working as an architect in Honolulu. I had to be separated from my family and children. And finally, I heard about this [Green Bank] in 2011, and as soon as I heard there was a place with no cell towers, I was here in 48 hrs.

"Sometimes when I'd have really bad reactions to radiation, I will actually sleep on the ground. I feel better on the ground and there may be a science to this because they say the ground emits 7.83 HZ and it's the natural rhythm for animals and humans."

In *WiFi Refugees*, Jennifer takes the filmmaker to a place in the woods she has found that is totally free from EMF interference, where she goes to recover from exposures.

"This is my spot where I come to recuperate. Sometimes I come for 48 hours straight. When I am here, I feel the same way I used to feel in the 70s and 80s. I feel completely normal. Every single symptom is completely gone. But you'd be surprised how many forests are no longer refuge. And it is affecting the trees—there is a lot of documentation on that.

"When I'm walking in those forests there is absolutely no relief at all. It's worse than parts of New York City. For example, I went to a beautiful, beautiful state park in Florida, and the town was quite good, you know in terms of EMR, meaning electromagnetic radiation, but the park—I was just dying. Torturous pins and needles going into your head, in your body. So, as I was leaving (I didn't stay long) I asked the ranger, or the lady at the center. I said, 'do you by any chance have WiFi here?' And she said, 'oh no, we have Zigbee wireless system, tracking everything.' I said, 'okay thank you' [and immediately left]."

With her bare feet in a stream running through the woods she says, "It feels so good. I can't put it into words, but it feels, it just feels like—YES, this is how it should be. It's as if my body knows this is how it should be. And it's been like this for thousands of years."

* * * * *

The following are excerpts from two more EMF refugees living in Green Bank, from the *Lo and Behold* documentary film by Herzog:

***Diane Schou***

"This place in Green Bank is wonderful, it's not perfect but I can go outside. I can see the trees. I can see the sky. I can see the stars. When I lived in a faraday cage, I had to live in a box and I only left the box when I wanted to go to the bathroom or wanted to shower. Otherwise, I stayed in that box day and night, for a couple of years, several years.

"I had a mattress and I didn't have a place to stand up. I had to stay on the bed the entire time. My husband went grocery shopping. He would cook the food. He would bring and serve me the food. We'd open the door; he'd hand me the plate and I'd eat inside the faraday cage."

### *Anonymous*

"I've heard it also called as a super sense, that we just have this ability to for whatever reason feel these frequencies. It's a very legitimate illness. At this point I don't consider it a gift. I would give anything to give it back."

In response to being called a refugee she says, "Yes, I don't know what to call myself. This is brand new. And I just want people to know that this type of illness is, and doctors, I really want them to hear that this is a legitimate illness and it affects our lives tremendously you have no idea... when you go home today after this interview you have the luxury of going home to your familiar surroundings. You have that luxury to go back to your families."

With tears welling up in her eyes she continues, "I haven't had that in four and a half years." Now crying outright, she further relates, "I haven't had any stability and I just have to impress upon you how serious this is for those of us that are suffering. And I am extremely grateful that there is a location here where I am no longer in pain. Or whatever type of irreparable harm may be being done to my body will be either suspended or temporarily arrested."

## *Canaries in Print Media*

### *VERONICA—My Life Six Feet Under Ten Cell Antennas*

This story was first posted to *magdahavas.com* and subsequently printed in the book *An Electronic Silent Spring* by Katie Singer (chapter two). Veronica's story began after November 2009 when an array of cellphone antennas was installed on her apartment building in Toronto, Canada. She did not learn of plans for the installation until the work crew arrived. I am including her story here because it so accurately embodies the feeling of shock and subsequent total upending of our lives, I think most of us who have been sensitized by EMFs experience in the beginning stages.

> *If you had told me three months ago not to hold a cell phone to my head or body, I would have listened politely. But I wouldn't have changed a thing.*
>
> *If you had told me to exchange my cordless phones for old-fashioned corded phones, I would have listened. But I wouldn't have changed anything. I liked my cordless phones.*
>
> *If you had told me to use an Ethernet cable with my laptop and to keep Airplane mode turned on, or to move the Wi-Fi router from my bedroom and to turn it off at night and when not in use, or to get rid of my Wi-Fi altogether, I would have wondered, why would I want to do any of that?*
>
> *If you had told me three months ago that baby monitors should not be placed near babies, or to ask my fourteen-year-old daughter to text more than talk and not to sleep with her phone or computer on the pillow beside her, or to replace every fluorescent light bulb with an incandescent one, I would have listened. But I probably would not have changed a thing.*
>
> *If you had told me that the microwave radiation emitted by cell phones, cordless phones, Wi-Fi, cellular antennas, and other wireless technologies causes people to experience all manner of symptoms from insomnia to high blood pressure and heart palpitations and anxiety and should be avoided completely or at least whenever possible I still wouldn't have changed a thing. After all, the government approves these devices, and the media tells us that they're safe.*
>
> *But before three months ago, I had not lived six feet under ten cellular antennas that were installed on the roof above my balcony. Before three months ago, I had health and vitality. I slept like a baby. I did not wake up with numb hands and feet, my body feeling prickly all night and tingling or vibrating almost all day. I did not spend night after night in a hyperactive state feeling like electricity was running through me.*

*Before, I lived in a home that I loved. It was my sanctuary. I did not have hissing, buzzing, or high-pitched ringing in my ear. I did not ever get tension headaches. I did not feel an invisible hand wrapped around my head creating pressure. I did not feel bouts of nausea on a regular basis, sometimes accompanied by a metallic taste in my mouth, and I did not get dizzy spells.*

*I was not afraid that I might have a heart attack as I slept on a makeshift floor mattress in my living room and felt my heart race all night. I was not without focus or direction or ability to concentrate. I had never felt shocks from touching my mattress, my light switches, my pots or my cars.*

*My daughter did not have inexplicable rashes that hurt 'in' her skin (as she described it). She did not have headaches or feel nauseous and dizzy in our home or experience the blood in her hand going cold. She never had sleepless nights.*

*Before three months ago, I had not talked with someone who could have sold me thousands of dollars' worth of products by convincing me that they would alleviate the situation, but advised me instead, 'If you care about your health and your daughter, get out of there. You have to move.'*

*I had never abandoned my home. I had never couch-surfed with my fourteen-year-old in tow while trying to maintain some semblance of a normal life. I had not spent fifteen days getting two hours of sleep each night because my body vibrated all the time. I had not cried for hours feeling like I was losing my mind from sleep deprivation and from feeling fight-or-flight twenty-four hours each day.*

*I had not researched everything I could find to educate myself about the dangers of exposure to human-made electromagnetic frequencies and microwave radiation. I was not fully aware of cellular antennas and the invisible wireless web that continues to grow around all of our heads.*

*I could not tell the difference between a Bell cell antenna Rogers, Globeavlie, Tellus, or Wind cell antenna. I had never heard of Industry Canada or Spectrum, Canada Safety Code 6 or the BioInitiative Report.*

*I had not spoken to Health Canada, Industry Canada, the Canadian Environmental Legal Association, Environmental Health Clinic, Environmental Health Association, the Environmental Protection Office, the Toronto Environmental Alliance, Canadian Association of Physicians for the Environment, or my city councilors' office, trying to find out whether it's safe to live six feet under ten cellular antennas. So far, none of them have told me, 'It's not safe.' But thankfully I have a body that tells me the truth, and the good judgment to listen to my body.*

*Before three months ago, I did not have clear and unpleasant physical reactions to my cell phone or the cell phones used by people near me. I did not get off of busses because half of the passengers were texting or talking and the RF signals bouncing around the bus were more than my body could handle. I did not react to the touch of my computer keyboard or from sitting close to the monitor for too long. I did not feel my legs tingling and going slightly numb if I spent too much time in a room with Wi-Fi. I didn't feel nauseous and have sharp pains go through my hand and up my arm if I made a call with a cordless phone or a cell phone. I did not feel nauseous if I sat for too long or too close to a television.*

*I do for now.*

*Before three months ago, I could not have told when I stood within four blocks of a cellular antenna installation. I never thought twice about leaning on walls or in close proximity to the electrical wiring in a room, or lying on the floor above a basement for the same reasons. I never considered the effects of my neighbors' Wi-Fi and cordless phone base stations broadcasting through the walls between us.*

*I had not heard the words 'electrosensitive' or 'electrohypersensitive.' I had not spent hours and hours on the phone trying to find a doctor who knows what a cellular antenna is and the effects of living six feet under ten of them. The few I found who knew what I was talking about and who offered treatment that would have been good for me cost an arm or a leg and a plane ticket.*

*Before three months ago I had not stayed in six places over nine days just trying to get a good night's sleep. Because even after friends and family unplugged their Wi-Fi and cordless phones and everything but the fridge my body reacted to their neighbors' wireless devices, which emitted through the walls to where I tried to sleep.*

*Before, I only knew the benefits of wireless devices. I had not heard of EMF Solutions, Earthcalm, magdahavas.com, the WEEP Initiative, the Electrosensitive Society, safelivingtechnologies.com, a Q-link, a gauss meter or an electro-smog meter. I had not read stories from hundreds of people around the world whose lives have been profoundly impacted by something we have come to believe is purely beneficial and harmless, something we cannot see: microwave radiation from a wireless device that emit at levels not meant for human absorption.*

*Three months ago, I was blindsided. My life got turned upside down by radiation from cellular antennas. Now, I'm looking for a new home. I'm challenged by the fact that I've got to consider my recently acquired 'sensitives' more than the home's location, size, cost or style.*

*Please, pay attention. Pay attention to the choices you make and the ones that others make for you. Pay attention now, before you pay dearly.*

*Chapter Eight*

# Falling off Perches

## *Bruce's Story*

AFTER HAVING NOW been introduced to some of the people living with the often-daily nightmare of EHS and reading some of their stories, it is not too difficult to imagine, after being so marginalized, misunderstood, disenfranchised, and suffering from the physiological and emotional trauma of EHS, how, many EHS sufferers could be driven to suicidal thoughts or actions when the future seems so entirely hopeless and the present circumstances too difficult to bear.

Earlier in this book, I made reference to Bruce Evans, a fellow EHS sufferer in Australia, whom I had met online while searching for EMF-free safe havens. At the time, Bruce had found a refuge on his father's farm in Wodonga, Victoria, Australia and had generously invited my partner Sam and I to come stay on the farm, free of charge. However, moving to the other side of the world coupled with difficulties obtaining visas proved an insurmountable task, so we had to abandon that plan.

Bruce is one of my heroes—of very strong character and mettle, who, with the support of other WiFi Refugees, pulled through and came out on the other side of a very difficult, dark period back in 2014 when he considered suicide as the only remaining answer to ending his EHS suffering. Always thinking of others, he had hoped to make his suicide public so that the world might learn of the truly devastating effects that our currently EMF-oversaturated environment has on a large portion of the population. Bruce also bravely consented to allowing the details of his story to be shared here.

But before telling his story, one thing Bruce wants made clear—he is not now, nor never has been any kind of wilting flower or snowflake. He understands that EHS sufferers are often stereotyped in the media and publicly perceived as, in his own words, "being weak, sensitive little snowflakes or conspiracy theorists with tin foil hats whose mummies clean up after them all day."

Bruce, in his 50s, is a highly intelligent computer programmer, with a quick wit and an attractive, muscular and stocky frame.

He is a former green beret in the Australian Army Special Forces, a former garbage man, concreter, builder's laborer, and furniture "removalist." He has done every "crappy job imaginable," and drives the point home by reminding me that he is definitely not a "snowflake." He also has a fellow EHS friend who is a former Regimental Sergeant Major and Vietnam War veteran, whom Bruce laughingly refers to as a "hard bastard." And he knows one EHS man who is an MMA fighter. He also feels it would be helpful to emphasize the fact that there are people who are affected by EHS "from all walks of life, from all nationalities, and all political persuasions." And that these people from all different walks of life, "are coming to together to fight as one against this injustice."

But even a fighter, someone as strong as Bruce, can be so devastated from EHS to consider throwing in the towel, and giving up the fight. And on December 2, 2014, Brue did consider just that, and sent the following heartfelt email to our EMF Refugee group, with the subject line, "Forced Eviction Pending from EMF Refuge; Suicide the Answer," which read as follows:

> *Hi everyone, sorry to bring the mood down with this post but this is the truth and I don't care who hears it.*
>
> *I had high hopes for establishing an EMF refuge where I am living because it is one of the few places on Earth that I can live due to my severe EHS. I was lucky enough to have a cottage owned by my father that I could retreat to where there is no mobile phone reception and few other EMR problems.*
>
> *This place has been the subject of a legal battle with the DEPI in Wodonga, Victoria, Australia where they have reneged on their*

*contract of sale without any valid reason. We paid a deposit on this crown land many years ago and then they immediately changed their minds.*

*This legal battle has been drawn out over many years and now it is reaching its end. I wanted to make a place where people like myself could come and be free from EMR and maybe enjoy what is left of their miserable lives thanks to corrupt governments and money-hungry Telcos* [Telecoms] *etc.*

*This really looked like it was going to happen at one stage and things were falling into place. Now the DEPI of Wodonga have sent a letter to us stating that I must vacate the premises and that they are going to demolish it.*

*So, my first question is simple—where the hell do they expect me to go?!*

*This same government has made it impossible for people like myself to live anywhere else. They have ignored all of the evidence and made the planet uninhabitable for people like me and now they are going to force me to move out of a safe zone into a world that I cannot live in.*

*So, I am issuing this notice to the public and to the regional manager of the DEPI in Wodonga whose name is James Stewart* 02 6043 7900, *if you follow through with your plans, expect to bring a body bag. As I have no other option and due to my condition have become broke, my only option is to kill myself. I am not joking and I am deadly serious. I have had enough of this. My life has been completely destroyed by my condition, which is created by the corruption of this government and by people like you. I no longer want to live if I am forced to move, as there is nowhere I can go. LEAVE ME ALONE!*

*It is people like you, James Stewart of the DEPI, that have made this planet uninhabitable for me and I no longer can live here. This is the only place I can live and you are forcing me to play the only card I have left, suicide.*

*I hope it haunts you for the rest of your life.*

After posting his email Bruce received a number of both private and public responses from those of us totally understanding of where he was mentally and hoping, of course, to talk him down off from the ledge. These were some of the responses posted to our group (I have deleted the posters' names to protect their privacy):

*"Many are living the same nightmare—you're not alone! There's no place to hide. EMF has a devastating effect on brain chemistry. How many are pushed over the edge as a result? Dr. Carlos Sosa, M.D. said that his neurological functions were so deeply affected that he thought about suicide every hour of every day. In addition to EMF affecting brain chemistry, it also alters the electrical activity of the brain. Dr. Olle Johansson said: 'it would take only one minute exposure (to cellphone radiation) to change for more than one hour the electrical activity of the brain.' Another horrifying effect of the wireless assault. Shielding may be the only option at this point. Please take a look at the Quiet Zone Retreat. Blocking EMF during sleep is key..."*

*"Don't give up. They want you to give up. When you give up, they win. And they don't care because they are usually a bunch of psycho/socio-paths."*

*"I would like to echo all the loving sentiments the people on this group have so eloquently written. I understand that it all may seem like too much to bear at the moment. Having your dream sanctuary shattered is truly a deeply traumatic experience. With that being said, I would like to say how much I admire you for creating that dream to begin with. You are a visionary, and heaven knows that that is what we ES folks need more of! WE NEED YOU, BRUCE! We need your vision to add to ours, to create a world in which we can not only survive—but thrive.*

*"Just take life one day at a time for now. Things will shift; they always do. Being ES has certainly not been a walk in the park (ha! really, because now parks are WiFi-d!), but I have had to learn to be more flexible than I ever thought possible, and also it has given me the opportunity to develop an even deeper spirituality than I had before.*

*"So it is your life, and ultimately your choice to stay or go. Just know that those of us on this little cyber group (who you have never even met) are rooting for you with all our hearts, and we would be very sad if you prematurely disappeared!"*

*"I am so sorry for your news. It's truly heartbreaking news, all of it. It does seem like this world is being made uninhabitable for us. As I try to find a refuge somewhere I am just finding how difficult that really is now and realize that if I do find somewhere it can easily and quickly be made to be uninhabitable especially with all the future long-term and near-term plans the industry has for us and the planet. For example, 'WiFi from the sky' as you mentioned in a recent email. I did in fact see something about a Google balloon for Australia ready to do that soon. And now I am hearing from this group, reports of plans for WiFi from Space!! It's all so much and can be extremely overwhelming and I understand that especially in your position right now with what you are facing that you would feel as you do, like it's a lost cause and why keep fighting and as you say you feel you have nowhere else to go. But I hope to somehow encourage you to stay with us and keep up the fight. Because the rest of us need you. We all need each other. And we need people like you who are willing to stand up to the governments and attempt to hold them accountable. Maybe we can help? Please let us know what we can each do to help. If you get enough solidarity, enough people to join the fight, we have a chance of winning.*

*"Can I ask you... Can you chain yourself to something on your cottage so they cannot demolish? At least this way if you force their hand to be violent and they end up killing you then it will be clear to all the world that they did this and you did not do it to yourself. The problem with the suicide idea is that likely it will be spun in the press as further validation that ES people have psychological issues as we are often told we are imagining our symptoms already. So that could just make it even easier to dismiss you as being disturbed and crazy and so by this not take any responsibility at all. I also hold onto hope that if this situation doesn't work out for you that we can find another one for you, and for everyone seeking a safe haven. But please don't despair and please let's all join together and help stop this.*

*"Hang in there Bruce. We truly need you to stick around if you are able. Many of us feel as you do, and often reach the brink of despair and feel the world as it is has no place for us and does not want us and we are too tired and unwell to fight. But we need to not give up as we may have a chance of stopping things from getting worse and ultimately saving many lives—those who are sick and don't know why, those who will be sick and will die from the EMFs. We KNOW what is happening, we FEEL it, but many others don't. We have to keep reaching out. I am afraid your suicide, though I understand how tempting that is, will not ultimate-*

*ly help and may make things worse. Plus, you would be dearly missed by many people and I am sure your father would be devastated. Thank you for your honesty, but please hang in there and please let us know what we can do."*

The following day, on December 3, 2014, with the subject line "I am still here guys" Bruce responded to the group, further explaining what had led to his first post and how forces since then conspired to keep him in the fight a while longer, with the following:

> *Hi everyone,*
>
> *It is good to hear from you. Thank you for your many kind and supportive posts it really helps me feel that there are others out there who feel the same. It is great to hear such support for one another at a time like this. I really appreciate the time people have taken to comment and offer their genuine support.*
>
> *I stand by my comments and am willing to play this out to the end.*
>
> *To answer some of the questions that have been asked of me:*
>
> *Yes, there is room on the other farm for me, but what happens if they introduce wireless broadband into regional areas?*
>
> *What happens when they introduce WiFi from satellite? This is happening soon.*
>
> *The other thing is I cannot live near my father and can only handle him in small doses this is why where I am is perfect. This may sound trivial, but there is more to this. Other people who go there to live will not have this problem with my father, they will think he is great. It is just me and my father that clash and always have. But we are in one mind about having an EHS refuge as he has EHS and MCS as well.*
>
> *So, to explain what I was feeling the other day, I am sick of this being treated like a loony by these assholes and now that my life has been completely destroyed I have very little to live for except more torture at the hands of these frigging phone towers and smart meters.*

*To add to my misery, I worked on a software program for a company for a while and they owed me $230,000. This is a fair amount of money that could change my life at the moment. I cannot do jobs like this anymore due to my condition. I am restricted to scraping through on scraps of work etc. to try and make ends meet.*

*So, I just got word that the company that I did the work for transferred the ownership of the intellectual property of the software program I built into another company without my knowledge and then put themselves into liquidation. I am now broke. I have no money left. I will soon have nowhere to live if the DEPI continues with their plans and no money to do anything positive to make a life for myself.*

*So, my main reason for stating that I intended to suicide was not because of the things that have happened to me that I have mentioned above. I have had much worse than this happen to me in my life. For example, I have just come out of a period of 40 years of chronic insomnia followed by a couple of years of sleep apnea. The last ten years of which I got by on between 5 minutes and one hours sleep per night. This was for ten years straight without relent, every single night. Right after beating the insomnia through self-studying health and nutrition for over twenty-five years, I developed EHS. Out of the pot and into the fire.*

*Insomnia was life destroying enough as I lost my sanity, my career and my entire social life and there was no chance of any relationships when you have insomnia. But EHS may even be worse. Aside from the headaches etc., it made me a prisoner in my own bedroom. At least with insomnia I could walk down the street without feeling like someone had stuck an ice pick in my head.*

Bruce then details the extreme challenges he faced growing up and how that led to a period of drug and alcohol abuse and continued with:

*I pulled myself out of the rut, cleaned up my act, joined the army and ended up in special forces, all this while having chronic insomnia. The funny part is, sleep deprivation is part of the training in Special Forces, ha! I earned the coveted green beret and have had many adventures, wins and failures since then including building my own successful website business using software that I taught myself to build. The story goes on and on.*

*The point I am trying to make here is not that I am awesome, or that you should feel sorry for me. The point is I am used to adversity and can handle almost anything that is thrown at me. But this problem I have is not going to go away unless the general public become aware of it in a big way and quickly—it is only going to get worse and more people are going to have their lives destroyed by this plague. They are rolling out new technology every day that is further complicating my own, and other innocent people's lives. This is treason and it must stop. Someone has to take action for all of us.*

*I did not plan to suicide in some quiet whimpering way that would have no impact and could be used as a means of saying that we are all a bunch of mentally deranged conspiracy theorists. I planned to film it and give a full explanation for my actions in a completely sane, factual, matter of fact way so that people could see that I was just a normal guy who had been pushed over the edge. I planned to stream it Live to the Internet using Bambuser and hope that it went viral. This way, it may have some impact and people will sit up and take notice. This does not scare me one bit and on the day fully intended to do it. Admittedly I was a bit emotional when I wrote that post and am a bit more composed now. But I stand by what I said.*

*But fate has intervened as it often does when you commit yourself. One of the things that was driving me insane was the fact that my uncle has a pump shed across the creek that is powered by three-phase power. I had made moves to have the power lines removed so that it would be more fitting for a EHS refuge by complaining to the power company and threatening legal action etc., and had heard nothing. (By the way, my uncle is the one that started this whole legal battle.)*

*This power line runs next to the cottage I am in. When it is running, like it is right now, I get deep pounding headaches and nausea that go through me like waves just like it is right now. I had moved out of Melbourne and found a refuge only to be subjected to that for 12 hours a day. This was no problem when I first moved here because it was winter and they were not irrigat-*

*ing. So, I was not aware that this could be a problem. But now that it is summer, they are on all the time.*

*This meant that it was almost impossible for me to work. All of this made me a bit emotional as it just seemed that the world was closing in on me and there was no escape, there was no help, I am alone in this fight, and that there was no one that I could get to take action on my behalf in a legal sense to stop this. I myself, although I have much of the necessary knowledge, cannot defend myself in court as I cannot walk into a courtroom without experiencing the usual symptoms of EHS and would be incapacitated within minutes. So, I felt like I was in a fight but I had been gagged and had my fists tied behind my back. It filled me with rage and anger, and eventually despair. This led to my post.*

*After I wrote the post, two things happened. A linesman from the power company turned up to inspect the pole that the line was on. Guess what? He inspects the poles and deems them unsafe as white ants had eaten half way through them. They will have to be moved to the other side of the creek, exactly where I wanted them to go. They will be no problem to me there.*

*Guess what else? The linesman and myself start talking about EHS... get this... he has it and does not know. He did not know the symptoms etc. I explained them to him, told him the origins of the problem (his fillings and the smart meters etc.), then we talk further about it. He is overjoyed that the insomnia and many other problems that he had been seeking help with for years were easily explainable. He was at his wit's end to find a solution to his health problems. In ten minutes they were all explained to him logically and methodically from someone who, although not professing to be an expert, had been through the exact same thing. This made my day and I am guessing, his as well. It definitely lifted my spirits a bit to see someone overjoyed at having the light shed on all of their persistent health problems all in one day. Nothing is by chance.*

*Then later that day fate intervenes again, but that is a legal matter and I cannot divulge that yet. Let's just say, help may have arrived in the eleventh hour.*

*So, it may just be that in my hour of desperation, when I was planning to end it all, the universe (or your choice of spiritual being whoever that is) has intervened and is step by step putting things in place. We will have to see what happens. Maybe there is light at the end of the tunnel.*

*Cheers everyone and my sincere thanks for your support. I have never had that sort of support in my life and it meant a lot to me.*

Bruce is one of the success stories, someone who was able to pull himself back from the abyss, regain some of his health, sanity and life to continue the fight in promoting awareness of the EMF issue and to help others suffering with EHS, which is why I still call him an EHS hero, although he would protest this and point to others whom he considers worse off than he is, especially those who suffer with both MCS and EHS. He has financially supported many of those worse off than himself whenever able and developed the Radiation Refuge website (*radiation-refuge.com*) where EHS sufferers can list their locations, share stories and solutions as well as list low-EMF accommodations around the world.

In 2015, Bruce was the first to post his EHS story onto the site. In the account he gives more details about how EHS has impacted his life and some things that have helped him recover:

*I did not ask for this to happen. I did not will it into my life. I did not do anything to deserve this. I was just unfortunate enough to be born in an age where profit is more important than people's lives and happiness. An age where mobile phone towers abound and a multitude of mobile phone users act as willing accomplices in this conspiracy. I am constantly hounded by these unseen high frequency waves of misery that penetrate my life and my brain.*

*I live like a rat in the maze. I cannot go outside without planning the trip to avoid the mobile phone towers and NBN. To take a trip outside I must set course through the maze that I know will minimise the after effects. I have had to map all of this out using an electro-smog meter and geographical radio-frequency maps. To complain about this attack on my person is to be labelled a tin-foil-hat-wearing lunatic and to be treated with a nudge and a smirk by the boys at the phone company.*

*They hide behind their so-called world standards for acceptable levels of radiation, but with a little investigation you will find that these are set to a level that allows the most amount of profit with the least amount of whining. It is the old-boys club regulating their buddies with the occasional tut-tut to give the appearance of actually doing something to keep the public safe.*

*My body is sovereign property. I own it. Nothing has the right to penetrate my body without my consent. This is the law. Yet, these microwaves bombard my brain and body day and night without my permission.*

*What gives them the right to do this to me?*

*I have complained to the phone companies to no avail. I have sought medical assistance and other help. But the world is full of new-age charlatans and peddlers of meaningless garbage that promises the world and delivers nothing. Doctors can offer nothing but drugs. And of all the products that I have spent thousands on to block the microwaves, not one of them has worked.*

*So why am I having this problem and other people aren't? (IN MY CASE – NOT EVERYONE'S CASE.)*

*Well, that is simple. Think about this, aerials, the type used to receive signals on your TV, radio, mobile phones etc., are all made of metal for a reason, they catch radio waves. Metal is very good at picking up and harnessing radio waves of all frequencies.*

*So, it stands to reason that if you have metal in your head, your head will to some degree act like an aerial. There is nothing mystical about it at all. It is pure physics. So how does this relate to me?*

*Well, I have had in the past many amalgam fillings and recently was the proud new recipient of titanium implants.*

*If the fillings in your head are deep and contact the root or nerve of the tooth, it allows for the metals to leach out of the filling and accumulate in your brain and your nervous system. Nothing mysterious about that either. It is quite logical, until you try to explain that to someone in the medical profession, like a dentist, then you will get all the sarcastic denials in the world.*

*Hair analysis confirmed the presence of ten separate toxic metals in my body in very high levels. Every single one of them is a common component of amalgam fillings. This was not too hard to find out. There is no other time in my life when I have come into contact with these metals in any way that they could have entered my body. So, there you have it, I am a walking antenna.*

*I am trying to leach the toxic metals out using coriander juice, DMSA and Chlorella. But this is having limited success and the symptoms are getting worse.*

*Prior to getting EHS, I had spent forty years with insomnia. This was also largely caused by these fillings. Ten years of that period were spent only having five minutes to an hour sleep a night. I lost my sanity and was a walking zombie ready to snap at any instance.*

*So, I have forty years of my life battling one thing and beat it, only to have another battle lumped onto me as a reward. It is very disheartening.*

*Thanks Malcom Turnbull. I hope you made another $10,000,000 out of your investment. I am pretty sure that you will sleep at night and have no pity for the misery that you have forced onto other people.*

*The fact that our communications minister has so much money invested in Telecommunications and is so closely connected to those that own all of these companies, as in, he is their former partner, just goes to show how corrupt this whole government is. It is run by proxy by foreign owned corporations. And the politicians will dance to any tune their handlers play.*

Bruce still sufferers the effects of EHS and it still severely impacts and limits his day-to-day life, but he has made some progress (since writing his post in 2015) via detoxifying heavy metals, and trying more EMF-shielding techniques and other ways to recover and protect from EMR exposures. He still lives on his father's farm but hopes to travel again and also find better EMF refuges than his current one. Recently (in

2019) he has recovered enough to handle some nights out on the town where he is able to play drums with a local band and regain some of the social-life he had lost due to the deeply impacting functional impairment that results from EHS.

He also has maintained his unique sense of humor and a very positive outlook on life in the face of all of the hardships EHS has bestowed on him, and has been able to use his humor to help educate others about the harmful effects of EMR (electromagnetic radiation). As a case in point, he recently (in 2019) related his latest approach in talking about EHS in an email to me:

> *I have a new way of introducing people to EHS. We all get sick of explaining EHS to people and we all get sick of the know-alls or the smart asses that try to make out you are a nut job. So, if I have to explain EHS, and most times I try to avoid it because I am sick of the whole thing, I now do it like this:*
>
> *I say something like 'yep, I am sensitive to stuff like phone towers, WIFI etc. When I get exposed to them, I start seeing unicorns flying around, and I start hearing voices in my head, etc.' All the time I am saying this I make an increasingly more insane facial expression with my eyes twitching and my head shaking. Then I add an evil laugh on the end as if to mock the whole thing.*
>
> *In a more formal setting, you can say the same thing, but with a complete deadpan facial expression. State it very seriously and matter of fact as if you mean every word of it, then stop speaking, look at them very seriously until they realize you are joking. It has the same effect.*
>
> *So far it has not failed to get a laugh. It has also disarmed the person I am speaking to. They have a laugh, but they have nowhere to go with any claims of you being crazy, because you have beat them to it.*
>
> *Because you have opened with a joke, making light of the whole thing, they now want the real information. So, they will often reply with something like: 'okay, now seriously, do those things really affect you?' Then I will parrot out my sermon about the metals in EVERYONE'S bodies responding to radio waves and how it used to be called microwave sickness. It does not matter if I am*

*100% correct about the metals. What matters is the person can find a logical explanation for the condition. The details of it at this stage do not matter. It just has to sound logical. No mysteries, no esoteric connection, no conspiratorial stuff, no lizard people who live in the arctic, just logic.*

*It works every single time. The person has understood it. After this explanation, it is no longer a mysterious disease that lies in the realms of tin foil hat wearing conspiracy theorists. It is something that is quite logical.*

*It really works well. And it works in all types of social situations. I have used this approach with the mayor of the town I live near and got a laugh with follow-up questions. I have used it with one of the richest women in this town, who actually owns half of the town, and she thought it was hilarious. She is also now considering going to Asia to get her fillings out because of what I told her. On the funny side, our standard joke now is that she also has a pet unicorn and when she sees me, she pretends she is taking it for a walk. That is the power of having a sense of humor. It breaks down a lot of barriers. Prior to this, this woman was known to be a coldhearted and shrewd investor. But now people see that she has a sense of humor and actually likes a good laugh. Amazing what can happen.*

*I have used this approach with truckers, bikers, bankers, real-estate tycoons, intellectuals and druggies. The delivery changes with each situation and setting, but the underlying humor does not.*

* * * * *

## *Canary Voices from Radiation Refuge*

The website, radiationrefugee.com, created by Bruce has served not only as a platform for canaries to post refuge/housing opportunities all over the world but has also served as a place for canaries to share their own stories. Here are a few samplings from the site.[113]

**ANTHONY—I worked with Smart Meters on a Daily Basis and Suffered**

January 10, 2015

*Hi There,*

*And wow, glad I found your website!*

*Eight years ago, I developed stress/anxiety symptoms, insomnia, fatigue, swollen glands, digestive problems, tinnitus, allergies, funny taste in the mouth. I couldn't smell properly. Brain fog. Finding it hard to make a decision—constant headaches, head pressure, high blood pressure, and uncontrolled hyperventilating on a daily basis. I did feel my body and/or immune system was under attack. These symptoms appeared to come out of the blue and I've never been the person to stress for one.*

*Anyhow a year later [after the symptoms first started eight years before] I started work for an electricity company and the symptoms were not abating. I tried going vegan. I tried yoga. Nothing was helping.*

*So, I began self-medicating with illegal and legal drugs to alleviate symptoms. The last eight years health wise, on a daily basis has been hell and I had no answer as to why I was feeling this way except maybe stress, up until two days ago. You see, after having these symptoms for the first year, I decided to get rid of wireless Internet and limit my exposure to mobile phones, though it did not help me. Anyway, two days ago, I decided to go back to wireless and ALL of my symptoms came back with a vengeance and the wireless was promptly disconnected. The last four months I've had off work. It was only up until six weeks ago I started feeling better and all symptoms had gone. That is up until two days ago. (By the way, I live in Dandenongs and the smart meter I do have installed can't transmit.)*

*So, with my work I worked with smart meters on a daily basis and suffered, unbeknown to me that they were the cause. Work suddenly dried up four months ago. They took my work car and wanted to downgrade my job. I didn't bother returning to work at that stage. I didn't care anymore. I wanted to feel better again as six months of this year I have felt heavily fatigued—never felt so run down in my life. It wore me down in the end.*

*So where to from here? I have a mortgage to pay and have been living on the little savings I have. Legal action is a consideration. My duration of employment has been a silent living hell and I was just trying a way to exist, to function. Anyhow, it's a weight off my shoulders knowing the cause and I too have not suffered alone from a perceived mental symptom and/or from imagining it.*

*Anthony*

## JODY WATKINS—Forced Exile

This story is especially heartrending given that Jody invested so much into building a home she could tolerate in a remote location only to have her safe haven destroyed overnight by the addition of a nearby cell tower after moving in. Her story aptly illustrates why most of we canaries strongly hesitate to financially invest in land purchases, knowing how quickly land can lose value, not just on the market (since cellphone and power-line infrastructures devalue home and land prices, in general), but fully lose all value to those of us in need of low-EMF refuges as priority above all else. But for those canaries like Jody (and myself) who are also sensitized to chemicals, finding any home, rental or purchase, is extremely difficult, so we do often feel the only option is to buy land and build our own homes—if at all financially viable. For Jody, who did feel the need to build, to have her custom-built home so immediately and irrevocably destroyed by the haphazard placement of a cell tower is truly devastating.

In the US, the 1996 Telecommunications Act signed into being by the Clinton administration, prohibits private property owners and local councils from protesting telecommunications infrastructure based on health concerns. Clearly this stipulation was included because the industry and government already knew that EMF radiation was damaging to health and that if they did not protect themselves via specific legislation in advance, massive amounts of lawsuits would be filed by those adversely affected by the added radiation from the same infrastructure. With the latest legislation drafted to protect 5G infrastructure, power is further taken away from home owners and local councils by declaring the "right" to intrude upon private property and place antennas on private land against the will of home owners if they wish, even forcing property owners to cut down trees that block the signals.[114]

Here is Jody's story, related firsthand and posted to Radiation Refuge on September 3, 2015:

> *I finally started to live in my home, after spending over five years homeless. This is Due to Multiple Chemical Sensitivities given the great difficulty accessing suitable housing for this condition. The impact of this means that I cannot live in a normal house due to the many building materials not being tolerated bringing about extreme loss of wellbeing and profound disability.*
>
> *I have endured years of stress to painstakingly build this personally handcrafted low toxic/healthy home. This was after years of enduring hard work collecting local air-dried forest timbers, researching and investigating the least polluting, select materials and then building in a remote and tranquil part of the states south away from pollution with pristine air quality.*
>
> *I had achieved my longtime vision of a healing environment where I was finally able to feel well and start living a normal life. This was the happiest time in my life, just knowing every day I was going to heal and get stronger. While also looking forward to completing this home and living the lifestyle I had created.*
>
> *Then an NBN fixed wireless tower was proposed to be built just 300 meters from my boundary. This threatened to take away my safe haven and everything I had achieved to feel well. This being the only place I know to manage my health.*
>
> *Among the many exposures, which I am sensitive to it has been recognized I have an outstanding sensitivity to wireless technology. I have never been able to use a mobile phone or tolerate Wi-Fi exposure. More recently I have found myself to be adversely affected by the new technology of NBN fixed wireless.*
>
> *After spending two and a half years of being consumed by overwhelming exhausting work to have the tower not go ahead next to*

*my home NBN Co. have just ignored and dismissed all my efforts on what impact this would have on my life and how detrimental it would be to my health.*

*In May 2015 the tower became active next to my home. The exposure made me intensely unwell, to where I was unable to perform even the most basic tasks at my home and I found I had never felt so unwell in all my life and never felt so degraded in all my life.*

*This resulted in my not being able to remain at my home and camping out on land many kilometers away, tucked out of the wireless emissions, where I felt the head clamp release, my body became lighter, my mind started to shift into clarity and I started to feel alive.*

*This has been my only option in one of the coldest Tasmanian winters, where I am only able to live from my car and cook on a camp fire, this means I only return home for water access, essential supplies and washing.*

*When I return to my home, I feel totally depleted like the life has been drained out of me where I become weak and exhausted. I am unable to concentrate or focus and feel disorientated. I become too nauseous to eat any meals and can't sleep due to a relentless buzzing stimulating effect through my head and body, which turns to tremors. I also develop the feeling of a clamp around my head that results in headache/migraine and the most intolerable agonizing and debilitating pain. I am left suffering a feeling of severe head trauma/injury as I am left with cognitive impairment, which I can only describe like a blow to the head and being left with a kind of brain damage. This felt like a massive oxidative stress.*

*I am devastated by this situation of being forced to leave my home, losing access to basic resources, losing basic dignity and not being able to plan my days or even know how I will get through each day. I am now left where I feel like my heart has been torn out and I have lost all reason to live. I have had everything taken away from me, not just my home and my health care but my whole world (everything I know to survive).*

*I feel strongly that EHS needs to be recognized and that sufferers need to be supported to live in the world, to live a normal life (just like any other disability) as everyone deserves to have a quality of life. Especially given we now live in an unprecedented world saturated in wireless signals from the many new technologies of pulsed microwave radiation.*

*The accumulative exposure from returning back to my home for access to essential resources has resulted in increased sensitivity and permanent injury. I feel I am now not recovering, where I am now profoundly unwell with relentless ongoing pain of headache/migraine, which I describe being felt like the most intense and debilitating pain accompanied by nausea and weakness, I feel there is lingering damage to my nervous system effecting mobility and coordination and I am left with permanent cognitive impairment like brain damage (making my life harder to manage). For this reason, my plans are to continue camping out to save my health.*

*I have investigated building a shielding wall designed to maintain ventilation that would allow me to temporarily return to my home, but this wall may need to be 20 meters high and success cannot be guaranteed given ambient levels and how sensitive I am. I have therefore been advised I would be better off looking to find suitable land out of the wireless.*

*Relocation is increasingly difficult from the ongoing NBN fixed wireless roll out. Where I live In the Huon Valley/Channel area from late 2013 I have found more and more places have been wiped off the map, with just driving out becoming more of a struggle.*

*It now looks like needing to go further remote making access in to such areas an issue. I miss my home and I desperately miss my garden, where I am able to grow my own food and live the life that I had. I am not sure how, but I will not give up on finding a place out of the wireless, where I can start again. It is my plan to relocate and finish my purpose-built house where I have a chance to look forward to healing and feeling well again.*

**STEVE's Story, IT Professional, 44 years old**

*Steve discovered he was sensitive to EMR well before he discovered through his own research that there was a label for the condition—electromagnetic hypersensitivity (EHS). He noted that, consistently, within minutes of using a wireless router he would feel a pressure in his head and chest, tingling sensations in his hands and face, and would be left with a headache that could last for days. He also experienced severe discomfort when using a mobile phone.*

*Subsequently he was able to function quite happily by minimizing his mobile phone usage and using only wired devices in his home.*

*In 2011 SMART meters were rolled out in his street. Although he resisted having one installed at his home, he has been severely affected by his two neighbours' SMART meters, which were installed three metres from his bedroom.*

*Soon after, he experienced a consistent pattern of waking up at night with a severe sharp pain in his head and great difficulty falling back to sleep.*

*He says, 'every morning I would wake up with a serious headache, which would make concentration and performing simple duties quite difficult. On a number of occasions, I would wake up with my heart beating irregularly.'*

*2012 was an ongoing health battle for Steve, with increased sensitivity to wireless devices and even things that normally would not have bothered him, such as, laptops, phone chargers, and light dimmers increasing his suffering from constant headaches and lethargy.*

*He underwent many tests with his GP and neurologist, with all results returning negative for any brain disorders or tumours.*

*Steve says, 'I know it is wireless that is causing these issues, because when I go to areas that have very low EMR and no SMART meters, I feel fine after several days.'*

*Steve has recently painted his house with shielding paint and installed RF blocking curtains. This action, he says, has helped immensely to reduce his sensitivity to manageable levels.*

*Although Steve has taken precautionary measures in his own home at great expense he is concerned at the lack of support, care, and understanding by his power utility and the various government departments that he has been in contact with over this issue.*

*He has effectively become a prisoner in his own home, because to venture out into the neighbourhood for extended periods of time leaves him drained and feeling unwell."*

## Rosemary and Vic's Story

*Rosemary and Vic had lived in their home for twenty years. Shortly after installation of a SMART meter they noticed ringing in their ears, dizziness, nausea, headaches, and heart palpitations.*

*When away from their home, their symptoms would subside, but would come back with a vengeance every time they re-entered their home.*

*'We phoned our power distributer to ask to have the SMART meter removed, but they refused despite our desperate pleas and having a doctor's certificate. The power distributer admitted that they were aware that 5% to 10% of the population are really sensitive to radiofrequencies.'*

*Out of desperation they moved into their van, away from the SMART meter. They lived in the van for six months while trying to resolve the problem by again contacting their power distributor, the Energy and Water Ombudsman and Energy Safe Victoria with no resolution. They have been forced to move interstate, leaving their friends and family, just to get away from SMART meters.*

## TIM's Story—Graphic Designer 22 years old

*Tim had just started a new job as a graphic designer in a small inner city terrace studio in Melbourne. Over a period of three weeks, he became quite unwell, suffering headaches, dizziness, blurred vision, heart palpitations, severe kidney pain and an inability to concentrate.*

*He found that after leaving work and having a few days off, he would gradually recover, only to have the symptoms reoccur each time he went back to the studio.*

*He then discovered that there was a smart meter on the studio wall opposite his head, hidden in a wooden meter box. To eliminate other causes, he underwent medical tests, ECG, blood tests, and a kidney scan, which proved to all be clear.*

*Still his symptoms continued and became worse the longer he stayed at work. Fearing permanent health damage, he resigned.*

*He feels fine when away from these pulsing emissions, but has now become sensitized to wireless radiation and becomes quite dizzy when exposed.*

## RFID—A Poem

*June 20, 2017, posted by Anonymous*

*This is a poem written by a person suffering from severe life-threatening electromagnetic exposure, which began when her name was registered to an RFID chip, which emits and receives signals to a server over distance. She believes that RFID technology may be one of the bigger electromagnetic threats, perhaps exceeding the effects of the SMART meter.*

*She is looking into the link between electromagnetic damage and spying practices in the Internet-of-Things Age.*

***RFID***

*I wander the streets without my head*
*I wander without a thought for bed*

*I wander without a thought in my head,*
*I wander blindly, just like you said.*

*I wander the streets, I look for air,*
*For gentleness, for decency, for care,*
*For warmth in someone's arms,*
*But beware, for arms are cold,*
*And men are sly,*
*Warmth is denied.*

*I wander the streets, RFID,*
*I see the skyline, RFID,*
*I know my mind, RFID,*
*I know my kind, RFID.*

*I wander the streets, listen to me.*
*I seek someone to speak to, listen to me,*
*Someone to tweet to, listen to me,*
*To meet to, to greet to, to preach to, listen to me.*

*I wander the streets, I know the text,*
*The scriptures, the books, the histories vexed,*
*The heroes, the failures, the conmen, the strange,*
*The mentally-split the upset, the deranged.*

*I wander the streets, not a sick one,*
*Not a shy one, not a dry one, a lost one,*
*I wander the streets, I know what I mean,*
*The horizon is close to the fingertips,*
*Yet by most so unseen.*

*I wander the streets, RFID,*
*Lock me away from society.*

*I wander the streets, listen to me,*
*The waves and the people and the words and the 1s*
*And the 0s, the heroes, the queer-os, the cheer-o-s,*
*I wander the streets, disconnect it from me,*
*I seek my own peace, Dear RFID.*

*The author is in need of assistance, as her life in currently in danger due to the electromagnetic threat of the RFID chip. She needs legal counsel, technology and privacy counsel, and general care and support. She has been hospitalised several times since the start of the illness, and spent long periods just trying to find people to understand and support what she's going through.*

**VIOLETTE's Testimony**

*To Everyone,*

*There it is: a letter from me. It is all that I could do to participate in this gathering. However, to be present, to meet other people, to exchange... I dream of this! To throw myself into the world, build with you... My 'fate' denies me this.*

*But is it really my 'fate' denying me this?*

*I am going to tell you the story of the disaster, which is occurring in my life. You may then form an opinion about responsibility for my fate.*

*I am now 14 years old. When this began, I was 10. Until that moment, my life had been normal; family life, school, my friends, my love for the mountains of Ariège... Great Happiness, very great Happiness: I was free (except when my parents restrained me!). The beginning of the catastrophe occurred during the start of the 5th year in school. I will forever remember this period as a nightmare.*

*I'm not sure I'm exaggerating when I say, it is like a person confined in the prison at Guantanamo: locked up, tortured...*

*Like all students, I returned to school, was with my friends, teachers... During the first week, I suffered an unbearable headache. 'Break' at home... Return to school, near-immediate return of the headache.*

*Unbearable, Torture. Return home for several days... during the month of October, I attended school only a third of the time. Very quickly, I realized that the presence of a cell phone, as well as an emitter of the relay antenna type, Wi-Fi, Wi-Max... violently set off the headaches.*

*Near-constant pain. Torture. I consulted Professor Belpomme. The result of the consultation came as a shock: harm to the blood-brain barrier, a very great deoxygenation of the brain...*

*After maximum protection, the damage lessened. Only, what did not subside was that each exposure to pulsed electromagnetic waves awakened the torture. It is violent, very violent.*

*I continued my education through the long-distance learning center (CNED).*

*Why did this intolerance occur so suddenly?*

*We noticed that a 3G emitter had been installed during the summer on an antenna near the school. Since then, the number of wave emitters has been increasing. I am more and more driven back into my 'protected' space, which is becoming my prison.*

*To tell you the truth, I have all the same found an activity that I am passionate about: Aikido. I can only go because a group of civilized persons practice it: they systematically switch off all cell phones when I am there.*

*It is also a place relatively protected from outside emissions. But I remain limited in possibilities for a social life: no concerts, no shows, no movies, no stores, no kind of outing in town... So, sometimes I snap. This year I did not resist the temptation to go to the music festival in St. Girons... I escaped my prison for two hours in order to do like everyone else, to have the same possibilities as other people. I rebelled, faced with my intolerance to electromagnetic fields. I do not want it; I do not want to live with it any longer. It has gained the upper hand! I was knocked out in bed for 8 days with an unbearable headache and since then, an intolerance that has increased. Do you know that even the mountains, my wilderness, are no longer accessible except in very rare places, the bottom of very deep valleys or small valleys where no one thinks of going, which turn their back on human conquest?*

*Because even in the mountains, there are antennas everywhere, which emit very powerfully... I let you form an opinion about responsibility for my 'fate' in the face of this disaster.*

*But I am going to give you my opinion: anyone who thinks my 'fate' is responsible for my situation, as for that of everyone, whose numbers are growing, who live with the same disaster, will sadly be giving up the values of Life, of the Earth, and of Humanity. I simply refuse without compromise to allow this disaster to spread. To all of you who read or are hearing my letter being read, please know that I reject all responsibility for my 'fate' and that, in my own way, I will fight.*

*Because I am convinced that I will not suffer the 'deviations of polluting technologies,' because in the name of freedom and equality, I will find the gifts that life has to offer. I refuse that they will be taken from me.*

*I am asking everyone to educate themselves and to act against the development of pulsed electromagnetic waves!*

*Violette*

## *Chapter Nine*

# Silenced Canaries

*"LIFE FOR AN ACTUAL canary in a coalmine could be described in three words: 'short but meaningful.' Early coalmines did not feature ventilation systems, so legend has it that miners would bring a caged canary into new coal seams. Canaries are especially sensitive to methane and carbon monoxide, which made them ideal for detecting any dangerous gas build-ups. As long as the bird kept singing, the miners knew their air supply was safe. A dead canary signaled an immediate evacuation."*[115]

Bruce, and many others who have had to battle with suicidal ideation due to their harrowing experiences dealing with EHS, have pulled through and continue to struggle to find ways of living with their impairment in our EMF-overpowered world. However, others were not able to continue facing the extreme challenges presented them and have flown this world in search of escape from the overwhelming pain electromagnetic radiation caused them. The following are only two such examples out of countless other forgotten canaries.

## *JENNY FRY*
### *October 3, 1999–June 11, 2015*

In 2015 in the UK, the suicide of 15-year-old Jenny Fry made headlines in the British newspaper, the *Mirror*. While news of suicides in teen populations may be unsurprising, the reason for this particular death *did* come as a surprise to most readers. Jenny killed herself because she could no longer endure the pain that the WiFi system at her school caused her. The *Mirror* claimed Jenny, "hanged herself from a tree following two years of crippling tiredness, headaches and even bladder problems."[116]

Jenny's mother Debra told the newspaper, "I believe she just couldn't take any more. She had overwhelming fatigue, headaches and ear pressure, difficulty finding words, itchy skin, dizziness and joint pains. She was getting into trouble at school because she couldn't concentrate and needed to urinate more than usual so was always leaving class. She'd always been a very good student and a very healthy child. I made sure she got the right nutrients, the right influences, the right education. I had no idea we were exposing her to something so dangerous."[117]

Debra Fry and her partner Charles Newman said Jenny's problems started when her school installed WiFi in late 2012, but her symptoms eased after Debra and Charles uninstalled their home WiFi router soon after. After personally contacting Debra to verify the facts of Jenny's story as presented here, she confessed to me that they actually had not had WiFi at home previous to the installation at school, but only signed up for the service at the time because most of Jenny's homework had then gone online. However only a few months after the WiFi installation at home, Debra figured out why her daughter was unable to sleep well, and so got rid of it, and hard-wired their Internet connection with Ethernet instead. To the *Mirror,* she said, "All her symptoms eased or went away when she came home, particularly at weekends and holidays. We wanted to take her to the GP [General Practitioner] but thought our fears would be dismissed."[118]

Turning the WiFi off at home not only helped Jenny but also her mother, who until that time had not realized that she too was being adversely impacted and she noted that her own symptoms of racing heart, sharp pains in her eyes and tinnitus (which she had initially thought to be age-related) all cleared up once they turned off the WiFi for Jenny.[119]

Jenny's mother did not stop at just turning off the WiFi at home, she did all in her power to have it disabled at Jenny's school as well, making multiple trips to the school armed with information regarding the dangers of WiFi radiation to health, leaving the head teacher with reports and websites to read, but was told that they could not turn off the WiFi for just one student. Debra made it clear that disabling WiFi would not only help *her* child but the rest of the children at the school as well, as it was not safe for any of them. But she was not allowed to take EMF measurements with her electro-smog meter to find safer areas within which Jenny could work.[120]

Debra also contacted, "the Governors of the school, the local Education Department, the Radiation Advisor used by them, the Chief Scientific Officer, the National Radiological Protection board, the Health Protection Agency, Public Health England, and my MP and GPs."[121]

Failing to find a school without WiFi, Debra offered to let Jenny stay home to work, but as she was a serious student in the middle of her GCSEs and also a popular well-liked classmate who wished to be with her friends, she opted to tough it out at school,[122] doubtless still underestimating just how profoundly she was being affected. Certainly no one had guessed, not herself, her friends, her mother or teachers, that it would drive her to take her own life.

Debra related her daughter's story in further heartrending detail, as printed in Olga Sheean's book, *EMF off! A call to consciousness in our misguidedly microwaved world*, as follows:

*To find your child hanging from a tree was not something I ever expected to experience as a mother. It is the worst nightmare imaginable, and one that I live with every day. I will never forgive those who allowed the use of this lethal technology [WiFi] to destroy my daughter's life, my family's life and the lives of other families affected by the unconscionable lack of safety and protection of children in schools. I will not rest until justice is found for what I consider to be the murder of my daughter by a state incompetence—a tragedy that is all the more devastating because it could have been prevented, had the authorities heeded my repeated warnings.*

*The school authorities refused to believe me when I warned them of the dangers of WiFi radiation and the problems they were causing Jenny. Because she was so severely affected, my daughter was treated as if she was suddenly a 'naughty' child, despite her school record of being a well behaved, hardworking child, with above-average intelligence... until WiFi was installed in the school. At the same time, the school uniform changed, with the introduction of blazers with pockets, which led to pupils carrying their cell phones and iPods in their pockets instead of in their bags, which, in turn, exposed Jenny to even more radiation.*

*When the school installed industrial-strength WiFi, Jenny started having difficulty holding her pen or pencil, concentrating, thinking and writing, which led to problems with many teachers. I was asked if anything had changed at the next parents' evening. I was initially unaware that the school had installed WiFi, and it was only when the school introduced a policy of most homework being set online that I realized what was happening. When Jenny started having nosebleeds, we became aware of the health dangers of children exposed to WiFi, cordless phone, etc.*

*Jenny tried so hard to do well and maintain her standard of work. When she became overwhelmed by the radiation and need to leave the classroom, she would find another room where she could work. However, this led to detentions when teachers refused to believe that she was being affected by the WiFi, which also caused urine urgency, which resulted in more detentions and in teachers refusing to let her go to the toilet, which created a lot of confusion and stress for everyone.*

*As a result of all this, her tutor gave up on her and no longer showed up at meetings. Incredibly, this tutor had told a couple of boys that they should go hang themselves and she would provide the rope. Is this what planted the idea in Jenny's head? Many children complained about this but were ignored. We also found that Jenny had written a paragraph in her English book, stating that it was against her human rights to not be allowed to go to the toilet. What torture for a teenager. Did the teachers expect her to wet herself in front of the class?*

*When exposed to WiFi Jenny's symptoms included headache, stress, nosebleeds, sleep problems, urine urgency, temperature control issues, fatigue, anxiety, ear noises and pressure, exhaustion, restlessness, confusion, skin rashes and irritation, joint pains, racing heart, hormone and cycle problems and difficulty concentrating and finding the right words.*

*As a dental nurse, I knew that cell-phone use was associated with brain tumors and parotid tumors, and that cancer clusters have happened around cell-phone masts. It was only when Jenny became ill that I became aware of the hazards of WiFi and cordless phones and I was not aware of the related suicide risk until after her death. I also discovered that there was a hidden phone mast behind a tree on a concrete water tower, which had 4G switched on the Sunday before her death. If I had known then what I know now, my daughter and possibly two other pupils might still be alive.*

*My advice to parents would be to home-school your child if the school authorities refuse to switch off WiFi, or to set up your own education groups and schools. I am not against technology, which I find amazing but WiFi Technology's decimating our children's health. We must avoid irradiating children and go back to safe, wired Ethernet connections. Take a precautionary approach and inform yourself of the dangers.*

*I will leave you with Jenny's words:*

*'I have no hope for humanity. We are destroying this beautiful Earth and I'm not good enough with words to stand out from the crowd and somehow help humanity. I'm insignificant. I'm an insignificant number on someone's screen, and so is my whole*

*life—just a tiny blip in the existence of the universe. I find it hard to be hopeful when I can barely enjoy anything anymore.'*

*I have just returned from putting daffodils and tulips on her grave.*[123]

The attitude of teachers in the face of Jenny's unrelenting pain and pleas for help, coupled with the inherent dehumanizing effect screen-learning has on children, and potentially topped off with a standard school curriculum, which all too often focuses on shaming children about topics like racial inequalities, climate crisis and environmental devastation, while using a storyline that blames the average human rather than industrial giants, billionaires, and politicians for these problems—all the while forcing children to use technologies responsible for some of these very same environmental degradations and humanitarian injustices taught in classrooms—potentially contributed to Jenny's feelings of insignificance while also denying her the justice she deserved.

All over the world children are forced into similar impossible situations, their voices silenced in the face of ignorant authorities, teachers, and sometimes even parents (but clearly not in Jenny's case whose parents were very supportive, with her mother going above and beyond to help her daughter) who refuse to believe their plight. In 2018 many children became ill and two children dropped dead[124] in a Simcoe County schools in Canada. Due to pressure placed on the schools by concerned parents, radiation levels were measured to find that the schools exceeded already inadequate radiation limits by 34%. Yet the school district board did nothing to remedy the situation and refused to remove the WiFi. The two deaths were the result of cardiac arrest while at school, in teenaged children. Two other children experienced cardiac arrest but were revived.

Radiation levels in schools can be enormous, much higher than in average public spaces as commercial-grade WiFi systems (such as "Wi-Max") are increasingly being adopted using several routers, or "access points" throughout the school, sometimes covertly placed in ceilings above student's heads. And these access points are never turned off, as Cecilia Doucette, technology safety educator explains in the 2017 RT Documentary *WiFi Refugees:*

"*Most schools now have wireless access points that are connected to their routers, and they are either put in every classroom—some class-*

*rooms have more than one, some do them in the hallways but they continually emit a microwave radiation signal broadcasting their address whether there is even anybody in the classroom. They continue all day, all night."*

Add to this, the hundreds of wireless devices in use all day long in these confined spaces, plus cellphone towers often placed directly on school grounds, it is no wonder so many children are experiencing headaches, irritability, ADHD, depression, anxiety, hormonal disturbances, nose-bleeds and ear-bleeds, cardiac arrest and increased incidence of suicide.

Patricia Burke, an electro-sensitive and EHS advocate, also while interviewed for the *WiFi Refugees* documentary lamented about the frighteningly high levels of EMFs at schools today:

*"I couldn't go to school in this day and age. I don't feel well just being in the building. I don't feel well in a lot of places, but in particular I can really feel myself reacting here [in the school]. I just feel wiped out."*

In 2018, Dr. Anthony Miller, World Health Organization advisor called for a ban on school WiFi networks explaining that:

*"Radiation from mobile phones and other wireless devices can cause changes in DNA and induce cancer in experimental animals. Children's skulls are thinner and absorb much more of this radiation. We ignore this at our future peril."*[125]

Jeanice Barcelo, in her book *The Dark Side of Prenatal Ultrasound and the Dangers of Non-Ionizing Radiation*, further explains (quoting Barry Trower) that, "Children absorb up to 60% more microwaves than adults because their bodies are more moist and their electrical conductivity is greater. So, children can expect 60% more danger from microwaves than adults, and one of the first symptoms for a child is suicidal tendencies."[126]

Jeanice points to several studies in her book related to the increase of these suicidal tendencies in children alongside the proliferation of EMF exposures in schools. "A [2018 Blue Cross Blue Shield] report examined major depression in millennials and adolescents. That study... reported 47% increase in depression diagnoses in millennials and a whopping 63% increase in adolescents—especially teenage girls—from 2013 to

2016." And "The number of children and teens in the United States who visited emergency rooms for suicidal thoughts and suicide attempts doubled between 2007 and 2015 according to a new analysis...43% of the visits were in children between 5 and 11 [years of age]."[127]

Ex Royal Navy officer and microwave-radiation expert, Barrie Trower says there is "no safety level of microwave radiation for children."[128] Having directly studied the adverse effects of microwaves as used in warfare while in the Navy, Trower does not hold any sympathy towards those still ignorant about the harmful effects of these frequencies and does not mince words when describing how those who continue to expose children to these fields should be treated. He says, "Anyone who puts WiFi into a school should be locked up for the rest of their life. They are not fit to walk on the surface of this planet, because they haven't looked at the research and, whatever incentive they have, it's not worth the genetic problems that parents are going to face with their children when they're born."[129]

Barrie has heard thousands upon thousands of testimonies from concerned parents and others worried about the harmful effects of microwave radiation from all over the world. Two poignant examples from Sheean's *EMF off!* are as follows, with the first a direct example of the "genetic problems" of which Barrie spoke, and the second an example of another death of a child due to WiFi exposures.

*"My child is one of several with cancer/birth genetic problems. These only started after the transmitter was turned on. My worries are two-fold and take every second of my life. Will my child ever marry or find a partner and be happy? What will happen when I die? I know I will die worrying. Regardless of who is to blame, it is me, the mother, who carries guilt and responsibility."*

*"My daughter has just died. I am holding her hand. She has just had her 11th birthday and she was number 11 to die since the transmitter for WiFi was put near her and others' desks."*[130]

In 2019, France banned WiFi from all kindergarten and nursery-school classrooms and also banned cellphones from all schools across the board. Perhaps eventually WiFi and similar sources of harmful radiation will be banned from all schools everywhere. But this ban[131] will have come too late for Jenny and others like her.

And for parents or educators seeking some sort of restitution, the task is nearly impossible given in particular that major insurance companies, like the Lloyds of London syndicate that underwrites for most major insurance companies, will not cover injuries in schools, "resulting from or contributed to by electromagnetic fields, electromagnetic radiation, electromagnetism, radio waves or noise."[132] And no matter what course of legal action is pursued, nothing can bring back a deceased loved one, and family and friends of the deceased have to bear the burden of the loss.

But some advocates are still trying and not giving up the fight. Professor David O. Carpenter of New York State University in Albany, New York, while filmed for the documentary *WiFi Refugees*, explains his advocacy role:

*"[What] I've been very much involved in has been electro-hypersensitivity in children, which is reflected primarily by headaches, by a sense of having brain fog, not being able to learn well. I'm presently involved in a legal case in Massachusetts of a nine-year-old child whose school refused to provide him with a room that was free of WiFi. When he is present for several hours in rooms that have multiple routers and multiple wireless laptops, he develops severe headaches and often will develop nausea and even vomiting."*

This child is one of countless complaining of the same symptoms, but for the most part these children, like so many other canaries of all ages, are being ignored. Kathryn Stauffer, a Green Bank, West Virginia resident and electro-sensitive, while also interviewed for the *WiFi Refugees* film perfectly summed up the desperate situation with, "What about the children and the animals that have no voice out there? They are suffering and nobody is really listening."

## *MARIA (SARGENT) AUGUST,*
## *Ani Thupten Tsondru*
*August 3, 1970 - March 12, 2019*

The following obituary was written by Maria in her own words before she ended her life.[133]

> *On March 12th, I chose to end my Earth Walk and reunite with the Mother of All Things. I wanted to free myself from the strait-jacket of electrifying pain and neurological debilitation of Electro-Hypersensitivity (EHS) and Epstein-Barr Virus (EBV). These hidden modern-day epidemics humbled me and connected me to the great suffering of others. Yet, they also shattered and dehumanized me. The constant tension of strange symptoms and crushing pain left me cut off from a life of embodied prayer and active altruism; I felt alienated from my true nature and impotent to be of service. I decided the greater good was to give my life with the aspiration to raise awareness and help others. As Thomas Merton said, 'Man has no greater love than that he lay down his life for his friend.'*

*If anyone asks, you can say I ended my own life. But it would be more accurate to say I died from Electromagnetic Field (EMF) poisoning. I am not ashamed of my actions. They were based on compassion for my own suffering and the desire to prevent more people from becoming sick.*

*I'm not the first person with Electro-Hypersensitivity (EHS) to die by their own hand, but perhaps I'm the first to publicize it. And that's my whole intention. Let me be the poster-child for this 21st century plague. Let me be the impetus for positive change. What sends a stronger message than death? The message is: seek the truth and learn how to protect yourselves. The message is: create housing opportunities for people with EHS.*

*Manmade Electromagnetic Fields (EMFs) at current astronomical levels are unprecedented in human history. And they are not safe. Human cells cannot function properly under this 24/7 radiation shower. That is why people are getting sick. I chose to die so you would know how important it is to reduce your exposure to this invisible toxin. I chose to die so the world would see that safe camps for EMF refugees are desperately needed.*

*EHS has all the ingredients of torture: cause someone intolerable pain, separate them from the community, and prevent them from sleeping. Is that a risk you are willing to take? Is that a life you want for your children? It's much, much easier to prevent than to cure. And those, like me, who get a severe case, find there is nowhere to live, detox, and recover. Let us care for this growing population of our society. Now is the time to start developing wilderness Safe Havens. Every person deserves a safe place to call home.*

*I do not want a funeral, memorial service, or life celebration. Instead, please help the living by honoring my three final wishes. Thank you and may you be well.*

*Take simple steps to lower your EMF exposure and stay healthy. Organize housing for people sick with EHS. Resources for this are: the fantastic book* The Non-Tinfoil Guide to EMFs *and the* Safe Haven *tab at www.HeartMind.info.*

*Cleanse your body of radiation, viruses, heavy metals, and pesticides, which cause EHS and other chronic illnesses, including lupus, Lyme, MS, RA, and cancer. It's fun and yummy! Read the beautiful book* Life Changing Foods *by Anthony William.*

*Take extra good care of each other. Spend time each week in loving service to a sick, injured, or housebound person. Let them know they are not forgotten. This is the true purpose of being human. Resources for this are: the books*: Everybody Always *by Goff, Peace Pilgrim, and the* TLC Book tab *at www.HeartMind.info.*

*May my death usher in a new era, an era in which EHS and EBV are taken seriously, diagnosed correctly, treated immediately, and prevented widely. May this be an era of permaculture eco-villages, bountiful food forests, abundant altruism and safe technologies. I gladly sacrifice my life for that purpose. May it be so!*

*Visit my blog, www.HeartMind.info, for more information and continued updates from friends. Look for me in the sun, the moon, and the stars. Pray that I've melted into Mother Luminosity and am helping all beings, in all worlds, in all ways. May the bodhichitta flourish!"*

PART THREE

# CANARIES IN FLIGHT

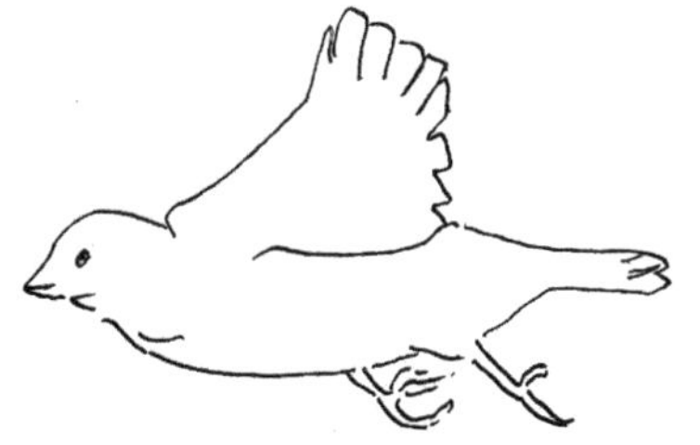

## *Chapter Ten*

# Exodus

### ***My story continued...***

IT IS SO HARD to think of moving again. It is autumn 2019 and we are now settled in comfortably, feeling fully integrated with our surroundings—the current state of affairs presenting in stark relief to our first encounters with our desert home—a home which had initially felt like a fascinating, yet harsh, inhospitable environment, making us feel almost like we had landed on another planet. In this unfamiliar terrain we were met with a wide variety of sharp and thorny plants that scraped at our legs and poked through our shoes, while various kinds of ants bit between our toes and made their way into our pant legs. (Yes, "ants in pants" is a real phenomenon!) At the mercy of relentless stings from scorpions, bees, and wasps—we let out little yelps or loud screams daily while trying to navigate our new world. Dodging the rays from the scorching sun, trying our best not to perish from heat stroke and dehydration in summer, or from below-freezing temperatures in winter, and while leaping away from the striking radius of venomous rattlesnakes, we often wondered why we had decided to move to this hostile land.

But just like others in our position, we had found it to be much more physically, mentally, and emotionally tolerable to live in an environment of natural harshness and extremes than to live in one of manmade artificial harshness and extremes. And so like the Hopis, the natives who inhabited this land long before us, we are also here as asylum seekers. The Hopis had learned they must live in the places other men disdain if they wished to live in peace.

But now, four years later, I cannot imagine *not* being surrounded by purple-green mountain ranges un-blighted by cell towers or tall buildings. Or being without the endless stretch of deep cobalt-blue sky filling the vast horizon, dotted with drifting puffy-white towering chimney-stacked clouds, or dark rolling thunderheads in summer, ushers of awe-inspiring lightning shows. Nor can I imagine being without the lights from thousands of sparkling stars by night as company overhead.

Yesterday a roadrunner was at my window, trying to balance on the narrow sill but failing miserably. She made several attempts before abandoning her pursuit, which had been undeterred by my peering at, and speaking to her from the other side of the glass.

Later, during my nightly half-mile jaunt to my homemade "sleep hut" tucked away in one of the several nearby canyons, I passed a herd of javelina (otherwise known as "skunk pigs"; an endearingly-ugly type of Peccary, with porcupine-like bristly coats, they are common frequenters of our compost pile where they indulge in regular all-you-can-eat kitchen-scrap buffets) and said goodnight to one of the many jackrabbits I often encounter on the path before bed. This morning I was greeted by a tortoise who had hidden in his shell upon hearing my approaching footfall only to shyly peek his head and legs out again as I spoke to him in soothing tones. He lifted his small, wizened head to look up at me awhile and locked my gaze with his.

Other days I wake to deer leaping over chaparral bushes and prickly pear cacti, impressive horns or white tails, the last and sometimes only traces of them glimpsed in their graceful stampede.

Some days we spend lazy afternoons in hammocks and watch what we have affectionately dubbed "Ant TV" looking down at the endless comings and goings of differently sized and colored ants scurrying across the many extensive networks of ant highways they have built for themselves on our property. Ants are amazing creatures. Being incredibly strong; they are capable of doing things beyond our imaginings when translated to human scale.

Overhead, a swarm of bees sometimes passes by, a mass of black haze accompanied by a nearly deafening sound of humming. The scene is powerful enough to send our cats ducking for cover under the nearest shelter of sticks or thorny foliage.

If we sit outside in the twilight of summer evenings, bats often swoop down from the night sky, sometimes grazing our ears as they capture mosquitoes buzzing near our heads. Or we might see the Great Horned owl who chooses to perch on the highest, most-impossible-to-balance-upon-looking tiny branch of a willow tree planted to the east of the house. The owl surveys the fields below from its' perfect, if precarious-looking, post and in a heartbeat opens its majestic wings and swoops down to within inches of the ground in order to snatch a small mouse in its great talons and fly off. Or if it misses, which it sometimes does, it flies back up to its lofty perch and patiently awaits the next opportunity. Some nights the full moon rises to rest just behind the owl, silhouetting the regal bird against its bright glowing spotlight.

Almost daily I find myself lifting a lizard, mouse, or snake by the tail and gently removing it from our house after one of our cats has carried it inside, only to release it so they can have a fun game of chase indoors. Just the other day, a mouse our cat Mina was chasing around in my office, brazenly scurried up my legs in its efforts to escape from her clutches. I admit to having let out a little shriek at the surprise, but still managed to compose myself enough to retrieve the little critter for release out of doors.

No longer do flies buzzing around my head much bother me. Nor do spiders hanging upside down from my sleep-hut ceiling positioned just over my pillow at night. I simply watch the spider in its moth-and-gnat insect-catching pursuits as I drift off to sleep. These creatures simply represent other facets of the vast web of life in its array of incredible diversity, dancing all around me.

During long summer nights, we go to bed to the sound of hundreds of crickets, lulling us to sleep. Some nights there are choruses of toads or frogs. Other nights, packs of coyotes sing their mysterious haunting melodies. Some evenings, bobcats saunter up to the house, and completely unperturbed, nobly ignore our shouting, jumping, and frantic waving of arms in our attempts to move them off.

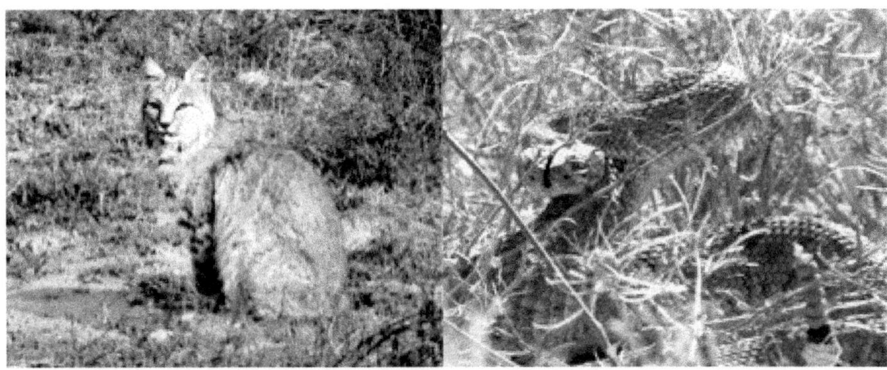

Over the past few weeks, since the summer rains started, we have been graced with shiny, plump toad visitors at our door. Each evening the gigantic Sonoran Desert toads come to sit in the puddles on our side-walk and catch bugs attracted to the house by the few dim lights turned on inside. The toads sometimes leap and crash against our glass door in their bug-hunting pursuits, which elicit pathetic meows of yearning from our cats, now shut indoors for the night, wishing to be on the other side of the glass so they can have a turn at catching one of the temptingly fat toads.

We have also been visited by, the fortunately very occasional, bear and mountain lion. But I can easily say that the morning I spotted the gorgeous, dark honey-brown, mountain lion prowling along the dry riverbed, from my hidden perch atop the twenty-foot-steep ravine overhead, was one of the most beautiful, unforgettable moments of my life.

We do have to watch where we step when the rattlesnakes are on the move. At least the noise from the shake of the rattle usually gives us ample warning to move away, and for this I am grateful. But even the dreaded rattlesnakes represent important members of our bio-diverse desert family. They help maintain balance by feasting on the overwhelming number of rodents, which also helps to keep the number of assassin bugs in check (bugs that leave nasty itchy bites and cause severe allergic reactions in some). It is also a pleasure to watch their seductive slithering movements and glimpse them asleep, curled up in self-contained cradles on the side of a hiking trail.

Springtime here, unlike in so many EMF-radiated areas experiencing "electronic silent spring," still means birdsong of all kinds, day and night. We live in a large bird-migration corridor and a lot of birders have made their homes here for this reason. But the birders tell us the numbers of winged visitors are falling annually. Some species they used to see regularly are not returning, because they are getting lost or dying off along the way.

This spring a pair of ravens built a large nest fifty feet up in a cottonwood tree by the river, right at our crossing. They circled daily, guarding the eggs and later their brood, after the large eggs hatched. Not long after, we heard the cries of baby ravens in the nest each day and were harassed by loud caws of the raven parents who sometimes came close to dive-bombing us each time we passed through—providing for an often-harried adventure walking to and from our car and the house. But it was also such a fascinating experience to hear the baby birds and finally watch them leave the nest on their first flights. These days it is a quiet river crossing, the family having since flown away. But the nest is still there, perched high up in that tree, serving as a snapshot reminder of our shared encounter from a few months back. Seeing the nest on our daily treks seems to us, a quantifiably better keepsake than browsing through photos on a computer screen via social media or other online platforms.

Moving means leaving all of these creatures behind, all of the living beings we have come to think of as family, and like Chief Seattle eloquently lamented in his famous speech, we fear suffering a great loneliness alongside most of "civilized" humanity, being separated from these wild creatures.

People from cities or suburbs, lands which have been tamed, subdued, poisoned and controlled, may look at our property on a map and smugly decree, "middle of nowhere," and "how lonely it must be" and have no understanding of just how lonely they themselves are and how much company we have by comparison. Diversity of life equates to richness of soul. Most humans are now soulless, and incredibly, unspeakably, and unidentifiably, *lonely*. It is a loneliness of spirit which increases with the extinction and suppression of every other plant and animal. Because we fail to see and understand our connection to all living things, so with each death of another life form we kill off some unknowable part of ourselves.

And we know that if we move it will be because we have been driven out, pushed out by the encroachment of "smart" technologies forced upon us by our utility company. It will be due to a huge increase in radio-frequency, microwave-radiation pollution with which we can no longer live. It will be because of an alteration in our natural electrical fields, which are fully supportive of life; fields, which will be suppressed and overtaken by artificial EMFs, manmade frequencies—fields which will harm, sicken and kill an increasing number of humans in our environment but along with them, and often before them, it will do the same to the rich abundance of life, which this unique valley is still able to sustain. It is one of the last intact wilderness areas in this part of the country.

Not only is the diversity of life being threatened with upset by the SMART grid, but it is being threatened by the encroachment of proposed new megalithic power lines, to be built upon our mountaintop and through our valley, which would irradiate the entire valley with harmful levels of high and low frequency EMR pollution and destroy wildlife corridors from the building of new roads and clear-cutting of large swaths of land to make way for the monstrous structures, solely to supply more power to southern Californian residents—a population which consumes four times as much energy and water as do residents of this part of Arizona. The locals here fought against this proposed assault for over eight years but lost by a narrow margin.

And it could be said that the stress of this battle cost one courageous and dedicated community member, who spearheaded the fight, a kindhearted, well-loved man, Mick Meader, his life, as he (a non-smoker) died the following year from sudden-onset lung cancer. The company, Sunzia, won its permit but is still meeting with resistance at other parts of the planned route. Meanwhile, the local inhabitants of this valley are holding their breaths, and crossing their fingers, hoping for a miracle.

Additionally, there are developers who are planning a twenty-eight-thousand-unit condominium complex in our nearest town of five-thousand residents. There is already a water shortage concern in our valley due to years of drought and these developers think it would be okay to somehow support up to eighty-thousand new residents along with an irrigated golf course to boot? The environmental assessments done so far state that this would not be possible and would be devastating to local ranches and wildlife, but these reports are being ignored and for some reason the Trump administration and the military are pushing this building project through undeterred. And although local residents and environmentalists are still trying to stop it, the future outcome is uncertain.

One of the more ludicrous proposals for dealing with the water shortage issues, which the development would surely cause, is to cut down the native cottonwood trees along our river, the San Pedro. Cut them down because they require "too much water" to sustain themselves. Cut down the only tall trees we have in this valley, the ones upon which the wild animals rely for shade, food, and shelter. The ones in which we hang our hammocks and watch the leaves dance and sway with the movement of ivory-colored branches, flickering in the dappled light, like thousands of pieces of celebratory confetti thrown to the breezes overhead. The same trees whose transitioning leaf displays signal the change of seasons with newly formed, bright, lime-green buds in the early spring, deep, dark emerald green in summer, yellow and browns in fall, and finally bare branches in winter. The same trees which throw-off packets of white cottony seeds to ride along the wind, filling the air like snow in springtime. Killing these beautiful ancient trees is their small-minded disconnected idea of a "solution."

If we have to leave, what saddens me more than leaving behind my new "family" is knowing that I may not be leaving them behind at all, that instead, they, along with countless other animals, birds, and insects, will have to leave as well, and they, unlikely able to find anywhere to go, will doubtless perish. I hope that if we must leave, we can find some other refuge. We still stand a slightly better chance than our animal friends, but somehow that thought does not give me a lot of comfort. It would be a heartbreaking exodus on so many levels.

Reports of other places now being devoid of insect humming, birdsong, and other signs of wildlife steadily increase. In cities it seems the only other life form able to survive at all is the common indomitable cockroach. Even the sparrows are disappearing from the skies and trees. Not just in cities, however (those places are especially empty of non-manmade sounds), but other countryside areas are no longer safe habitats for wildlife either.

One of our visitors who had been traveling around the country as an EMF remediation consultant, helping to make the EMF exposures in homes safer, had remarked upon the vast quantity of ants in our area, the likes of which he had not seen elsewhere, even though he himself was an Arizonan desert native. He also noticed a pronounced difference in the number of ants[xxiii] present near the power lines along the road versus on our property. Near to the power lines they were almost non-existent.

Given what is known about the harmful effects of manmade EMR on wildlife, there is an alarming trend among biologists and environmentalists to tag or inject wildlife with RFID chips in order to track their movements. They reason that this practice will help them to discover what is going wrong in the environments of these animals; to discover what is serving to contribute to the steady decline of these species in many areas—all the while, fully ignorant to the fact that the chronic microwave exposures to which these animals will now be subjected 24/7 (much like giving them cellphones to carry on their bodies at all times) is very capable of causing severe impacting damage to the animals' DNA and thereby likely to cause cancers and other illness in the very animals scientists and conservationists are trying to protect. The accompanying

---

[xxiii] Ants, like bees are especially sensitive to manmade EMFs as these fields seem to serve in confusing them by interfering with their EMF sensitive homing abilities.

wireless infrastructure implemented and needed to track these same animals also contributes to the animals' demise by irrevocably degrading its' environment and additionally makes such would-be safe havens for electro-sensitives, such as myself, off limits. Similarly, wireless-transmitting "critter cams" placed in sensitive wildlife habitats for remote picture-taking has the potential to harm these very animals that the photographers and scientists wish to study and save.

The criticalness of educating biologists, conservationists and environmentalists about the dangers of wireless and other forms of manmade EMF radiation cannot be overstated. This must happen quickly. The blind spot these well-meaning people have on this topic is as huge as it is devastating. All of our lives hang in the balance.

## *Nowhere to Run, Nowhere to Hide*

Thanks to the megalithic multi-trillion-dollar wireless industries and industry-funded government plans to "serve" the rural areas with these ever rapidly increasing new "smart" technologies (heralding these plans with media taglines like "rural first"), those desperately in need of a refuge are left with few-to-no options and must as a result, simply suffer incredible daily torment from the effects of this so-called "smart" grid, fully knowing that if they could only get away from wireless radiation they would feel better than fine and no longer be "functionally impaired."

As Gary, whom I introduced at the end of chapter five, so perfectly explains, "Once I was on the land and away from the field sources, I started recovering my full function. Just like Olle Johansson from the Karolinska institute says, this is full functional impairment. It's an environmental problem, you get the environment fixed and you are just fine. It's very hard to believe when you see someone in my condition or in Ruth's condition, that there is nothing wrong with that person, even that person thinks there is something wrong. But I had been on the land long enough to realize every single day I'm out here in the wild by myself I get better. I feel incredibly good, there is nothing wrong with me, my mind is completely clear, I have more energy than I've ever had, I don't get sick anymore. Then I go back into town, in a really short period of time, it's right back again."

The obnoxious attitudes of news writers offend we rural folk with their self-proclaimed hero stance written up in their own biased mainstream-media articles, which presume to know the desires of rural people and pretend we do not already have reliable (and *safe, wired*) high-speed Internet connections. They also use us without our consent as their "brave-new-SMART-world" poster child, and inundate the public with ceaseless propaganda, all the while, failing to mention the millions worldwide protesting 5G and SMART grids, or that many rural residents do not want the new tech, and were never asked if we wanted it in the first place.

And with Elon Musk's plans to fill Space with a staggering 20,000 5G satellites, intended to blanket the Earth with millimeter wave radiation, by 2021 (and up to 40,000 and possibly more in the years after) these mighty forces hope to leave no stone unturned and no corner of the Earth spared this harassment.

Leaving us to ask the same tired question, "*Where can we go?*"

*Chapter Eleven*

# Addendum 2022—Three Years Later Can You Hear Me Now?

*"Every creature on earth returns to home. It is ironic that we have made wildlife refuges for ibis, pelican, egret, wolf, crane, deer, mouse, moose and bear, but not for ourselves in the places where we live day after day. We understand that the loss of habitat is the most disastrous event that can occur to a free creature. We fervently point out how other creatures' natural territories have become surrounded by cities, ranches, highways, noise, and other dissonance, as though we are not surrounded by the same, as though we are not affected also. We know that for creatures to live on, they must at least from time to time have a home place, a place where they feel both protected and free."*

–Clarissa Pinkola Estés, PhD, *Women Who Run with the Wolves*

AFTER A YEAR-and-a-half battle to stop the SMART grid from destroying one of the last remaining white-zone outposts for electro-sensitives, my partner and I, along with a few other neighbors, had to admit defeat and seek (yet another) refuge, elsewhere. We simply did not have the funding (a minimum of one-hundred-thousand dollars as retainer for a lawyer) for legal representation to continue the fight against our electric utility company–SSVEC. Also, the pandemic of 2020 quickly diverted the attentions of other members of our desert community from the pressing matter of the coming, forced radiofrequency grid to what most considered an even greater threat–that of the mysterious new virus wreaking havoc across the world.

The news of the novel coronavirus emergent from its deep slumber in the bat caves of China unleashed upon the world after a bat-eating Chinaman allegedly became infected in a Wuhan wet market hit the headlines on nearly the same day (rather "conveniently" for the wireless industry) as the first worldwide protest against 5G involving 200 participating nations, on January 25, 2020. Many of us coping with EHS and engaged in decades-long battles with the wireless industry and industry regulators, strongly feel that this viral pandemic has served as a "necessary" scapegoat to protect vested interests in both tech and pharmaceutical industries, whose financial positions were previously threatened in the face of mounting lawsuits and outrage at corrupt practices and lack of accountability and oversight into human health and environmental protection.

Since the hallmark symptoms of Covid-19 (this latest added to the long list of new diseases of civilization started with the first installations of electrical grids and wireless-communication systems in the late 19th century) demonstrate as insufficient oxygen levels mimicking high-altitude sickness; the predicted fallout described by Dr. Erica Mallery-Blythe in her 2014 "EHS—A Summary" paper (presented in chapter one), if nothing were done to slow or halt the rapid expansion of wireless systems—in the extrapolation of EHS statistics increasing to fifty percent by the year 2017—seems to have manifested itself not under the label "EHS" or "Microwave Syndrome," but as "Covid-19," and also given other labels like, *chronic fatigue syndrome, fibromyalgia, MS, hypothyroidism, anxiety,* and both neurological and psychological or psychiatric labels.

Oxygen starvation is likely a hallmark symptom of this new disease due to the direct effect of millimeter-wave exposures from the new 5G-communication grids on oxygen molecules.[134] So concerning is this effect that NOAA[135] has publicly declared its opposition to Elon Musk's 5G SpaceX satellites as, due to millimeter-wave impacts on water molecules, they pose a direct threat to accurately predicting weather patterns and warning people of hurricanes and typhoons in a timely manner.[xxiv]

---

[xxiv] 5G millimeter waves also threaten to interfere with airport/airplane navigation equipment and major US airlines are calling for a halt to installations near and at airports. https://stevekirsch.substack.com/p/80-of-airline-pilots-arent-going

5G millimeter waves operate at between 30-60 GHz (Giga-Hertz), but oxygen molecules oscillate at between 10-60 GHz. These molecules are diatomic, meaning that oxygen (O2) represents two oxygen molecules bonded together by shared electrons. According to Dr. Robert O. Young, DSc, PhD, "When the oxygen molecule is hit with 60GHz 5G waves, these waves affect the orbital resonance properties of those shared electrons. It is those shared electrons that bind to the hemoglobin in our blood. When the oxygen is disrupted, it will no longer bind to the hemoglobin and myoglobin (oxygen-carrying molecules) and therefore the red blood cell will not be able to carry oxygen to the cell's powerhouse 'membrane.' Without oxygen, the liver becomes congested, and the body, and brain, begins to break down due to slow suffocation. Because the brain is the bodily organ most sensitive to the lack of oxygen, not getting enough oxygen to the brain will result in brain hypoxia."[136]

In other words, 5G frequencies affect water molecules by "exciting" them—meaning that they cause orbiting electrons to "jump", and sometimes even jump out of orbit from around the nucleus of a cell (an effect normally attributed solely to ionizing radiation, however this can also happen in non-ionizing radiation exposures—effectively blurring the line between the two, with biological effects often identical with each type of exposure).[137]

For blood to transport oxygen, it has to bind one molecule of hemoglobin with four molecules of oxygen to form oxyhemoglobin. This binding happens as a result of the concentration of oxygen being greater in lung tissues than oxygen concentration on hemoglobin molecules. "Partial pressure" due to this oxygen imbalance affords the means whereby oxygen can transfer from the lungs onto hemoglobin molecules elsewhere in the body.

But microwaves and millimeter waves can cause interference in this process by creating permeability in the cell membrane, which allows for the entrance of other molecules like excess calcium ions (as related by

---

Dr. Martin Pall[138] with his studies on EMF and Voltage Gated Calcium Channels) and the displacement of others (such as carbon dioxide molecules).

If the hemoglobin molecule is occupied by other elements, its structure can change and if the structure changes, so too can its function—resulting in an inability to bind with oxygen and carry oxygen, leading to oxygen starvation in the body, resulting in this "high-altitude sickness" or "suffocation" related by Dr. Kyle-Sidell[139] in early 2020.

Heme is also adversely affected in people with porphyria and as mentioned early in this book, porphyrics can be born, but also made as a direct result of injury to heme from exposures to wireless radiation.

In fact, the Chinese government's initial response to reports of "unique" respiratory illness presented in patients at the Wuhan hospital, was a warning that their symptoms may have been radiation-induced. But this information was quickly buried as it threatened industry interests, and was traded in favor of the Germ Theory model of disease. However, the government's first reaction is likely the more correct one, since CT scans of Covid-19 patients compare directly with those of lung scans of patients injured by radiation.

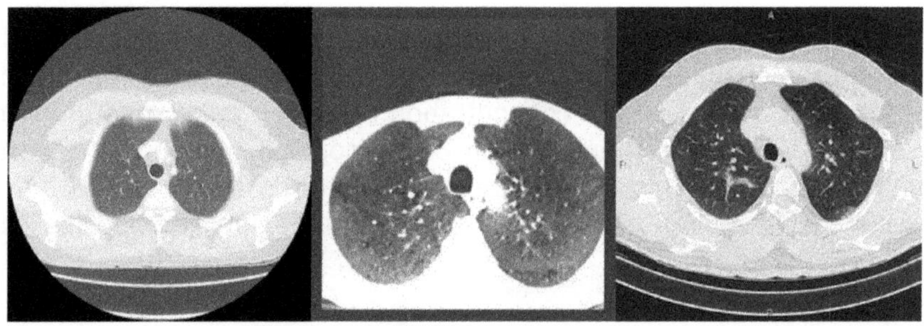

**CT Scan of Healthy Normal lungs**
Case courtesy of Dr Andrew Dixon, Radiopaedia. From case 36676.

**CT Scan radiation damaged lungs called Radiation Pneumonitis**
Case courtesy of Dr Bruno Di Muzio, Radiopaedia. From case 150530.

**CT Scan of COVID-19 patient lungs**
Case courtesy of Dr Mohammad Taghi Niknejad, Radiopaedia. From case 75684.

***The above images show CT Scans for healthy, normal lungs (far left), radiation damaged lungs aka, "radiation pneumonitis" (middle), and for a Covid-19 patient's lungs (far right).***

***The scans for radiation damaged and Covid-19 damaged lungs bear striking similarities, even as described in the case notes.***[xxv][140]

Apart from respiratory distress, many other symptoms now attributed to this "new disease" have also been linked with microwave/radiation injury, all having been reported by those suffering with EHS or from microwave-radiation damage—symptoms like: cognitive dysfunction ("brain fog"), short-term memory loss, difficulty concentrating, trouble sleeping, chronic headaches and migraines, and neurological issues. These symptoms have been, in particular, listed as chronic in "Covid-19 survivors."[141]

And so, it is easy to see how anyone already adversely affected by wireless radiation from wireless-communication grids, like canaries such as myself, and for those who have studied the biological effects of EMFs for decades before the alleged "outbreak" of a new plague that just happened to coincide with the "outbreak" of 5G systems[142] all over the world, may understand the new viral pandemic to be acting as scapegoat in an industry attempt to cover-up the ill and *known* effects the new systems would doubtless have on a percentage of an unwitting public.

This might seem a farfetched idea until one understands just how powerful this multi-trillion-dollar industry actually is—one that has already bought out world leaders and news media agencies long before. Couple with this the power of a similarly grossing pharmaceutical industry poised to further line its pockets with obscene profits[xxvi] from government-mandated vaccines as *the* "solution" to ending the pandemic—AND

---

[xxv] The case comments accompanying the second image for "radiation pneumonitis" are as follows: "*In cases of early or subtle radiation-induced pneumonitis,* ***areas of ground-glass opacity*** *may be evident on CT despite a normal chest x-ray.*" And the case comments accompanying the third image of "Covid-19 patient lungs," are as follows: "*Multiple patchy* ***ground glass opacities*** *and consolidations are scattered in both lungs predominantly at subpleural regions. This patient had positive RT-PCR testing for COVID-19.*" (emphasis added)

[xxvi] In 2021, pharmaceutical companies selling COVID vaccines, collectively earned approximately $100 billion—far more than grossed in any previous year to date for ANY drug. (See NY Times journalist, Alex Berenson's, *Unreported Truths about COViD-19 and Lockdowns, Part 4: Vaccines*)

add to this the fact that the virus (SARS-CoV2) alleged responsible for Covid-19 has yet to be properly isolated[xxvii] and identified, with all models of the virus presented thus far based solely on computer renderings and conjecture—it is not such a stretch to suspect that we might not be getting the full truth from our elected and unelected leaders.

As it happened, the pandemic, real or not, has served to effectively drown out voices of canaries and their allies, while at the same time enabling massive installation of 5G systems into empty schools,[143] government buildings and open public spaces while the people of the world were forced to stay at home and prohibited from gathering and putting a stop to the encroachment. Lawsuits against the FCC and telecom companies, years in the works, screeched to a standstill, as government offices shut down with all "non-essential" workers forced on hiatus. I wonder at what has become of our world when the pursuit of justice is no longer considered "essential," or feeding one's family, or touching and hugging one's children and friends as well.

Telecom and medical tech employees continued on their path of destruction and devastation unencumbered, freely roaming the streets and cutting down trees to make way for 5G antennas, and in my own community, busily installing SMART grid repeaters on our rural dusty road, as they, like gods among us, carried the invisible "essential" pass bequeathed upon them by our "powers-that-shouldn't-be" but sadly still are.

---

[xxvii] No alleged pathogenic virus to date has been properly isolated or purified—all such claims to virus isolation are fraudulent and refer only to isolation of cell cultures from sick patients not of virus isolation from cell cultures. And all images of viruses show only what is assumed to be "virus particles" (really just strands of genetic material), not full virus genomes as no such photograph of a virus has been taken, even though the modern-day electron microscope is more than capable of photographing viruses at their alleged sizes—if they exist at all. The Germ Theory of disease itself, first popularized by fraud and plagiarist Louis Pasteur in the 19th century, that views all illness as the result of exposure to microbes, has been disproven many times over. Blaming microbes for illness has served the purpose of making immense profits for pharmaceutical companies that rely upon public belief in one microbe = one disease = one cure. It is high time we reviewed our acceptance of this flawed theory. A statement signed by Dr. Andrew Kaufmann and Dr. Thomas Cowan, detailing the lack of evidence for isolation of the SARS-CoV2 virus alleged responsible for Covid-19 can be found here: https://andrewkaufmanmd.com/sovi/

While still under "stay-at-home" orders, I, my partner, and our EHS neighbor **Amy and her family** had to seek new homes for ourselves, far from the encroaching grid. Amy's family was fortunately able to sell their home and farm in Cascabel and buy another off-grid property several hours to the north in a more exposed landscape surrounded by public lands and sprawling cattle ranches. They bought 40 acres of scrub brush, red rock and sand dunes with no structure or power, but with a well (a survival essential in the intensely arid desert). Incredibly, only a year and a half later, she and her family were able to build for themselves a lovely comfortable-but-cozy home, a root cellar, acres of productive vegetable gardens and beautiful landscaping. After several months of having had to first live in a tiny, but beautifully decorated, metal box of a trailer in very challenging conditions, they have succeeded in creating a stunning oasis in the desert, where fortunately Amy and her own are thriving.

However, there was a worrying time in the spring of 2021 during which unusual radiofrequency activity caused Amy and her nearby neighbor (another canary who also bought a property in the area in order to escape SMART grids) intense head pains and other debilitating symptoms. Unable to locate and discover the source of the new activity in a previously quiet, low-EMF zone, they had to wait and hope it would prove an anomaly and not a permanent new source of RF pollution. Fortunately, it turned out to be short-lived and likely related to covert military testing, known to happen on public lands from time to time, as the military is allowed use of such lands for testing radiofrequency communications systems and even EMF weapons. But the worry was that testing for the latest satellites launched into low orbit, thanks to the continuance of Elon Musk's SpaceX plans, had occurred, and as such was serving as a harbinger of worse yet to come. We are still hopeful it is not the latter but the former hypothesis that is the correct one.

As for myself, Sam, and our two cats, we decided against joining Amy and her family, although the generous offer stood for us to bring a trailer of our own and camp out on the land—mainly because I could not bear the thought of having to live so much further from the coast than we already were, nor to move to an even harsher, more arid, and colder desert environs. Now that the world had also closed off its nations' borders, including the land border between the US and Mexico (Mexico being where I went to get my regular "water fix" at the Sea of Cortez), we decided it would be best to move to the west coast if we could, so that I would not be kept away from the sea or ocean indefinitely.

To this end, after first having explored several properties in the north-eastern portion of Arizona, within a hundred-mile radius of Amy's new home, we scoured housing opportunities in the northern-most, sparsely populated, western portion of California. We primarily sought work-trade situations, as our finances, after having bought a 6x10 trailer for the move and as a back-up housing plan along with a truck for hauling it, were quickly dwindling with little means available to us for replenishing our accounts after the death of my employer in January 2020 and subsequent death knell rung to that business thanks to subsequent pandemic lockdown policies.

Landing upon a farm work-trade position in a densely wooded forest with spotty cellphone reception, we hired a cat sitter and set off on a 10-day-long adventure to see the place in person before committing, so as to determine whether or not it would really feel good and prove the safe haven we hoped.

It is necessary for me to back up a little in my story here and tell of our thwarted plans to emigrate from the US before seeking refuge in the northwestern states. Prior to pandemic lockdowns, grounded flights and border closures, feeling a stark lack of remaining and affordable options for us in the US and knowing that the US, on the whole, was showing no signs of rolling back wireless technology anytime soon, we spent several months seeking refuges in a variety of overseas countries from Belize, to Panama, the Azores and Canary Islands[xxviii], to the Philippines in South East Asia, French Polynesia in the South Pacific, remote parts of Ireland and Scotland, Uruguay, Chile, Bolivia, the Maldives and even Pitcairn island, one of the most remote outposts in the world—a 2-mile-wide island with only 50 inhabitants.

---

[xxviii] We thought it would be particularly fitting had we found refuge in the Canary Islands, with myself being a "canary" in need of a home. However, even on the sparsely populated island of Las Palmas, the off-grid farming homesteads all seemed to have 4G cellphone reception and/or WiFi signals. Apart from possibly buying a cave house and living in it (a real option on those islands), it did not seem the islands would be a viable solution and hiding out in a cave is not exactly the lifestyle we were hoping to find.

We left no stone, country or island unturned in our search. As others in our situation can relate to, we were willing to go anywhere in the world provided it was somewhere we could live free from the torment artificial EMFs cause us. But for the more ambitious of us, we also hope to find other things that feed our souls. As in my case, I hoped to be near the sea as I have a strong passion for the ocean and had my fill of time spent away from it (excepting the aforementioned excursions to Mexico) during my desert retreat. Also, many with EHS find that they do not well tolerate living at high altitude, and so while possibly gaining some relief in high-elevation locations with low RF interference, they may not fully recover and thrive due to the stress imposed by lower levels of oxygen at these higher altitudes.

This was true for me. I did recover quite a lot at 3,500 feet due to the complete absence of radiofrequency interference in our location, however I suffered shortness of breath, muscle weakness and fatigue due to the thin air. While not medically confirmed, I could be one of those with EHS who also has porphyria and as such fares better at, or nearer sea level. But finding safe locations by the sea can be extra challenging as these areas tend toward being more densely populated, which also means more densely populated by wireless and electric grids.

All of the places we searched abroad posed some kind of insurmountable obstacle to us, either in the form of financial hurdles or logistical trouble with migration, either physically or legally or both. Also, many of them did not present us with the low-EMF options we sought. During these investigations it was not uncommon to be met with the sometimes-scathing suggestion to go live in the Amazonian wilderness, as that seemed the only place left for us. Sadly, this assumption proved to be false. The Amazon is actually now one of the most EMF-contaminated places on Earth.

As such, indigenous low-tech people living in the Amazon have been "infected" by "Covid-19" due to the massive amounts of EMF-radiation pollution forced upon these primitive peoples in the form of "twenty-five enormously powerful surveillance radars, ten Doppler weather radars, two hundred floating water-monitoring stations, nine hundred radio-equipped 'listening posts,' thirty-two radio stations, eight air-borne state-of-the-art surveillance jets equipped with fog-penetrating radar, and ninety-nine 'attack/trainer' support aircraft,' [all of] which can track individual human beings and 'hear a twig snap' anywhere in the Amazon."[144]

When considering a move (to wherever) we also had to consider industry and government plans for the deployment of 5G, or expansion of 3G and 4G systems. We hoped to, at the least, find a place where these systems were either being blocked or whose deployment, for whatever reason, might be delayed. After exhaustive searches and weighing all of the options, we finally found a low-EMF refuge in the mountains of southern Ecuador in an expat community called Vilcabamba.

After having had several online chats with expats living in Vilcabamba about our need to escape the encroaching SMART and 5G grids for my health, and having been reassured that the community was actively fighting the same and were determined to keep the area unpolluted by additional RFs, we decided that an exploratory visit was merited. I personally doubted how well I could endure such a long international flight after having avoided air travel for several years due to inflight WiFi and radar systems, along with ubiquitous chemical pollution in the form of disinfectants and passengers' perfumes. I certainly did not feel I could survive a round-trip passage within a two-week period on top of an additional flight for the move. So, it was decided that Sam would make the trip for both of us and meet up with another friend of ours who happened to be visiting neighboring Columbia at the time. Sam planned to arm himself with various EMF meters and track exposure levels at a number of available low-rent houses in the area while also getting a feel for the community socially.

We had tickets in hand plus a number of Airbnb rentals booked. All of the travel itinerary details were printed out and Sam and friend were ready to go when we got news of the first Covid-19 cases announced in Ecuador alongside the first wave of flight cancelations. Only two weeks before Sam's travel dates, did all of his flights list as canceled. Soon after, Ecuadorians found themselves under extreme militant lockdown conditions with absurd curfews and police- and military-enforced mask and glove wearing, disinfectant use, and severe travel restrictions which, at the worst of it, allowed only one person per household out one day per week for grocery shopping while limited to only 25 minutes inside a store. Thousands were forced not only to remain on their properties, but inside their homes, for weeks on end.

Residents of Vilcabamba helplessly watched as teams of biohazard-suit-clad men sprayed their yards with unknown oily chemical substances. One person was able to get an answer of "petroleum-laced ammonia" from a worker when questioned as to the contents being sprayed on her garden. Many similarly watched their properties and tourist-rental businesses, built upon years and decades of hard work and significant monetary investments, crumble to ruin overnight. What has followed in the wake of the early, extreme pandemic measures in spring and summer of 2020, has borne all the hallmarks of police-state behavior, more so even than witnessed at the same time and since in many other countries including in the United States.

And as in the case of so many other places where locals were fighting against 5G and SMART-grid installations, the people's efforts were rapidly steamrolled as utility trucks swarmed their towns installing more antennae as the residents were made to stay at home, unable to resist the tech takeover. Excuses made for the "need" for increased surveillance systems in order to "protect" citizens in contact-tracing efforts under public-health directives has paved the way for the increased proliferation of such wireless systems the world over, even in direct opposition to public sentiment and town council rulings.

Emergency public-health laws have served to override all others, allowing for unprecedented industry and political power-grabs and overreach, all without proper oversight. Our personal freedoms and liberties have been made to take a backseat to "public-health" concerns. But where were the concerns over the health of the public when it has come to dangerous radiation exposures, particularly to vulnerable subsets of the population like those coping with EHS and MCS? It is a hard pill to swallow that all of a sudden, the same leaders responsible for consistently turning a blind-eye to the plight of the canaries all of these years, in favor of lining their pockets with colossal profits made by these same harmful, health-destroying industries, wish now to safeguard public health. It is highly unlikely given who has benefited by this current crisis—namely tech and pharmaceutical companies and Big Box stores—that our best interests are really in mind as decisions are being made to increasingly erode personal liberties in favor of increased police-state dictates.

How can further polluting our environments with surveillance-enabled wireless systems and with increased use of sterilizing chemical agents *safeguard* the health of the public, the next generations, and of the environment? And how can recklessly injecting billions with toxic, genetically-modifying, experimental drugs similarly safeguard public health, especially when the same drugs, wrongfully labeled as "vaccines," do not make claims to providing immunity to any disease, but simply state that they *might* help to reduce the severity of symptoms *if* one were to contract this new disease? The logic behind the administration of this "vaccine" along with all other traditional vaccines, is that by deliberately sickening those who are still well, those who are already sick will be protected from getting sicker. It has only been the already-sick (with a number of other chronic and acute illnesses, including late-stage cancers) who have been "at-risk" for death or physical impairment due to contraction of the alleged virus, *not* the already-healthy. Instead, it has been this latter group being made sick and even killed by way of injection of these so-called prophylactic drugs, rather than anyone benefitting, from poisoning with these drugs.

Further burdening an already over-burdened populace with pumping more chemicals into the environment while forcing more into our bodies and adding more EMF exposures to our everyday lives, never has been a recipe for health and never will be. And it is no accident or coincidence that the first alleged outbreak of the virus being blamed for Covid-19 happened in a 5G mecca—Wuhan, China, the world's very first "smart" city, which boasted over 10,000 5G antennae[145] by the end of 2019. All other outbreak centers, following the first, have also boasted 5G networks[146].

The red-covered outbreak maps depicting the spread of the new "plague" has borne a stark resemblance to the red-marked cellphone coverage maps my partner and I consulted so many years previously when searching for areas untouched by red—a color, as I wrote when starting this book three years ago, that reminded us of spilt blood and actually, quite ironically, of a *plague* outbreak. It has been natural and logical for us to link the two together—the red of the outbreak with the red of the radiation coverage in these maps.[147]

Many Covid-19 hotspots were also polluted by factory smog or biofuel fumes (biofuel is a derivative of GMO plant-based oils containing glyphosates, especially dangerous to the lungs when burned). As I know from personal experience, and a fact, which is otherwise well documented, the combination of both chemically polluted air alongside EMF pollution makes for the perfect storm when it comes to disease outbreaks. These are classic disease conditions creating disease.

If we do not look to the true cause of illness,
we can never hope to find its cure.

Ever since nonsensical pandemic policies have been allowed to rule our nations, the already marginalized canaries have been forced into further retreat. We already had lived lives of intense social isolation, with many of us the only ones wearing masks in public—the freak anomalies donning these masks, worn, not out of fear of contracting a germ, but from our inability to cope with incessant VOCs (volatile organic compounds) polluting the air in all towns and cities, and often in chemically sprayed farmland countryside as well. And we were already well versed at practicing "social distancing," repelled from others due to colognes, perfumes, fabric softeners and always-on, always-transmitting, cellphones, stowed away inside pockets and purses.

The few remaining safe havens for us have rapidly dwindled, as the excuse for 4G and 5G networks to encompass the planet has been announced, and for the most part it seems, readily accepted. No stone, no forest hamlet, no desert island must be left untouched by this technology, so that all humans, we "disease-carrying vectors" can be consistently tracked, monitored... and controlled.

More is at stake for humanity than the overlooked needs of the fleeing canaries, as we are backed into corners and rounded up for drugging and monitoring. [xxix]

But for now, I will go back to my own story, that of my latest, and hopefully last exodus.

After our Ecuador relocation plans fell through and all other international travel suddenly became off limits, we re-shifted our focus back onto domestic options. In a way, the timing of the manufactured crisis (aka "pandemic") worked out for us. Had it started a few months later we might have found ourselves stuck in Ecuador, once a golden refuge for those seeking escape from New World Order dictates, it proved just another pawn in the pocket of world leaders bent on executing their nefarious control agendas over all nations of the Earth—showing just how easy it is to buy off world leaders, especially those of smaller "Third

---

[xxix] For more on this topic, about which I have already written extensively along with co-author John Hamer, see our book, *Welcome to the Masquerade—Prelude to the Coming Reset,* available on Amazon.com

World" nations, who will readily sell their own people off as guinea pigs, if the price is right.

It seems clear today that Ecuador is one of the nations chosen as a test bed for the NWO surveillance society, as even beaches are not off limits to 5G-enabled real-time "contact tracing" where beachgoers are now alerted by drones for breaches in social distancing—something also tested in pockets of the US, but has not (yet) gained a foothold as it has in Ecuador, where the will of the people is being fully ignored. And in December 2021, the country has announced that only "fully vaccinated" international visitors may be allowed entry into the country, while still required to wear facemasks, social distance *and* subject themselves to invasive Covid-19 testing. And now in January 2022, vaccine mandates are being forced upon the entire populace—an unconstitutional move that is being challenged in the courts by concerned Ecuadorians, including residents of Vilcabamba. (Let us hope they win.)

In the end, it turned out to be quite a good thing that we were denied entry into a country, turned from democracy to dictatorship overnight. Also, some of the other countries we had explored enforced similarly frightening measures to the point many expats in Panama (as one example) decided to repatriate back to the United States after enduring extreme restrictions on liberties where ridiculous measures, such as only allowing women to travel for essential purposes on certain days of the week and men on others (as if mixing genders in public has anything to do with viral transmissibility) prompted this repatriation.

Unsure of whether or not I would be up for such a long car ride for our exploratory visit to the northern California coast, having to travel over highways dotted with cell towers—we packed the car to the brim with camping gear and food enough for the ten-day and nearly 2,000-mile adventure—to the ancient redwoods and wild, windswept beaches of the Pacific Ocean. We traveled a longer route through the Nevada desert to avoid the more crowded highways and so that we could camp in low-EMF wilderness areas along the way. Even though driving through the most remote and desolate of desert landscapes, it was still difficult to find campsites free from EMF radiation, but we managed to find enough for the three nights required for the trip.

After the long four days' journey and arriving in the forests near the coast, we were instantly entranced by the constant presence of water after so many years confined to the desert. Rivers, creeks, lakes, the ocean and dense green forests met us at every turn and we feasted upon the abundance of life in this new setting.

Camping next to a creek in the woods on the farm we were investigating felt nourishing and sleep came easily in this low-EMF haven. The farm itself boasted several head of goat, chicken, duck, and an abundance of fruit and nut orchards alongside beautiful vegetable beds packed with rich, black soil. In early summer the weather was perfect, the sky a deep blue and wildflowers graced the hillsides overlooking ponds teeming with darting golden koi.

The property owner offered us a tiny house—a cute 20x10-foot cabin without electricity and equipped with a bedroom loft and small kitchen in a private wooded area where no cellphone signal reached, along with food in exchange for part-time hours of work on the farm. The situation seemed perfect for us, just waiting for our arrival. After a few days visit, we agreed to return and join the newly forming community, even though joining meant not knowing who else would arrive to be part of the group and also meant serious "roughing it"; with compost toilets (mere buckets outdoors), limited solar power and shared communal areas with limited-access hot-water showers and daily hiking up and down very steep terrain plus hard manual farm labor milking goats and shoveling animal waste as examples of some of the many tasks we would have to regularly perform.

Having returned from the trip, the next few weeks seemed endless. After giving notice to our landlady about the move, we had to pack and clean—no easy task given the size of the house and property, the number of things we had accumulated in our five years there, the extreme heat of late July and early August, constant threat of encroaching wildfires and no direct drivable access to the house (meaning we had to carry the bulk of our things half a mile to the truck and trailer parked at the road). But somehow, we managed, and finally got underway, to set out on our latest adventure. So, after packing the cats into our already over-packed vehicle, we watched as Cascabel faded from view in our rearview mirror, likely for the last time.

We had planned four days for the journey west—the amount of time it had taken on our exploratory visit, but in the end, it lasted eleven days. We had made it to our first campsite in northern Arizona without a hitch until I threw out my back repacking the trailer at the site. This injury cost us a large setback in our move, as I was unable to get out of bed (the air mattress on our tent floor) for six full days.

Once I was finally able to sit up without terrible pain, we continued on our journey. By this time the wildfires had spread all over the western states so that it seemed we journeyed hundreds of miles unable to put the black and gray smoke-laden skies behind us. Instead, it felt as if we were constantly traveling deeper into the dark clouds with every alternate course we mapped in our attempts to flee from it. Our campsite intended for our third night, we discovered to have burned to a crisp, only hours prior to our arrival. As we passed the charred landscape and pressed onward, we found that our truck battery was suddenly running dangerously low.

The battery finally drained completely just as we rolled into the nearest town across the California border. Fortunately, we were able to get help just as a hardware store that sold car batteries was closing for the night. The new battery proved only a temporary solution as we discovered the alternator to be the real problem. It being a Friday night, all mechanic shops in the next larger town (there were none in the small town where we stopped) would not reopen until the Monday following.

We had just enough charge on the newly purchased battery to make it to a safe campsite outside of town in the mountains. There we rested easy for a couple of days and enjoyed relaxing next to a stream. However, the smoke that it seemed we had finally escaped, soon moved into the town and increased to an alarming level prompting our need to move on again soon, as the whole town was poised for potential evacuation. Monday came and Sam was able to make it to the next town and have a new alternator installed in our truck, after which, later the same day, we continued onward, finally arriving at a campsite near our destination—but only after I managed to break a toe, after stubbing it on a large rock while carrying one of the cats after taking her for a potty-break walk.

Two weeks after settling in at the farm, and after a glorious reprieve from smoke-filled skies, fires encroached on our area as well, requiring a speedy and immediate evacuation early one morning. Just after our cats had adjusted to the new home and surroundings, we had to pack them along with a few essential belongings into our truck, and head to the coast. It was not until ten days later that we were able to return to the farm. In the interim, we had to sleep in the truck at an entrance to a park where the radio signals were lower than in town, and once again camp with the cats, who at that point were more than fed-up with the camping lifestyle, so that one of them cried most the night through every night, leaving us all extremely sleep deprived.

Those ten days slowly dragged out. Each day we watched the fire maps and made inquiries with local firemen and police about the situation on our mountain. The fire had come very near to burning through our part of the forest. Had it done so, the whole farm would have been destroyed along with many of our possessions, and our new home would have been burned to ashes. Miraculously, the weather conditions changed enough to slow the speed of the fires and allow for firefighters to begin containment efforts when the fires had reached a mere four-and-a-half-mile distance from the property.

The weather also cooled and we enjoyed a few days of light rains. Our hopes for being able to return to the farm and our new life were renewed with the rains and the change in wind direction—so that at the end of the ten-day evacuation we were able to return and begin cleanup efforts. Ash from the nearby fires covered everything, and in our haste to help move out valuable farm equipment during our departure, the place had been turned upside down.

We stayed on ten more months at the farm, but somewhere in the middle of the stay we realized it was not going to work out for us as the long-term solution for which we had hoped. While it was a good low-EMF location, the farm also rented out a few cabins to Airbnb guests—visitors who shared communal spaces with farm residents; most of whom often had a plethora of wireless gadgets with them. Worse still for me (since I was able, awkward as it often was, to ask guests to turn wireless devices to airplane mode when really bothered) was contending with the level of perfumes and fragrances many of the guests wore. On too many occasions, I had to flee the communal kitchen due to the room being oversaturated with artificial scent, so that I often had to postpone or skip meals all together or delay my work. Ultimately the guest situation proved too much of a "wild card" to make things comfortable for me. And often the volunteers who came to work on the farm posed the same problem.

Additionally, we also were put in a position, due to the lack of consistency with the volunteers and other temporary residents, of taking on extra work, far beyond the agreed-upon hours for the work-trade arrangement—a mere hand-shake deal without signed contracts. After months of the situation not abating and understanding that it may never do so, the extra physical work burden was taking its toll on us both, robbing us of much-needed rest and downtime, as well as time to pursue our own interests or to earn extra income to cover our many debts, most of them newly acquired related to the move (having to buy a truck and trailer).

Again, we were met with the daunting task of finding, yet *another*, suitable home. While we often felt strongly that what we sought in a refuge was very basic and simple, as we were willing to live off-grid with no hot water, indoor plumbing or power, and since we only required a very small living space—but because we also needed to be free from EMFs and chemical pollution, our basic needs proved quite a tall order in today's industrialized world.

We met a few others who had stopped at the farm during that year as both guests and work-trade volunteers, who also were wrestling with EHS or with both EHS and MCS, who had likewise sought refuge in that forest for the same reasons we did. They too felt that their needs were simple but so hard to meet, and they too were seeking permanent homes, but struggling to find suitable options.

When I first became sensitized to EMFs in 2014 and told others of my plight, I was mostly met with disbelief and shock. Almost nobody I met back then had heard of the condition. But by 2020 I noticed a big shift in awareness. Now when I tell others of my EHS label I am rarely met with surprise and, more often than not, others have either heard of it, know someone with it, or even have it themselves. Or when they have not heard of it, it no longer shocks them as before.

By the summer of 2021, after many months searching for other work-trade housing opportunities in the area, we got lucky and found something that is (touch wood) working out well so far. As I do not wish to disclose our exact location for reasons of privacy, I can say that it is the best situation we have found up to this point, but has not been without its challenges. And we still have to live in a way most people would find incredible in the 21st century and in this industrialized "First World" country. Not having access to indoor plumbing, hot water or electricity—things most people in these times and this part of the world fully take for granted, is not the usual state of affairs for most. But the sacrifices are worth the benefits of living in an environment far removed from chemical and EMF pollution.

We are fortunate that we are not averse to roughing it, to living such a lifestyle. For those other canaries who are not physically or psychologically up to living in such rudimentary conditions, finding safe havens is even more challenging and generally impossible. This should not be so. These extremely marginalized groups should be helped. Safe places need to be carved out for them. Is it too much to ask for governments to create "white zones," places free from radiofrequency and low-frequency EMFs? And as Professor Olle Johansson so astutely pointed out to me in our phone conversation back in 2016, accommodating electro-sensitives benefits everyone. Making low-EMF environments in public spaces would serve to lower overall incidences of disease and prevent chronic illness, cancers and premature deaths.

It is possible that governments do not want to create such safe havens simply because offering other options for world populations as retreat from the harmful effects of modern technology, may prove to tip the scales to such an extent that these industries lose their power-hold on societies. If people were allowed to choose between lifestyles and have the opportunity to see what it feels like to live and sleep in a low-EMF environment, they may find that they do not wish to go back to the EMF-polluted ones, once they are able to feel the difference.

But by not giving the choice, industries cannot be blamed for harming the masses. They do not want an obvious control-group to exist. The same holds true for vaccines. Drug companies never test unvaccinated populations to see how their health compares with the vaccinated, even if they are quick to blame these same groups for making others (who are vaccinated and so by their own logic should be "immune" and protected from disease regardless) sick.

And this is why we have seen increasing pushes, as pharmaceutical companies gain more financial and political power, to vaccinate entire populations. If everyone is equally poisoned, there will be no one left to prove that the drugs are causing deaths and disease. If people are allowed total free will and free choice, too many of those who choose to remain un-medicated and remain in perfect health as a result, while the injected others fall ill, will count as evidence against the drug manufacturers. The best way to hide the evidence is to get rid of it—this holds true for any corrupt, greedy corporation vying for power. Those of us who suffer horrendously as a result, whose lives are totally destroyed, are merely considered "collateral damage" by such power oligarchies.

As it stands, the same industries degrading our natural environment use the "sustainability" label in order to manipulate the noble sentiments of a people increasingly feeling the sting of separation from the natural life-giving sustenance nature provides. We have to be wary of falling prey to this manipulation as it threatens to further remove us from wild lands in the name of reclaiming them for "re-wilding" initiatives under the political guise of "saving the planet." What we really need is to allow the people to do the saving themselves, to "re-wild", with themselves a direct part of that wilderness. Since when were humans not part of what we call "wild"? It is our birthright, alongside every other creature born into this world, to live in the wild, directly sustained by our mother, Earth.

What humanity does *not* need is to be further driven from the land and forced into the prison of overcrowded cities, trapped inside four walls of small apartments, made to live out their lives virtually through screens. If we allow Big Tech and Big Pharma to continue dictating public policy—lockdowns, for all kinds of reasons including "sustainability" efforts, will be forever imposed upon us and the loneliness and loss of soul we already suffer in varying degrees, directly related to the degree of removal from the land, plants, other animals, and from each other, will continue to haunt and torment us.

We need to put trust back into ourselves, our families and our communities and stop giving power away to outside authority figures. We do not need more "management" from industry-controlled, corrupt governments. We already are suffering from the effects of intensive micro-management by a small group of powerful elite (the billionaire's club members) who fancy themselves masters of the universe and of the people whom they call the common "herd"—viewed as such, we are slated to be micro-chipped, vaccinated/drugged and tracked under "management" and the aforementioned "sustainability" pretenses.

Education needs to move from placing undue burdens of guilt and blame on the average person for the destruction of the environment and shift it back to where it belongs—onto the very leaders who pretend to care and offer more devastating "solutions" to the problem they themselves started through their desire to seek total control and "full spectrum" dominance of all life.

Limiting the average citizens' access to wilderness and catch of fish or foraged food is not addressing the true problem of industrial-scale fishing and farming—a model which allows for foods to be shipped out to faraway lands, taken from the tables and mouths of the people who need it, who worked the land and sea for these same big industries. In California, crab and other catches are being shipped overseas to China, while locals pay unaffordable prices for seafood, often shipped there from elsewhere, frozen and chemically preserved, as local fresh foods are increasingly off limits to locals in the name of corporate global-empire "needs."

During the manufactured worldwide financial crisis of 2020-22, using stimulus money meant to help a deliberately tanked economy, billions were allocated to furthering 5G infrastructures in Trump's Secure 5G and Beyond act[148], and billions more given to aiding biotech interests in Trump's $2 Trillion-dollar stimulus package[149], adding more surveillance tech in hospitals, schools, and public spaces, enabled by 5G, to aid contact-tracing efforts so that trillions in total went back to the very interests that have gotten us into such a terrific health and environmental crisis to begin with, while small "mom and pop" businesses were destroyed in a matter of days and weeks as these locally supported and owned businesses were deemed "non-essential" and Big Box stores raked in obscene profits while all competition was instantly wiped out, possibly for good.

The canaries of the world, the WiFi refugees, already caught between a rock and a hard place long before the new crisis, have been further marginalized as access to safe havens in public and state parks has been intermittently and unpredictably cut off, and travel severely restricted as people were forced to "shelter in place" and risk fines or arrest for disobeying orders. While thus under "lockdown" many refugees, including Ruth and Gary were prevented from moving to safer locations and many of the once-safe locations were destroyed as "essential" utility workers installed more SMART grids and 5G antennas throughout the land.

The world's technocrats, bent on dominating and micromanaging every last corner of the Earth alongside every last creature, plant and human, seeing all equally as resources or commodities to be managed and sold, are in positions of power all over the world, and as such, are most assuredly orchestrating this current unending "public-health crisis," designed to collapse cultures and economies in order to rebuild the world in the way they deem best.

In this world no one will be free, no one will be allowed to live outside of oppressive totalitarian societies; ones kept under constant surveillance using health-destroying microwave and millimeter-wave frequencies.

We can rest assured that true public health is the last concern for those posturing as health officials and caring leaders. The voices of those who suffer the most from exposures inherent to the new prisons, enclosed by invisible EMF fences, will remain silenced and the plight of the canary lost to a world held captivate by an unmerited fear of microbes, as their attentions are diverted from the *real* health threat and allow their manufactured fear to rule over reason and give away further liberties and rights to our controllers.

## Canaries Still Singing—an Update on our EHS Friends

But the canaries are not all silenced yet. As I already touched on early on in this chapter ***Amy*** and her family and well and thriving at their new remote desert outpost.

**Bruce Evans,** who had hoped to move away from Australia to Southeast Asia pre-pandemic, has been stuck on his father's farm in rural Victoria, but has far from lost his fighting spirit. Now, in addition to continuing to fight for WiFi refugees, he has taken on the corrupt local police who have been instrumental in enforcing unlawful restrictions on Australians under bogus "emergency orders" during never-ending lockdowns witnessed in his country over the past two years. Bruce has received reports of and seen firsthand, the brutality of the police against the people where crowds of peaceful protesters against the lockdowns have been fired into with often-*lethal* "non-lethal" rubber bullets where women, elderly and children are present. They have bludgeoned, beat, and strangled innocent people and force "vaccinated" native aborigines and taken an unknown number of protestors to camps. Bruce himself has been arrested multiple times, but this is not stopping him and others from continuing on in the fight for freedom and justice and from filing lawsuits against the police.[150]

**Liz Barris** true to her Viking goddess nature and in spite of impossible challenges, is still fighting the good fight. Prior to the pandemic Liz had already needed to take a break from lobbying, as she was no longer able to tolerate the EMF exposures for travel or time spent in government buildings. Yet in spite of these newly imposed limitations, she has managed to file a number of lawsuits and join in as a petitioner (along with Theodora Scarato of EH Trust) on the successful lawsuit against the

FCC[151] for failure to review RF safety standards for the US, as she was able to join in remotely.

In 2020, Liz even ran for Congress after, as she related to me in a recent phone conversation, "catching our incumbent Congressman Ted Lieu" lying to them "about his knowledge of 5G and health effects," in order to try and replace him. While she did not win, she came in second with the Democrats, and third overall, with 15,180 votes in her favor. Liz did not expect to win—she simply wanted to use the opportunity to raise awareness to, as she says, "this covered up issue and also let our Congressman know that we will not stand for his lies and corruption on this issue and [that] it could cost him his career."

Liz also filed a lawsuit in regards to 5G, non-thermal effects from wireless radiation, Constitutional violations, the American Disabilities Act, Fair Housing Act and the utility company Edison and LA County. After raising $65,000 for the lawsuit on GoFundMe[152] Barris and her group were de-platformed[xxx], she assumes due to their newsletter, "exposing the Covid-19 fraud and vaccine side effects including severe disability and death."

All the while, Liz has contended with severe health effects from continuing wireless radiation exposure. As a result, she underwent surgery for her cancer after her 2017 diagnosis, which left her clear of cancer for two years, but recently she feels it has come back and spread to other areas. Currently in 2022, she is suffering a myriad of debilitating symptoms, primarily affecting her kidneys and she suffers chronic migraines—all since the 5G was activated[xxxi] in her city in 2020.

For the last two years, Liz has been living out of her SUV parked near her apartment (from which she runs cables for utilities to her vehicle) since

---

[xxx] This has sadly become common practice to de-platform or withhold funding for causes relating to anything that may threaten profits of powerful industries since companies like PayPal and GoFundMe are owned by these same powerful players.

[xxxi] Liz noticed a big change in radiation levels two years ago and that is when the local wireless telecom said they activated 5G, but she has also noticed the levels increasing even further (alongside her worsening symptoms) in the past year and in recent months. She believes that the 5G may have been turned up to a more powerful level and/or new installations turned on, in addition to more powerful WiFi systems turning on.

she is no longer able to endure the increased levels of EMR in her apartment. As her health deteriorated, she finally has felt she should flee the "mine," at least until she can recover her health enough to keep fighting for her rights to live free from EMR assault in her home and in her hometown. An EHS friend living in the wilds of the Arizona desert had offered her a shielded house for purchase—something Liz seriously considered until intensification of signals from the nearby highway broke through the previously well-shielded house, after the 5G was activated. (It seems the new 5G frequencies are not being blocked by the shielding that had worked well for previous generations of wireless radiation.) This friend had put tens of thousands of dollars into shielding, to have these extremely expensive measures be totally nullified, but at least she has another house in a different location still working out—for now. But all of this has left Liz still in need of safe housing.

Liz now feels she needs to focus on making enough money to afford a home elsewhere and successfully shield it. She tells me she is now in survival mode and needs to address her growing serious health issues. But it is extra challenging for her, because she is not willing to undergo another round of toxic radioactive injections for diagnostic imaging. The symptoms are serious enough that Liz is certain she is at stage 4 with her cancer and is sure the EMF exposures (specifically from cell towers), which caused the cancer and continue its progression, will ultimately kill her, since she has not been able to get fully away from the exposures.

But Liz is determined to live out her dream of finishing a documentary film[153] about EHS and related topics she started back in 2007. While she is not planning to succumb to invasive, toxic and highly dubious conventional cancer treatments, she still hopes to film her doctor visits to get a conventional diagnosis for the film. Liz is just praying she can live to see the film to its completion, and document her life as a canary to bear witness to these crimes against humanity after she is gone.

She also hopes that the latest attorney she is now working with, who has been injected with what she calls, "the COVID death shot," lives long enough to help her win her current lawsuits against the FCC, FDA and another agency (to remain undisclosed at this time), as two of her other lawyers recently died, one directly from the "COVID death shot."

Liz tells me that after the 5G turned on, she could no longer do anything or go anywhere. She could no longer go to her gym for even an hour, as the exposures make her sick in less time than that. She says she never fathomed her life could ever had turned out like this.

**Tim** is still living in the wilds of Arizona and still living a rough and simple lifestyle from a tent. However, he has upgraded from a regular tent to a canvas one equipped with a woodstove for heating. And he feels fortunate to have found a community within which to pitch his tent. While not specifically a community for electro-sensitives, they are understanding and accommodate his needs.

**Mario** is feeling good and fully engaged in living, in the slice of paradise he has created for himself in Bolivia, and certainly glad that he is no longer living in Canada, which, in its current state, he refers to as a prison. He still conducts ayahuasca ceremonies once per month offering them free of charge, and he still feels the effects from microwaves when exposed, but he is managing his condition and continues to view it as a gift.

Much of his family continue to view him as a bit "special" or crazy, but they do not see him as being quite as crazy as they once had, since the pandemic has served to further expose some of the control systems to them, under which we are all subjected. Mario, instead of being bitter about the worldwide lockdowns and billionaires getting richer, sees the drama as an invitation to greater awakening for the many people suffering under the effects of mass hypnosis and microwave-radiation-induced mind control. "If you don't wake up now," Mario observes, "then you are just not ready. Maybe in the next life or in 1,000 years, in the end they [those still asleep] will wake up."

Mario has compassion for those not ready to awaken as it is, "very hard for an individual to accept that he's been lied to by authority. Belief systems get shattered—belief in democracy, etc. But that's not the *real* world. If you want to know the truth, to see the real world, just accept that you were lied to and be open, without judgment. At least now you will have an open mind. My thanks to Bill Gates and everyone else, they help us to wake up. The world you think you live in, maybe one percent or less of that is reality. There are so many dimensions and inter-dimensions. When you accept the reality as it is, you begin to learn how to manage that reality, to live a special life for you and the universe. That's how you begin to understand stuff when you stop judging and this is what I call love. It needs love, the capacity to love. To accept what is happening and observe without judging and it's difficult because we are programmed to judge... All this manipulation [related to the pandemic] and confusion, humanity needed [it] to wake up. So, in a sense, the plan to control us is backfiring and our controllers are self-destructing... We can create and spread our own reality so that it becomes 'viral' and wipes out the other. You can have an effect on others and they can have an effect on others."

**Olga Sheean** is now living in the countryside in France with her husband Lewis and continues to educate others on how to heal their bodies, minds and relationships while holding a vision for a better world for all. During the 2020 pandemic lockdowns, she used her unique talents to write a suspenseful and incredibly moving novel, called *The Parents; How Far Would You Go to Save Your World*? The story is about quantum magic, enduring love, family bonds that break the barriers of place and time, and about creating a vision for a new world—one free from the harmful effects of wireless radiation and driven by a fierce desire for love, harmony, true connection, healing and realizing the highest levels of our human potential. As ever, Olga enchants readers with her wit, wisdom, creativity and undying optimism to captivate readers while challenging us to radically change for the better—something she is certain we can all do, even in the face of loud and crazed media voices telling us the opposite. And, in 2022, Olga, launched a series of empowering blog posts titled *Rebooting the Species.*[154]

The series is designed to help others find a way forward and help heal locked trauma we all harbor within, especially as we have found ourselves *locked*, not just in enclosed spaces, but in the grip of fear that has been greatly amplified since the beginning of 2020, engendered by a manufactured pandemic.

Olga says that it is no surprise that, "we now live in fear, fueled by generational trauma that prevents us from understanding the deeper truth, processing our pain, taking assertive action or bravely creating the world we want. Instead, locked in the illusion of subservience, we wait to be granted freedom by some external authority, while adapting, conforming and complying... *to stay safe.*

"With the crescendo of corruption and violation of our rights, we are so fully in the grip of global PTSD that we often can't think straight, act rationally, feel compassion, or focus on a future we once dreamed of creating. In reacting to what is coming at us, we cannot see the power of our pain or how we are being pushed to heal it.

"Whether we're dealing with EMFs, conflict or a loss of personal freedom, it's a form of abuse that can only be resolved when we understand the deeper drivers.

"Healing our personal and collective trauma is the most powerful and effective way of setting ourselves free—physically, emotionally and spiritually. It is what we are being pushed to do. When we witness the growing conflict, separation and us-vs.-them divisiveness in our world, we can see what is missing: unity, harmony, togetherness, collaboration, community and the collective healing that will bring us all back together, like nothing else on earth."

Olga asks if we are ready for the REAL reset, since, "The so-called 'Great Reset' will never take us back to being fully human, only further away. We need a human-greatness reset, to reboot the species so we can get back on track with our human evolution of empowerment, enlightenment and conscious creatorship."

**Ruth** is still searching for safe land to purchase so that she can have the option of settling, but has yet to find that perfect place unpolluted by EMFs. And **Gary** and Ruth continue to dance the Tango Libre whenever they come together, but they too have noted a further decline in their environment due to the manic devil's dance the tech industry, along with the masses of tech addicted, like the girl in Hans Christian Anderson's story of the *Red Shoes,* unable to take off the red shoes of their addictions, will not cease this dance of destruction until the shoes (with feet enclosed) are cut off. In the meantime, the dance of the red shoes continues its path of destruction into wilderness areas, home to WiFi refugees, like an out-of-control dust devil, it continues to adversely affect the health of electro-sensitives and other wildlife. Gary's last update to me going back over and confirming some details about his story for this book, contained the following lament:

> *I moved from LA to Colorado in 1993, extremely sick, but not yet aware that MCS and EHS were the problems. This* [knowledge] *came gradually from an acupuncturist and Bau-Biology trained EMF practitioner Roland Holzwarth in the period 1993 to 1996—all* [his symptoms previously] *massively misdiagnosed with $15,000 worth of tests/consulting and 4 years with the Veteran's Administration. They said it was psychosomatic, take Zane and see a shrink. If I had* [followed their advice] *I would be painfully dead.*

*Now thirty years later I am back to the VA* [Veteran's Association hospital] *with emphysema and asthma,*[155] *both increasingly looking to be EMF induced. The VA still hasn't caught up with the rest of the world in EHS recognition and diagnosis. Until we have physicians and nurses trained competently in Environmental Allergencis, Western medicine and this culture will continue its sickening and now rapid disintegration into device addiction, wireless radiation induced immune suppression pandemics, and depression/inflammatory illnesses.*

*They do, however, now recognize the broad-spectrum chemical allergenic reaction but fail to see it as simultaneous if not identical to EHS; known and well documented by the German Bau Biologie academic medical research community 70 years ago. This is like paying millions of dollars to learn how to harness and ride pterodactyls and ichthyosaurs.*

*The experienced and progressive deteriorations in all wildlife, insectary, aviary, flora and fauna on our wild lands became obvious after the change from analog to digital transmissions in 2009. The impacts on urban, power-line radiated trees, etc., were and continue to be disastrous and also apparently invisible to the device denial protecting the wireless stimulus-addicted public—now showing evidences of debilitation (including immune suppression basing the COVID pandemic) approaching or exceeding 90% of the global population. This stimulus addiction deterioration should be obvious to all the clean and sober alcohol/drug dependency recovery community adequately versed in Environmental Allergencis of wireless debilitation and equipped with metering equipment capable of indicating whether or not EMF is present and to what power level.*

*To this day, out of thousands of contacts we have yet to find one licensed MD or psychologist in the Tabauche corridor equipped to make this scientific determination in spite of their remaining and historic insistence that whatever it is and to whatever extent it pervades it [EMF radiation] cannot possibly have any effect. The school children and parents in Hiroshima and Nagasaki Japan in 1945 prior to the blasts harbored exactly the same theories regarding EMF and radiation effects. Intelligent science terms this convenient and expedient technique: 'Contempt prior to investigation,' which has apparently not become the methodology of preference for government regulation, medicine, psychology, school systems and the CDC.*

And Professor Olle Johansson is still dedicating himself to helping canaries all over the world. In a recent court case, in which he was directly involved, with a ruling dated November 26, 2021, Nadiane, an electro-sensitive 60-year-old French woman, was permitted to continue living in her makeshift house (a mere shed) on her land. Nadiane had been living as a recluse with only two donkeys for company, on her 13-hectare property in a forest in Haut Vallée in Narbonne, France, where she had resorted to building a shed for herself as shelter because she had nowhere else that she could retreat from EMFs. In September 10, 2021, Nadiane had been tried by the Carcassonne Criminal Court for having "built a makeshift house on her land, without the consent of the Departmental Directorate of Territories and the Sea (DDTM)." Fortunately, the court ruled to allow her to continue living in her shed due to her unique circumstances.

Even though this was a positive outcome of sorts, it still means that this woman has to live in a cold shed totally isolated from the world. As Olle correctly observes, "It would be much more honorable for the French authorities to allow Nadiane an equal life in society already based on equality (and *liberté* and *fraternité*)... but this is a first step." Olle is involved in other similar court cases he hopes will be ruled in favor of WiFi refugees.

* * * * *

I sometimes miss the desert, the open spaces, endless expanse of sky, the purple mountain peaks, the pinks and red of boulders and sand, the smell of the creosote-scented rains, the company of desert wildlife, and especially my corner of the canyon in which I slept undisturbed for so many years. It saddens me when I think that I will likely never return, like so many other places I was fortunate to at one time have called home. If I were able to return, even for a visit, I fear I would be crushed at the changes—ones of which I have already received reports by neighbors who stayed behind after the encroachment of the SMART grid.

One neighbor related on the community forum, the strange behavior of ants on her property, whose movements appeared drastically changed from before—something which distressed this particular woman. However, she did not seem to draw a correlation between the strange behavior and the new radiofrequency pollution coming from the SMART meter placed just opposite her bedroom wall.

And in December 2021, at the Oasis Bird Sanctuary in Cascabel, tropical birds are dying in alarming numbers and totally out of the blue. Whenever a bird at the sanctuary dies, it gets an autopsy at a special veterinarian in Tucson. The autopsies for these recent bird deaths are showing heart disease, stroke, and in some cases, respiratory distress. These are all in keeping with symptoms of excessive electromagnetic radiation exposure. But the head of the sanctuary is baffled by these deaths and cannot understand what could be the possible cause.

Some members of the community died within months to one year from installation date of the SMART meters. One died following her second stroke in a matter of weeks. Another death was oddly attributed to "West Nile virus." And another, only in his late 60s, died from a final bad bout of dementia, after a struggle with it for a number of years. None of these deaths will be blamed upon the utility company and its horrendous meters, nor upon dangerous vaccinations, but I am at least certain that both of these factors must have and will continue to play an important role in the inevitable degradation of health of my former neighbors and to the changing behavior and ultimate destruction of surrounding wildlife. So, my memories of my desert haven will have to remain as such, filed away in vivid color, texture, and sound under a mental folder of "once was."

It is my sincere hope that the spirit of the people will win out over the oppressive rule of the military-industrial complex that has overtaken our governments and continues to tighten its stranglehold on the world. And I do believe that the fight is far from over, and that the wild human spirit and that of the Earth itself will not be fully vanquished and that we canaries will not have to forever flee, but instead, one day be cherished and aided alongside all life equally—and that our plights will not have been in vain.

# PART FOUR

# CROSSROADS

*Chapter Twelve*

# The Road Not Taken

*Two roads diverged in a yellow wood*
*And sorry I could not travel both*
*And be one traveler, long I stood*
*And looked down one as far as I could*
*To where it bent in the undergrowth:*

*Then took the other, as just as fair,*
*And having perhaps the better claim,*
*Because it was grassy and wanted wear;*
*Though as for that the passing there*
*Had worn them really about the same,*

*And both that morning equally lay*
*In leaves no step had trodden black.*
*Oh, I kept the first for another day!*
*Yet knowing how way leads on to way,*
*I doubted if I should ever come back.*

*I shall be telling this with a sigh*
*Somewhere ages and ages hence:*
*Two roads diverged in a wood, and I—*
*I took the one less traveled by,*
*And that has made all the difference.*

—Robert Frost, *The Road Not Taken*

## *Stalking Cyborgs*

Not long ago I had a nightmare. It was a vision of the not-too-distant future. In the dream I had been living somewhere in hiding with a small group of WiFi Refugees and was tasked with going into a town or city for supplies. The environment I next entered was unrecognizable to me. I was seized with a terrible apprehension as I attempted to navigate this high-tech dystopian world, in which I was totally disoriented and lost. I did not understand how to engage in the most basic transactions and I did not know how to travel and move about in this strange place. The people around me were using electronic devices I did not recognize or understand. When I tried to utilize a system of public transportation that resembled a steeply inclined moving walkway or escalator, I did not know how to engage with it. It seemed to make use of some kind of mental link, like artificial telepathy to communicate with AI. Unable to make sense of the contraption, I became trapped within it, unable to disembark or to slow its ever-increasing ascent. I felt that the machine might kill me if I did not quickly learn how it worked.

In a more recent dream, I was working in an office building in which the constant presence of surveillance drones was the norm. These drones did not hum. They were quiet and flapped thin black-metal wings and moved about like butterflies. I became increasingly distressed and threatened by their incessant hovering. I attempted to swat at them but I was not quick enough to catch them. I thought to myself, "I need a baseball bat so I can bash them to the ground." Senior colleagues noted my "odd" behavior and quickly took action to "correct" my dissenting thought processes. With arms weighed down by stacks of books containing propaganda about how much safer societies and the workplace were thanks to surveillance drones, two women cornered me. Their eyes were wide and fearful. They simply could not understand how I was incapable of seeing how much better off we were now in this "new normal" society than we previously had been. Did I not know how much more *dangerous* it used to be when we were not constantly monitored by "intelligent" robots? I wanted to scream at them, to tell them they were wrong, that it was NOT more dangerous before the AI eye-in-the-sky came to dominate the planet, with this new artificial Big Brother floating in "the cloud" ever present, always near and at the ready to ensure our "safety", and above all, our obedience.

In yet another nightmare of recent years, I was sleeping in a small bedroom in which my large, high-standing bed took up one half of the room and my monstrous computer the other half, with its ridiculously large flat-screen monitor filling most of the wall on top of a heavy desk. In the dream, I had been sleeping in my darkened room with the computer turned off, when suddenly I awoke (within the dream) fully alert and aware of being watched, only to discover that my computer had turned itself on and seemed to have been engaged in spying on me. I had caught it in the act and it, having sensed this, quickly turned itself back off. I finally awoke in reality, heavily breathing and sweating. It is difficult to describe just how much that simple dream scenario affected me. It had felt so real.[xxxii] The machines, it seemed, were turning from tools into our masters and this idea left me reeling for some time after.

*"Whether created by the use of hypnosis, drugs, behavior modification, electronic or sonic brain stimulation, or through a combination of these tools of psycho-science,* ***the cyborg is stalking us in our dreams.*** *And just as life imitates art, men live out their dreams in their waking state. The dream, expressed by the prophetic visions of men from all walks of life, is of a time when the machine or the drug will take over and relieve man of his difficult burden of self-responsibility.*

***For better or worse, self-responsibility—***
***where each individual acts consciously,***
***and accepts the consequences of his own actions—***
***is the stuff of which freedom is made.***

*"The prophecies of poets, writers, scientists, and futurists express what can be considered a regressive, devolutionary myth. Sprung from the complexity of technological life, where* ***self-responsibility is largely directed by propaganda and indoctrination, where an ignorant rather than an enlightened public is desired****, the majority of responsible actions can result only in cultural disaster. This, in turn, adds to the frustration of the individual who, weighing all the facts—or what were presented as facts—thought he had made the best choice possible. When these decisions, based on false information, are shown to result in negative effects, the frustration of the individual grows. Weariness eventually*

---

[xxxii] Sadly, this nightmare is not far off from actual recent reports of AI bots not doing as they are told and turning on or off when not instructed. For examples Internet search: "Creepy things Alexa does".

*sets in, and* ***the individual becomes willing to surrender his self-responsibility and eagerly awaits his liberation by some authoritarian figure.***"[156] (Walter Bowart, *Operation Mind Control*, emphases added)

As Bowart correctly observed in his 1978 book, *Operation Mind Control*, the cyborg is stalking us in our dreams. We are being carefully groomed for a technological takeover of humanity. Once merely tools and communication aids, our devices have become *divisive*—that is, they are dividing us from our true powerful selves as we relinquish our inner authority and knowing to artificial "intelligence." Just like *HAL*, the thinking computer from the 1968 Kubrick classic film, *2001; A Space Odyssey*, our machines are beginning to have minds and intentions of their own, and their "desires" (really just coding programmed into them by their masters) are not aligned with our best interests.

Digital addiction, once confined to small subsets of our populations (the "nerds" of the world), has spilled over into the population at large at terrifyingly dangerous levels whereby our very lives, and the lives of others, are put in real danger as people glued to portable screens lose touch with physical reality and move through it like a phantom with one foot teetering over the abyss as they are pulled into other virtual realms while still having to navigate this one.

In this disconnected state while "connected" to devices, people stumble into fountains, fall off of bicycles, step over the edge of cliffs and derail crowded trains, killing dozens of passengers—as tragically happened in Los Angeles in 2008 when a Metrolink transit train conductor got distracted because of texting while driving the train, subsequently running a red light and colliding head on with a freight train, killing himself and 25 passengers and injuring 135 other commuters, making it the worst US train accident in 15 years.[157]

These incidents have risen dramatically in number over the years alongside growing social acceptance of texting while driving (or texting while doing *anything*), in spite of (what should be) the obvious risks inherent to this behavior. Of course, the *solution* offered us by the perpetrators (the designers of addictive devices, the ones who caused the problem in the first place), is to hand over authority to "intelligent" driverless vehicles, "freeing" us to dedicate even more of our attentions to the tiny screens, so we can text, game, or stream videos to our heart's content while we let the machine take full control of our transportation. However, autonomous vehicles are not infallible, nor can they even be proven to be *less* fallible than human drivers, and already have killed a number of pedestrians[158] in crosswalks or on streets, even though they are not "supposed" to do this. In one of these incidents the pedestrian was hit because the car lacked "the capacity to classify a pedestrian from an object unless that object was near a crosswalk."[159] A person holding a driver's license or not, would never make this mistake, why would an autonomous vehicle even be released to market if it cannot differentiate between a human and an object?

Tech researcher and author of *The Glass Cage*, Nicholas Carr writes in his book, "Artificial intelligence is not human intelligence. People are mindful; computers are mindless."[160] And, "Whatever it is that goes on inside our brains... can't be reduced to the computations that go on inside computers."[161]

And he points out that...

***It's impossible to automate complex human activities without also automating moral choices.***[162]

In the case of driverless vehicles, moral choices, and liability, he says, "Many technical challenges remain to be met, such as navigating snowy or leaf-covered roads, dealing with unexpected detours, and interpreting the hand signals of traffic cops and road workers. Even the most powerful computers still have a hard time distinguishing a bit of harmless debris (a flattened cardboard box, say) from a dangerous obstacle (a nail-studded chunk of plywood). Most daunting of all are the many legal, cultural, and ethical hurdles a driverless car faces. Where, for instance, will culpability and liability reside should a computer-driven automobile cause an accident that kills or injures someone? With the car's owner? With the manufacturer that installed the self-driving system? With the programmers who wrote the software? Until such thorny questions get sorted out, fully automated cars are unlikely to grace dealer showrooms."[163]

Asking similar questions about robots also taking over human jobs, Carr postulates, "Who determines what the 'optimal' or 'rational' choice is in a morally ambiguous situation? Who gets to program the robot's conscience? Is it the robot's manufacturer? The robot's owner? The software coders? Politicians? Government regulators? Philosophers? An insurance underwriter?"[164]

There will always be a ghost in the machine and it is to this ghost we are relinquishing our power and liberty.

The road less traveled in today's world is that of the one that leads into the woods—that takes us back to nature.

When someone, almost anyone, is asked to describe paradise, he or she tends to imagine natural settings–waves lapping gently upon a white-sand beach, birds singing softly, or imagining him or herself lying in a meadow of wild flowers and tall grasses, walking through an ancient forest, sitting on a rolling green hillside, or looking out over a vast ocean. They do not generally imagine AI-powered machines ruling their lives. They do not think of drones flying to and fro, filling the sky like so many menacing, swarming insects. They do not call to mind the sound of screeching bus brakes, crowded elevators, people walking around wearing masks, or driverless vehicles failing to stop for them at a crosswalk.

In 1975, Russian president Brezhnev sought a UN treaty on electromagnetic weapons, but the US blocked this move, refusing to allow it. Brezhnev warned that electromagnetic weapons were a far greater threat than nuclear weapons, calling the latter, "old technology."

And Tom Bearden, warning about the same[165], said about the potential for use of artificial electromagnetic fields:

***"…will give us the capability to engineer reality itself—both physical reality, life and mind—and we will be able to engineer it to be either a heaven or hell: the choice is up to us."***

What kind of world does it look like is being created today thanks to the introduction of these new technologies–a *heaven* or a hell?

The world being quickly shaped and imagined for us is what the average human would consider to be a dystopian nightmare–not a paradise. Why are we letting our world be turned into a cold, heartless cyberspace fit only for robots and cyborgs and not instead, creating a paradise for ourselves?

The futurists are running the show. They are the billionaires, like Elon Musk and Bill Gates, who have no interest in living in harmony with nature and living connected to it. Musk is busy making plans to colonize Mars just in case "something goes wrong with Earth."[166] That "something going wrong" *is* happening, thanks to Musk and his development of AI, brain-based microchip implants (Neuralink)[167] to connect our brains with this same AI, his development of autonomous vehicles, and his launching tens of thousands of 5G satellites into our low-orbit stratosphere. With only 6,000 visible stars in the night sky, Musk's 60,000-plus reflective satellites are not only poised to blanket the planet with harmful millimeter-wave frequencies, and as such threaten all life on Earth with annihilation (hence his prophesying about our possibly having to move off-planet), but they also threaten to obliterate the very stars of our night sky from view.[168]

Instead of looking up at the same constellations our ancestors have for eons—beautiful glittering patterns above, we will see a grid of lights in straight lines crossing the sky in ugly streaks, similar to what we are seeing with the polluting of our sky in the day by persistent contrails from planes spraying poisons with the express intent to block out the sun in "geo-engineering" attempts to "cool" the allegedly overheated planet—even though we are also told that we have now entered (as of 2020) a grand solar *minimum* and that our Earth has already cooled a number of degrees, more than global-warmists claimed we needed to halt climate change decades ago.

Musk says that AI (Artificial Intelligence) is a threat to mankind (so why is he developing it again?), and so we must join it, as we cannot ever hope to beat it. This is his rationale for linking our brains to the AI "cloud" so that we will not be outdone by this other "intelligence"—one, he says, that is set to outpace us in the job market.

Futurists like Musk pretend that it is not possible to stop the runaway train of technology. They tell us it is a "natural" and "evolutionary" event—that merging humans with machines is the next step in this evolution. The "singularity" (this merging of man and machine) is also being referred to by Ray Kurzweil (prominent futurist and current engineer director at Google) as "Human 2.0." Kurzweil sees the chance to become a cyborg as a potential means to prolong human life—maybe indefinitely.[169]

The futurists are chasing the Holy Grail, the fountain of immortality. But what is "human life" if one is no longer *human*? Kurzweil says we may be able to download the contents of our minds into machines and continue life in semi- or full-metal form. But the mind cannot be located, nor can the location of our memories within our brains be found either.

This plan is doomed to fail, as have so many other similar promises made to us before—promises that machines will make our lives easier, that they will free us from the need to work all together, that they will fix our broken parts—have also failed. (Kurzweil, also rather significantly states that in the near cyborg future, humans may not even "need"[xxxiii] a heart.[170] In the futurist's or technocrat's mind, our biological organs are merely parts of a "biological machine," and can be replaced with synthetic parts or discarded all together. They ignore the aspect of spirit, of the vital force that truly defines and gives us life—*the divine spark*.)

But machines are not perfect. And neither is technology. They can, and have, both failed us repeatedly. Today our lives are not easier—they are more complicated. And we are not healthier. A metal hip or metal plate in any part of our body is not superior to our own bones as we are made to believe. And as we allow machines to takeover and run our lives, we, "expect more from technology and less from each other," as author and tech researcher, Sherry Turkle, astutely observes in her book, *Alone Together*.

And we are now truly "alone together" as we sit in rooms next to one another while, instead of making eye contact and speaking to the person beside us, we distract ourselves with portable screens, preferring the "company" of virtual "friends" to those nearby. And nowadays we not only stop making eye contact, we move farther away as we "social distance" (really "anti-social distance" as there is nothing remotely "social" about it) and cover our facial expressions behind masks or stay locked in houses, apartments or single rooms, conditioned to fear

---

[xxxiii] In Kurzweil's *The Singularity is Near,* under the subheading "Have a Heart, or Not" after describing the many "severe" problems with the heart and why it could use "enhancement" or upgrading, Kurweil writes, "Although artificial hearts are beginning to be feasible replacements, a more effective approach will be to get rid of the heart altogether." In order to accomplish this task of ridding us of our "problematic hearts" our blood stream would have to be incessantly riddled with nanobots. Nanobots, he says, can also help us to also eliminate our "lungs, red and white blood cells, platelets, pancreas, thyroid, and all the hormone-producing organs, kidneys, bladder, liver, lower esophagus, stomach, small intestines, large intestines, and bowel." And, he proposes that the skeleton can also be replaced by self-replicating nanobots. Kurzweil promises that we will become Cyborgs, what he calls Human 2.0 by the year 2030 and that we will become "more non-biological than biological." (p. 306-309)

"disease-carrying humans" to such an extent that we avoid physical contact all together in favor of socializing solely through screens.

And Turkle, who spent decades studying human and robot interactions and tracking the changes in attitudes of humans toward robots over the years, when addressing the topic of robot "companions" and noting the steady decline in humans' ability to empathize and engage with one another as they devote more of their social time to machines, observes, "Perhaps when people lose this ability [to empathize], robots seem appropriate company because they share this incapacity."[171]

In April 2020, Ahunja Sonalker, CEO of Steer Technology, autonomous parking technology, described the effects of a world under quarantine on the increasing acceptance of this technological takeover thusly:

***"There has been a distinct warming up to human-less, contactless technology. Humans are biohazards, machines are not."***[172]

Today we are "warming up to machines" because they are not considered to be carriers of disease. However, they *do* harm us by bathing us in unnatural frequencies. Robot "caregivers" are increasing in popularity as humans are demonized and presented as threats to individual human safety. Robot nurses are fast becoming the norm in Chinese hospitals where they "helped" deal with the pandemic by exposing patients to 5G frequencies—frequencies that can literally take one's breath away through triggering asthmatic attacks and hypoxia, or at the least exacerbating the problem of respiratory and other illnesses, and at the worst by killing people outright.

The big irony is that robots and machines are the *only* things capable of being infected with actual "viruses," when they are attacked with malware and malfunction or are destroyed as a result. *Viruses,* as mentioned already, have not been viewed in humans, plants or animals. They have only been found in sea algae[173], and even then, understood to be benign and possibly beneficial, as the algae with the largest number of viruses appears to be healthiest. So yes, "viruses" can be found in machines, but not in humans.

Our world is being reshaped for us by those holding the world's purse strings. AI and wireless-enabled technologies are only being developed so quickly because they are being financed by the world's billionaires. *Power corrupts and absolute power corrupts absolutely.* It is the drive for even greater wealth, more power and control—the full-*spectrum* dominance of the planet, by those bent on being its masters that is driving the push to overtake our environments and bodies with invasive surveillance technology. It is also the force behind the reeducation agendas that tell us to call digitally controlled environments; digital "ecosystems," and to refer to fields ruined by acres of buildings filled with computer servers; "the cloud," and computer software; "windows," and machines that monitor us and mimic intelligence; "smart."

It is those behind the control agendas hijacking our language and re-"programming" our thinking and behavior—making us redefine what it means to be human and to redefine life itself. In today's world, run by technocrats hell bent on labeling, tracking, tagging, chipping, drugging, genetically modifying and commoditizing every human, plant, animal, insect and microbe on the planet, we are told to assign genders and human names to robots while humans no longer are permitted to be given genders or even names. (Futurist and billionaire Elon Musk's own unfortunate child[174] is named **X Æ A-12**, pronounced X-Ash-A-12, but he just calls him "Baby X", telling us just what he thinks of babies, similar to how he views space, something to be *X*-ed out and replaced by technology.) And genderless, or *trans*-gender humans now call themselves "non-binary" as if to say it is *they,* not the rest of us, who are *not* being reduced to binary code, like nothing more than ones and zeros used in computer programming.

## *The Internet of Things and You*

The push for "smart" cities, phones, computers, and appliances, and what is being called the *Internet of Things* (IoT) is, again, being orchestrated by our futurist billionaire overlords. This frightening *Brave New (SMART) World* is only being presented as beneficial and sold to us under the guise of "sustainability", when in reality it is anything but that. SMART grids require a great deal of electricity to power and also increase levels of harmful EMFs in our environment. There is absolutely *nothing* "smart" about it, unless you refer to one of more than twenty military acronyms for SMART, then it starts to make more sense. "Secret Militarized Arms in Residential Technologies," "Self Monitoring And Reporting Technology," "Satellite Monitoring And Remote Tracking," and

"Simulation Modeling Anchored in Real-World Testing," are a few of those listed on acronyms.com.

Planned obsolescence of *non*-SMART electronics is making it increasingly more difficult for people like myself and other canaries to opt-out of wireless-enabled devices—ones which also increasingly do not allow wireless features to be switched off. Many new electronics cannot even be powered down fully or turned on and off without an app loaded onto a smartphone to engage it. New computers are designed without an Ethernet port, making it more challenging for people who wish or need to plug into the Internet via Ethernet cables. Similarly, wired "mouses" or keyboards are disappearing from the market, being replaced by wireless counterparts. And it is harder to find household appliances not equipped with surveillance chips—even though the inclusion of such *should* be illegal, as it constitutes wire-tapping. Prior to these intrusions finding a "bug" in one's home would have been considered alarming, but now the "bugs" are just a part of life, slowly but surely replacing real *bugs* or insects in our environment.

The world is quickly resembling one that mimics that of my worst nightmares—one within which I feel certain I will not be able to navigate in the very near future. It is already near impossible to function in society without a SMART phone—the only thing capable of powering the many apps now required to engage in basic commerce and travel. And the 2020 pandemic only made this situation worse by requiring people in many countries to carry a microwave-emitting phone with them at all times so that they can be "contact-traced" for reasons of public health and safety (so we are told). These phones cannot even ever be turned off in many cases, if one does not want police showing up at one's door, as happened to an American student living in Taiwan when his phone battery depleted one morning while in quarantine at home (after having traveled abroad) at 7:30 am. By 8:15 am the police were at his door. [175]

***"The internet will disappear. There will be so many IP addresses...so many devices, sensors, things that you are wearing, things that you are interacting with, that you won't even sense it. It will be part of your presence all the time. Imagine you walk into a room and the room is dynamic.***

—former Google CEO, Eric Schmidt, at Davos, 2015

*"What's electricity? It's a socket in a wall—you plug in, you get electricity. You don't get how it's made. It's not a complicated interface. It's invisible. The internet is yet to evolve to that goal I was hoping for—of being invisible."*

—Bob Kachov, Internet founding scientist, interviewed for the 2017 documentary film, *Lo and Behold*

Google's former CEO, Eric Schmidt and Internet founding scientist Bob Kachov, say that the Internet should disappear and fully integrate into our surrounding environments. They say that it should be like electricity, something we use all of the time but cannot see and something of which we no longer question its origins or existence. (But I personally feel it is not wise to not question the existence or origins of electricity, since we *do* need to be aware of electricity and its effects on us, and from whence it comes.)

This is the world the futurists are creating for us. We did not vote on it. We did not get to have a say in it. If we had been told only two decades ago that our governments wanted us to carry with us a tracking surveillance device that would record our every move, conversation and written correspondence (and ultimately even our thoughts)... and not only would we be required to do this, we would be made to *pay* for the privilege ourselves, *and* that it might destroy our minds, our emotions and bodies and turn us into pharmaceutical-dependent zombies—I doubt many people would have been on board with that mandate. Likely there would have been massive protests against it.

But because this same health-robbing surveillance device has been sold to us as first a convenience and then a necessity, we have fallen for the bait. And certain world events like the 9/11 World Trade Center attacks in New York served to "warm up" a cellphone resistant public to the idea of "needing" a cellphone for "emergencies," in case of terrorist attacks,[xxxiv] just as we are now being made to "warm-up" to the idea of contactless everything, constant surveillance and robot caregivers after the world's populations have been turned into a race of germaphobes.

---

[xxxiv] Not that anyone's cellphones did them any good during that attack, as the overwhelmed system created a large-scale radio blackout. The same thing happens during storms—these vulnerable systems easily go down. Landlines were made for the purpose of providing communications during emergencies as they work when the power is out.

Now a once-resistant-to-the-Internet-of-Things public, is fast being made to see the "benefit" of doing everything remotely from one's home (in some cases these homes are merely four walls—essentially a box) and now "live" one's life on *Zoom*. And so, we have been groomed for *life* on Zoom and are being set up for the next stage—acceptance of "life" as an avatar or cyborg. Even Japan's government has said it expects half or more of its population to be "living as avatars"[176] by 2050 and is gearing up for just such a scenario.

SMART machines are not *intelligent.* They can only be made to *mimic* human intelligence. Human intelligence is dependent upon traits like empathy and compassion—human traits not possessed by machines, no matter how advanced they might be. Algorithms are devised by humans to create the illusion of intelligence and decision-making capacities in machines. But computing machines are only ever doing just that—*computing*. They are only able to make calculations based on data entered. You may have heard the expression "crap in—crap out?" Well, that is part of the problem. What data is entered is going to determine the results calculated, and this is an inherent problem in trusting computer models and simulations—they all too often miss the mark—sometimes by quite a lot.

Nicholas Carr perfectly sums up the difference between computing "intelligence" and human intelligence with, "The replication of the outputs of thinking is not thinking. As Turing himself stressed, algorithms will never replace intuition entirely. There will always be a place for 'spontaneous judgments, which are not the result of conscious trains of reasoning.' What really makes us smart is not our ability to pull facts from documents or decipher statistical patterns in arrays of data. It's our ability to make sense of things, to weave the knowledge we draw from observation and experience, from living, into a rich and fluid understanding of the world that we can then apply to any task or challenge. It's this supple quality of mind, spanning conscious and unconscious cognition, reason and inspiration, that allows human beings to think conceptually, critically, metaphorically, speculatively, wittily—to take leaps of logic and imagination."[177]

I remember a time in the early 1990s when I was dealing with the daunting problem of my university scholarship funds having not been processed by the school's admissions department, making it impossible for me to sign up for classes at the beginning of my third semester at school. I visited the office daily to try and understand what went wrong and get the problem resolved. A different person helped me each day—a student administrator standing behind a computer who would read me something from a screen in answer to my questions. When I contradicted the information that they had accessed, I was simply told that this was what the computer said. This statement was given to me as the final word since the computer could not, it would seem, be wrong.

The idea that machines were "smarter" and less fallible than humans has been deeply imbedded in our psyches for some time and it was certainly driving the decision-making at my university's admissions department. When I told the student employee that someone had entered the data into the computer and had clearly entered it incorrectly, I was met with blank uncomprehending stares. Ultimately, I had to fight to see the head of the department before I was helped and it was finally confirmed that yes, someone, a *person*, had screwed up. The computer was just spewing out the incorrect data that had been entered. It was just "following orders." And the school staff were also just following orders, those, they believed, that were given to them by the computer.

The trouble with increasingly putting our faith in and becoming reliant upon machines to do our thinking for us is sometimes called "automation complacency," which Carr tells us, "takes hold when a computer lulls us into a false sense of security. We become so confident that the machine will work flawlessly, handling any challenge that may arise, that we allow our attention to drift. **We disengage from our work, or at least miss signals that something is amiss**."[178] (author's emphasis)

This reliance upon "smart" machines also has a tremendous and potentially disastrous dumbing-down effect upon tech-dependent human populations. As Carr also says, "As machines become more sophisticated, the work left to people becomes less so." And, "As their programs become adept at doing our thinking for us, we naturally come to rely more on the software and less on our own smarts. We're less likely to push our minds to do the work of generation. When that happens, we end up learning less and knowing less. We also become less capable."[179]

And tech historian George Dyson warned:

***What if the cost of machines that think, is people who don't?***

This problem of disengaging and checking out from our work as we become monitors of screen monitors rather than active participants in our tasks, becomes more dangerous and life-threatening as the stakes increase for the kinds of tasks for which we allow computers to take the helm. In instances of airline pilots literally handing over the helm to computers for the duration of flights as they are set to "auto-pilot," has too often resulted in loss of lives when pilots are needed to step in and once again take the reins when computerized systems fail, as they inevitably sometimes do. In these cases, the checked-out pilots tend to respond too slowly and many alarmingly admit to having forgotten how to fly altogether.

Cognitive scientist Donald Norman wrote, "Society has unwittingly fallen into a **machine-centered orientation to life**, one that emphasizes the **needs of technology over those of people**, thereby **forcing people into a supporting role**, one for which we are most unsuited. Worse, the machine-centered viewpoint compares people to machines and finds us wanting, incapable of precise, repetitive, accurate actions... **When we take the machine-centered point of view, we judge things on artificial, mechanical merits.**"[180] (author's emphases)

The lines between tool as servant or as master are becoming increasingly blurred, as we hand over more jobs and important decision-making tasks to machines—ones that not only require basic data-analysis capabilities but also the ability to make moral choices, as pointed out in the self-driving vehicle scenario previously described.

Hannah Arendt, author of the 1958 book, *The Human Condition*, writes, "Unlike the tools of workmanship, which at every given moment in the work process remain the servants of the hand, the **machines demand that the laborer serve them**, that he adjust[s] the natural rhythm of his body to their mechanical movement." (author's emphasis)

"The ease with which we make technology part of our selves can also lead us astray. We can grant **power to our tools in ways that may not be in our best interest**. One of the great ironies of our time is that even as scientists discover more about the essential roles that physical action and sensory perception play in the development of our thoughts, memories, and skills, **we're spending less time acting in the world and more time living and working through the abstract medium of the computer screen**. We're **disembodying ourselves**, imposing sensory constraints on our existence. With the general-purpose computer, we've managed, perversely enough, to devise a tool that steals from us the bodily joy of working with tools.

"Automation can narrow people's responsibilities to the point that their jobs consist largely of monitoring a computer screen or entering data into prescribed fields... The apps and other programs we use in our private lives have similar effects. By taking over difficult or time-consuming tasks, or simply rendering those tasks less onerous, the software makes it even less likely that we'll engage in efforts that test our skills and give us a sense of accomplishment and satisfaction. **All too often, automation frees us from that which makes us feel free."**[181] (author's emphases)

And the Internet of Things in which, you and I as humans will merely be seen as another "thing," is set to dramatically increase the number of wireless devices polluting our environments with exponentially greater levels of EMR than we are already witnessing. Projections from 2018 predicted that 31 billion wireless transmitting devices along with 4 billion people would be connected to the Internet by 2020.[182] These new "smart" devices are being added to our new digital "ecosystem," and rapidly overtaking us, as the dream of the Internet disappearing and our surroundings predicting our moods, thoughts and desires, is fast becoming a reality—a dream, a fantasy, and maybe even paradise for the futurist but not for the last of the humans, for which this dream is a devastating nightmare—especially for WiFi refugees.

## *Jeanice's Story*

I met Jeanice Barcelo on the EMF Refugee forum in early 2019. Jeanice is an inspiration and unstoppable force—as an incredible researcher, author and educator, she helps to raise awareness about EMF hazards, birth trauma, abusive medical practices and industry corruption, while also offering life coaching in the areas of relationship, natural birthing, and is an ordained minister and a Jin Shin energy healing practitioner.

Jeanice's interest in the EMF topic in particular was first sparked when researching into the dangers of ultrasound radiation used as a regular diagnostic tool in tracking development of babies in the womb, but later extended into other aspects of EMF-exposure harm after Jeanice herself began to suffer from the effects of microwave exposures and became electro-sensitive. Ever since she was injured by microwaves in 2018, her life has been turned upside down and she has been made a refugee multiple times as a result.

Like me, Jeanice has also searched the planet looking for a safe environment in which to live and call home, but has run up against multiple unexpected obstacles in the process.

At one time Jeanice and I had discussed sharing a home on the west coast of Ireland after Jeanice traveled there in the autumn of 2019 thinking she might have found the perfect place, as it had at first seemed to be a very radio-quiet location. However, when signals of mysterious origin (possibly from radar used for military applications) haunted her on her exploratory travels, she had to abandon that particular plan and head back to Long Island, New York. Complications with bringing her cat[xxxv] and the cost-prohibitive nature of the current economy in Ireland served as further deterrents.

After her return stateside we discussed other possibilities abroad; including, Costa Rica, Panama and Belize, but found that in each case these countries did not offer rental properties in low-EMF areas. Even the jungle proved polluted with radiofrequencies as wired cable or landline Internet options were not available in remote locations—instead people in these locations used radio towers for their Internet access. The new radio towers were not part of the wireless phone service grids, but independent, and it seemed they were now littered throughout these remote jungle regions. Jeanice and I both were hoping to find or form a community of like-minded people and live off of the land, growing our own organic foods and living harmoniously with nature in a mild climate near the sea—one totally free from EMF pollution—our own version of paradise.

Jeanice's recent ordeal, at the end of 2021, of being made repeatedly homeless, illustrates the growing problem of the encroachment of the Internet of Things into our world. Her story is told in her own words

---

[xxxv] Now that no airlines traveling to Ireland allow pets to travel with passengers but must instead be stowed away along with baggage in the undercarriage of the plane. Jeanice says her cat is not a piece of luggage and I agree one hundred percent with this sentiment.

below. (Note I have left in her excellent footnotes and endnotes—please see each for more information, especially about the devastating effects of EMR on wildlife and our environment.):

### *A Month in the Life of a Wireless Radiation Refugee*

*By Jeanice Barcelo*

*The term 'electro-sensitive' is a misnomer conjured up by the medical establishment to cover-up the fact that a growing number of people are becoming ill from exposure to manmade, technologically produced, 'non-ionizing' radiation. Such radiation is being emitted by all wireless technologies, ultrasonic devices, many medical 'diagnostic' and 'therapeutic' machines, solar systems, wind turbines, and more. The number of humans, animals,[183] [184] birds,[185] [186] [187] insects,[188] [189] [190] trees,[191] plants, etc., becoming sick and/or dying from exposure to this radiation is growing. Cell towers, microwave antennas, transmitters, boosters, satellites, solar arrays, wind farms, etc. now blight the landscape all across the world. On almost every street corner of every city in the United States, and especially in front of schools, retirement homes, and hospitals, you will find a cell tower or microwave transmitter blasting its deadly frequencies at the innocent. There is no doubt the dark side is targeting the most vulnerable.*

*Radiation exposure, and in particular, exposure to pulsed microwaves, has led to a plethora of illnesses and deaths.[192] [193] [194] Yet very few people understand what is causing their sickness and fewer still are willing to make the necessary changes to keep themselves healthy and well. Combine this with the 'warp-speed' rollout of 5G and the uptake of people receiving COVID jabs (many of which appear to contain nanotechnology that can self-assemble into a wireless operating system within the human body),[195] and we are now facing a health and environmental crisis of gigantic proportions. Before too long, we are likely to bear witness to millions (perhaps billions) of people losing their minds and/or collapsing into severe illness and sudden death. In truth, it is already happening.[196] Young people,[197] [198] [199] athletes,[200] children[201] and even newborn infants born to vaccinated mothers[202] – are developing severe neurological problems, blood clots, strokes, myocarditis, and fatal heart attacks. These are all symptoms of radiation sickness caused by exposure to pulsed, modulated microwaves. The creatures that run the medical establishment have known about the real cause of these illnesses/deaths for decades but they have chosen to cover it up for at least 100 years. Today they are blaming the illnesses and deaths on a non-existent virus called 'Covid.' In 1918, they blamed it on the 'Spanish flu.' This alleged 'flu' was neither Spanish nor a flu, but rather symptoms of radiation sickness that first occurred on military bases where wireless telegraphy was being used and then spread around the world on military ships that were also using wireless telegraphy.*[203]

*Like the so-called 'Spanish Flu,' the term 'Covid' is a cover-up for radiation sickness.*[204]

*In 2018, I became a victim of radiation sickness (otherwise known as 'microwave illness'). After experiencing two severe bouts of 'the flu'*[205] *during the last half of 2017, in February 2018, my body finally went into full system collapse. I have honestly never been so sick in my entire life and I did not know if I was going to live. I was so weak that I did not have the strength to make it from my bed to the bathroom without collapsing to the floor. I was losing my hearing, my eyesight, and my sense of taste and smell. I lost my ability to stand up without falling down. I was running a low-grade temperature for several weeks, indicating my body had given up fighting against whatever was causing my malaise. My ears were ringing incessantly so loud that I could not sleep. I developed autoimmune hyperthyroid disease (Grave's disease), terrifying heart irregularities, muscle cramps, headaches, night sweats or 'hot flashes' (often blamed on menopause but actually the result of being cooked by microwaves), and more. It was one of the most frightening times*[206] *of my life.*

*Sadly, once someone becomes 'sensitized' to the radiation (i.e., once their body reaches saturation), there is no turning back. That person will feel sick whenever they are exposed. Hence, the symptoms I just described above persist to this day whenever I am exposed to wireless radiation or devices. My extreme sensitivity to the radiation makes living in this wireless society very, very difficult.*

*My illness was brought on by exposure to smart-meter radiation combined with a solar system that was placed on our home and was being powered by a deadly WiFi router that pumped high frequency pulsed microwaves into the house 24/7. Within one year of placement of these devices on our house, everyone in my family became severely ill. My mother experienced respiratory failure, unrelenting noise in her ears, chronic, itchy skin rashes, chronic insomnia, and illnesses so severe that she came very close to death on two separate occasions. My stepdad experienced rapid-onset memory loss, severe muscle cramps, balance issues (he was constantly falling down), thyroid cancer, dangerously low blood pressure, and more. It was during this time I finally figured out what was causing all the sickness. I hired a building biologist who showed me the myriad sources of wireless radiation in my house which included two WiFi routers, several cell phones, a wireless printer, several cordless phones, smoke detectors, iPads, and much more. He advised me to turn off the solar system, which I promptly did, and it was never turned on again. I then discovered how to remedy the myriad radiation problems in my home and I demanded the utility company remove the smart meter (which they finally did). I hard-wired everything in the house and strictly prohibited exposure to wireless/microwave radiation and devices. I lived inside of a radiation-shielded canopy and only came out to use the bathroom, kitchen and laundry or if I needed to go to the store for food.*

*Once I made the necessary adjustments to my life and my home, everyone in the house began to get better. Our symptoms all but disappeared.*

*Fast forward to January 2021 when I decided to move to North Idaho because the radiation levels on Long Island had become unbearable and the 'Covid' situation in New York was not something I wanted to participate in. By visiting a website called Radiation Refuge, I was able to find a place to live in North Idaho. I agreed to rent a small cottage, sight unseen, because it was protected from radiation and already set up with wired Internet and telephone. I knew I could get right back to work. I then traveled across the U.S. to a living situation that was extremely small and way overpriced, but it was radiation-safe, so I paid the price. A few months later, I ended up living in a larger house in a rural area that also had very low radiation levels. Unfortunately, I end-*

*ed up sharing this space with two extremely toxic people that had an addiction to using their cell phones (and other serious addictions as well). I erected an 8x8 radiation-shielded canopy around my bed and also brought in a comfy rocking chair, a small table that I used as my workstation, and my cat's kitty bed. I spent most of my time living and working inside the canopy and I always felt better when my cat was safe in the canopy with me.*

*Things at the house took a serious turn for the worse in October/November of 2021, when I let a nasty creature masquerading as a human woman stay in my spare bedroom. Suffice to say, I was forced to move out of the house on very short notice and on December 4, 2021, my time at the rural house was complete. Since I was unclear about what I should do next, I found an Airbnb rental close by that turned out to be a well-equipped house with WiFi and several other 'smart' appliances.*

*On Day 1 at the house, I noticed that it was secluded enough to avoid WiFi from neighbors. Hooray! I also noticed that the smart meter was placed several dozen feet away from the house instead of right on it—another very good thing. So, I entered the house and immediately unplugged the WiFi, hoping all would be well inside the house. Sadly, when I took out my radiation meter, I could see that the radiation levels in the house were still exorbitantly high. I quickly looked around and began unplugging every appliance I could, one at a time, in the hopes of discovering what was causing the radiation problem. I learned that it was not the refrigerator or the microwave, but the stove that was radiating pulsed microwaves. This new stove/oven was a wonderful 'smart' appliance that turned the entire house and the surrounding area into a microwave oven! So, I pulled out the plug and watched radiation levels drop, but still, they were much higher than they should be. I then discovered, upon closing the door of the dryer, that the dryer was also 'smart,' but after trying multiple times to pull the dryer away from the wall, I realized I was not going to be able to get to the plug behind the dryer to unplug it. I became frantic and started to search for the breaker box so that I could shut off the circuit that controls the dryer. Finally, I found the breaker box outside the house and then turned the breakers off, one at a time, until I found the ones that controlled the dryer and the stove. Still, there was a problem in the house and at last, I discovered the 'smart' thermostat—another wonderful 'conven-*

*ience' that allows the dark side to spy on those inside the house while irradiating them with pulsed microwaves 24/7. I pulled out the plug to that noxious device and quickly carried in my portable electric heaters which are radiation-safe and which I used to heat the house for the duration of my 2-week stay.*

*Finally, radiation levels in the house dropped into the green on my meter and my body was able to relax. I felt relieved to know that I could use the fridge without being pummeled by pulsed microwaves, and I could also use the washing machine without putting my life in danger. I did not use the dryer the entire time I was in the house, but rather brought my wet laundry to the Laundromat in order to use their dryers instead. When I had to cook, I put on radiation shielded clothing first, then went outside to turn on the breaker to the stove, and I limited my cooking time to 20 minutes, tops. I brought myself and my cat to the farthest corner of the house away from the stove while it was on and I immediately turned the breaker off when I was done cooking. Only then was my body able to relax once again.*

*When I needed to use the Internet, I was forced to plug in the WiFi. But before I did so, I took out a large piece of radiation shielding and wrapped it around the WiFi router multiple times in order to trap the radiation inside the shielding. I also plugged an Ethernet cable into the back of the WiFi router and ran the other end of the cable into my computer so that I would not need to turn on the WiFi on my computer. I sat very far away from the WiFi router when it was on even though it was shielded. All of these measures made the use of WiFi in the house manageable.*

*The WiFi router was unplugged immediately when I was done using the Internet. It was unplugged every night while I slept in order to protect myself, my cat, and all the creatures that live in the area. I am consistently hyper vigilant to doing everything I can to avoid hurting the birds, insects, plants, and animals in my vicinity with wireless radiation. It disturbs me greatly to see humanity carelessly using wireless devices for their own convenience even though these devices are deadly to every living thing. I refuse to participate in this satanic 'civilization' which normalizes the killing of innocent living things. I will not use wicked technologies that have no business even being on the Earth. I work overtime to try to avoid hurting innocent creatures. Sometimes, my efforts exhaust me.*

*After almost two weeks at the comfy house, I realized I was not going to find a long-term, suitable home in North Idaho. There were a number of reasons for this, but one of the most significant involves the fact that Airbnb has gobbled up all the rentals in the area and those who own the rentals have become greedy. Local people blame the lack of housing on 'transplants,'*[xxxvi] *claiming that newcomers to North Idaho are gobbling up the rentals and causing the housing crisis. My experience, however, showed me that there are actually many, many rentals available in North Idaho. It is just that the majority of them have been listed on Airbnb or VRBO at very high prices so that the owners can rake in maximum dollars. Therefore, local people who are desperate for affordable, long-term, housing, are out of luck. Similar to the way Amazon and Wal-Mart have caused millions of small businesses to go under, Airbnb has completely altered the housing market in the U.S. (as was probably the intention all along).*

*While the controllers of Airbnb have made millions (perhaps billions) of shekels for themselves, they have simultaneously turned everyday landlords into greedy 'vacation rental hosts,' rendering an untold number of low and middle-income people unable to find affordable housing.*

*In addition to this stark reality, I also had to acknowledged that the weather and environment in North Idaho were not something I really enjoyed or felt cut out for. For example, having to lie down in five inches of snow while simultaneously getting soaked and filthy in order to put chains on my car, was not my idea of a good time. And so, after months of resistance, I finally accepted the fact that I was not meant to remain in North Idaho. I surrendered to the idea of moving back to the east coast, which move necessitated a 40-hour drive across the U.S. on highly irradiated interstate roads, with several overnight stops along the way so that I could sleep and let my cat get out of the car, eat, and calm down after the stress of the day's drive. I was not looking forward to the journey. I knew it was going to cost a small fortune and I*

---

[xxxvi] "Transplants" are those who flee democratic tyranny in "blue" states by moving to "red" states like Idaho.

*feared the radiation levels that I would encounter along the way. I tried to plan ahead by booking vacation rentals where I would have an entire house to myself. After my experience living in a house with two cell phone addicts, I had to acknowledge that I cannot be around other people. I must keep myself separated from other humans because the majority of them are addicted to their noxious devices. This is a very sad fact of our current reality.*

*So, after giving away virtually everything I owned, including dozens of beloved plants, several boxes of stored organic food, a brand new Dutalier rocker/glider that I paid a lot of money for, a new djembe drum, several electric space heaters, and much, much more, I packed my van to the ceiling and headed out on the road. I was suited up with a radiation-shielded hat, shirt, and pants*[xxxvii] [xxxviii] *and also had a head net that completely covered my face, neck, thyroid, and upper shoulders in case things got really bad. Once on the road, I quickly discovered that I was literally taking my life in my hands since the roads were horrifyingly treacherous with ice and snow everywhere. Nevertheless, I felt I had to go.*

*With white knuckles, I gripped the steering wheel of my van, trying to hold the vehicle steady on the road. My stress levels escalated every time I noticed a cell tower and they absolutely skyrocketed when I was forced to sit in traffic in Couer d'Alene, Idaho where the roads were so bad, people were hardly moving. Couer d'Alene is a HIGHLY IRRADIATED city and I do not like being in that town because of it. Being forced to sit in traffic and be cooked by microwaves for over an hour was truly unnerving. Fi-*

---

[xxxvii] Jeanice used to wear radiation-shielded socks too, but—she says, she left them in New York, wrapped around a noxious medical alert device that her mother was told to wear 24/7 in case she fell down. The fact that this radiation-emitting device is likely to be the *cause* of her losing her balance and falling down is never acknowledged by the medical establishment. Instead, they just keep giving ignorant and vulnerable elderly people more radiation-emitting devices (like hearing aids) that will ensure their rapid demise.

[xxxviii] Because, Jeanice explains, she had no shielded socks to wear for the trip, she developed an itchy radiation rash around both of my ankles. The rash healed up within days of her arrival at a radiation-free house.

*nally, I got past the traffic and onto I-90 which interstate also had patches of ice and snow everywhere. Driving was extremely treacherous.*

*After hours of stress, I finally reached my destination at a vacation rental in Montana. To my horror, I discovered that the studio cottage I had rented for the night was in direct line-of-sight to a cell tower. Definitely not good. I panicked and felt I wanted to crumble to the floor and die. My host, however, was fantastically kind and understanding. She allowed me to turn off the electric breakers in the cottage save for one, which gave us light from one lamp and hot water. She then offered to help me carry in my things and set up my radiation–shielded canopy. We spent over an hour moving furniture around to make room for the canopy and putting it up. Radiation levels were still somewhat high in the canopy but much more manageable.*[xxxix] *My super kind host, who lived in the building next door, also agreed to turn off her WiFi while she slept. Radiation inside my canopy then dropped to levels in the green. My body could relax then, especially once my cat joined me inside the canopy. I slept OK. When I stepped outside the canopy to use the bathroom or kitchen, I became dizzy and lightheaded, with tension across my chest and heart area. I was looking forward to leaving the rental, which I promptly did the next morning.*

*On Day 2, I headed out on 1-90 in Montana where there were cell towers every few hundred feet. From my vantage point, it appeared the entire state had been turned into a radiation death trap. My observation was confirmed by my host who said that, 'every town in Montana has a cell tower.' Frankly, I could not wait to get out of that state and get myself into Wyoming, where I knew things would be better. But Montana is a BIG state and the-powers-that-be suddenly decided to close I-90 due to extremely high winds. Not good. After an hour of utter frustration sitting in non-moving traffic and surrounded by people in highly microwaved vehicles using their cell phones, I somehow found my way*

---

[xxxix] Jeanice would discover three nights later that there was a tear in the shielding they had placed on the floor, which allowed some of the radiation to leak through. This is why radiation levels in the canopy were still somewhat high.

*around the road block and continued on my journey. I finally reached my second destination later that day in Wyoming— a place I have stayed once before. I looked forward to arriving there because I knew that once the WiFi is unplugged at that location, radiation levels would be in the green inside the cottage. My only fear was that the dark ones might have erected a cell tower close by. But much to my happy surprise, there were no new cell towers in the area and I felt deep relief being able to spend a night at that beautiful location in a radiation-free environment. I did not plug the WiFi in at all. I simply did not want to go near it.*

*On Day 3, I headed out on the road again to a rental in frigid South Dakota. The roads in South Dakota were free of snow and ice and there were very few cell towers as well. This was a relief. When I arrived at the rental for the night, however, I realized it was an ENORMOUS house, with many TVs and other potential 'smart' problems. Quickly I found the breaker box and turned off the smart meter that was located INSIDE the house. Lucky for me, some of the electric in the house still worked after that, including the fridge and stove (not sure how that happened, but I'm glad it did). I turned off a few more breakers until I achieved radiation levels in the green. I slept downstairs where levels were the lowest. Things were OK until the morning, when my car wouldn't start. My host gave me a jump and I travelled to the nearest city to a garage to have them look at the car. I was forced to sit in a highly irradiated garage for about one hour and then told that the car was fine and that it must have been the cold or some loose wiring. All was well with the car after that.*

*On Day 4, I continued travelling east to my next destination and was concerned when I realized the place I had rented for the night was in a neighborhood with houses that were fairly close together. I entered the house and immediately unplugged the WiFi. Still, radiation levels were high and I seriously did NOT want to put up my canopy again, especially by myself. So, I unplugged the dryer, washer, TV, and a few other things. I could not find the breaker box but noticed that radiation levels were highest by the windows on the side of the house facing the neighbor. I figured it must be their WiFi. I chose the bedroom with the absolute lowest radiation levels and put shielding on my bed both under me and over me. I slept OK and felt grateful that I was somehow surviving the trip.*

*On Day 5, I got on the road again and was thrilled when I got to Missouri and felt the sun shining brightly on my face with temperatures around 65 degrees. Driving felt much more pleasurable at this point. The days were getting longer as I headed south. The sun was no longer setting at 3:30 but rather between 5-5:30, which felt like a welcome change to me. Driving is much easier in the daylight. I stopped in Columbia where I knew there was a good health food store (I lived in Columbia for a whole year in 2016). I walked into the store and immediately my eyes met with those of another as if we had known each other before. As it turned out, I had not met her before but she had recently seen my interview with JC Kay and somehow 'knew' me. We talked for many minutes about many things, but mostly about the radiation issue. She got it. I then drove around Columbia for a few minutes and noticed there were no cell towers in the area. How could this be, I wondered? My mind began racing. Should I move back to Columbia? It is a college town, after all, with lots of young people ripe for TRUTH. But I quickly came to my senses and realized that today was not the day to be thinking about relocating to Columbia. I knew I needed to continue my journey eastward, even if I did not understand why. I had to go where the opening had occurred and my spirit was urging me on.*

*I then travelled way out of my way in to get to a rental in the country, imagining this would be a safer location for me than in a town or city. However, I grew concerned as I got closer to my destination and saw that there were wide-open fields in the area, and miles and miles of expansive horizon, with absolutely nothing to block cell tower radiation at all. Fretfully, I arrived at the rental, which was beautiful and in a fantastic location, on a dead-end street and on the water. It was such a sweet spot. But damn. There was a cell tower about two miles away in direct line-of-sight of the cottage. Immediately, my inner voice screamed... 'Those bastards are destroying everything!' I despise the wireless industry and everything it represents.*

*As a result of the cell tower, radiation levels in the cottage were high. I immediately unplugged the WiFi router and turned off all electric breakers, but radiation levels were still high. I noticed the smart meter was pulsing away. And the cell tower was just sitting there, beaming its Luciferian*[xl] *'light' directly at the cottage and for miles around. Naturally, all living things were/are being negatively affected by that hideous and evil structure. I visualized it going up in flames.*

*With much regret, I then surrendered to the fact that I would have to put up my bed canopy if I wanted to survive through the night. What's worse is that I would have to do it by myself. I felt exhausted and stressed. But I had no choice. I got to work.*

*The struggle of erecting the canopy was almost too much to bear, especially because I was already beginning to feel ill from the radiation. Dizziness and a strong headache were coming on and tension gripped my heart area. Each time I tried to bring the pieces of the canopy together, one side inevitably collapsed back to the floor. I was overwhelmed with a sense of doom and frustration. Finally, I figured out that I could use an open dresser draw to help support one of the legs of the canopy while I pulled the others into place. At last, the structure was up, and then I had to put the shielding into place, which was another interesting chore. About two hours later, I was able to enter the canopy. With great relief, I saw that radiation levels inside the canopy were in the green!!! I began to cry with relief. I seriously did not want to ever come out of the canopy. I was feeling so stressed and overwhelmed that all I could think was that I do not belong in this nightmare 'civilization' and do not want to be a part of it. I begged the Creator to help me get to safety. My body was longing for a feeling of safety and grounding. I was not sure how long I could go on feeling this sick and stressed. Finally, I fell asleep.*

*At 5:30am, I bolted out of bed and got ready to high tail it out of that place and face yet another day on the road. I wondered if*

---

[xl] My own understanding of Jeanice's reference to Lucifer here is that Lucifer represents artificial light, as in, artificial EMFs. You do not have to be religious to understand that if what we call God is the creator of our natural world, then Lucifer, as a light source/ or "bearer of light" that seeks to replace God with its own light or energy—one that also seeks to control man, and rob man of free will and agency, then it makes sense that this Lucifer may really take the form of health-destroying and mind-altering non-native (and so not from God), "light" or EMR.

*my next destination would prove harmful to my health. I was seriously not looking forward to the drive. Still, I packed my belongings and put them into my vehicle, making sure my cat was well fed and watered first. Every few minutes, I jumped back into my canopy to recover from the radiation levels outside the canopy. Finally, it was time to take the canopy down, which I did quickly and then jumped in my van and drove away. Once on the road, I stopped at McDonalds to use their WiFi and ask the host if there was a cell tower near the rental that I had reserved for the night. He assured me there was not and offered to turn off the WiFi in the unit before I arrived. I felt relief over this and continued on. As I got closer to the east coast, I remembered why I so urgently wanted to leave. There were cell towers* ***everywhere*** *and traffic was a nightmare. Congestion, overcrowding, toxicity, and frustrated people surrounded me. I wondered why people continue living that way. I simply don't understand it. It's not how I want to live. I'm ready to return to nature.*

*My next destination was a one room, no frills cabin that I thought would surely be a reprieve from wireless radiation. I was WRONG!!! In this no-frills cabin was a smart heating system and several other smart devices causing high radiation levels. I panicked once again and asked the host for help. He agreed to help me by turning off all electric. Radiation levels immediately plummeted into the green. Sadly, he refused to turn off his cell phone, even for 5 minutes, since he felt that every interaction with a guest must be recorded. This is the insanity of our world. Obviously, I could not allow him to stay in the cabin because (a) he was irradiating me with his phone and (b) I would never have been able to figure out which circuits were causing the problem if he kept his bloody cell phone turned on!!! Finally, he left and I got busy figuring out which breakers I could safely leave on and which ones needed to be shut off. I was able to get radiation levels down into the green by keeping the smart heating system turned off and hauling in my electric space heaters to heat the cabin. I got a good's night rest. My host greeted me the next morning with questions about how he could learn more about the wireless radiation issue. I directed him to my book[207] and website.[208] Then I packed the car and was on my way.*

*A few hours later, I arrived at my final destination, exhausted from the trip and ready for a much-needed break. But I did not like the looks of the house I was about to enter. Although it was huge and in a really nice location, it was in decrepit condition. When I stepped into the back porch, I was overcome with the smell of mold. I then entered the house and almost cried out in horror. I had never before seen such a dirty house. It looked like the house had not been cleaned in 25 years. I was horrified by what I was seeing and began sobbing uncontrollably. I wondered what the heck I was supposed to do next.*

*Frantic, I drove into town to a nearby McDonald's to use their WiFi in order to make a few calls. I poured my heart out to two friends, one of whom is only 4 hours away and offered to let me come and stay at her place. I contemplated this idea but knew it was not the right thing because, as much as I love this sister, she is addicted to using her cell phone. I cannot be around these devices. I* ***will not*** *be around them. There has to be another option.*

*Exhausted and broke, I surrendered to the fact that I had no choice but to stay in the dirty house. I toughened up and mustered up the courage to begin cleaning the house. First, I cleaned the bedroom, then the bathroom, and then I moved onto the kitchen. It was more disgusting than anything I could have ever imagined. I suddenly realized I had not yet checked the radiation levels in the house. I stopped what I was doing and turned on my meter to discover that radiation levels were in the green. This was a tremendous relief. Then, with great trepidation, I walked around the house to see where the smart meter was. Happily, I discovered it is located on the opposite side of the house, far away from my bedroom. There appeared to be no microwave pulsing going on in my bedroom. This was a big relief since the pulses negatively affect my heart and my ability to sleep. I felt grateful to realize I would not need to put up my bed canopy here – or at least not right away. I then went back to cleaning the house and continued doing this for the next 3-4 hours before taking a shower and collapsing into bed. I slept well and continued cleaning the next morning and throughout the duration of my stay.*

*At my new location, I had no Internet service or ability to communicate for the first 10 days. Since I do not own a cell phone, I usually make all my calls from Skype on a wired computer. But in the new house, I could not do that yet and so I had to travel into town to a McDonald's to use their public WiFi in order to check email and make calls. Each time I had to do this, I dreaded turning on the WiFi on my computer. Immediately, I would feel pressure in my chest and across my heart area – as if something were gripping my heart from the inside. I also quickly developed headaches. So, I limited my time using WiFi to 15 minutes and then turned it off and busied myself doing other things. I've noticed that there are not many cell towers in this new town but still, there are days when I feel decidedly unwell in the stores. I figure it must be their WiFi or being around so many people with cell phones.*

*After spending a few days at my new location, I also notice that, because I am living in a house with a smart meter attached to it, the 'ringing' in my ears is loud and incessant. The noise I hear is metallic and electrical. It is a high-pitched, mosquito-like hum that sometimes escalates into a tortured screech from hell. This sound is worse than any annoying mosquito I have ever heard. From all of my research, I know that what I am hearing is ultrasound. The lying medical establishment will never tell you this. Instead, they call this phenomenon 'tinnitus,' but it is not tinnitus[xli] at all. Rather, these noises are the result of what scientists call 'microwave hearing' and/or 'ultrasonic hearing.'[209]*

*According to the science, millions of people around the world have the ability to hear the microwave frequencies and/or the ultrasonic frequencies that now permeate our air space and are being emitted by wireless devices.[210] Babies can hear the frequencies;[211] children and teenagers can hear them too.[212] Many adults can hear the frequencies (but not all), and pets, especially dogs, hear the frequencies as well.[213] Some people hear the high-pitched, metallic noise. Others hear 'clicks' or 'hisses' – which are the microwaves. Being forced to listen to these horrid sounds 24/7 is a form a torture and some people have commit-*

---

[xli] I too have suffered with tinnitus/ringing in the ears for decades with increasing persistence. I have also noticed this horrible high-pitched ringing increase in the presence of EMFs. But some level remains even in low-EMF or seemingly no-EMF locations, which suggests that either I have suffered permanent damage from microwaves or that there are no truly RF-free locations in this world anymore. But I think, at least in my case, that both are true since the tinnitus is much worse in my left ear—the ear against which I used to hold my cellphone.

*ted suicide as a result. The only way to stop the noise is to shut down the wireless grid. Sadly, this solution will likely take some time because humanity is currently addicted to their wireless devices. It is not clear what it will take to get people to snap out of their addiction and let go of these technologies, but let's hope something happens soon because, if we do not get these technologies off the Earth, life on this beautiful planet is likely to collapse.*

*On the final day of 2021, I decided to travel to the nearest city to see if I could find quality health food. The city is about an hour away from where I am staying. As I neared the city, cell towers began appearing everywhere. The most grotesque and offensive tower was sitting approximately 100 feet away from a grammar school. I wondered how it is possible that people do not see how evil this is. Do people really send their children to that school? Do they not fear for their children's lives? Don't they know that children who attend schools near cell towers are becoming ill and dying?*[214] *Have they not heard that WiFi in schools is causing children to have heart attacks and die?*[215] *And what of all the people who live in the neighborhood who are now being massively irradiated by the tower? Are all these people prepared to die in exchange for the 'convenience' of using wireless devices? I just do not understand what is wrong with humanity that people do not acknowledge the danger. I grow dizzy and exhausted just being around the radiation and I do not understand how other people do not feel it.*

*I briskly walk toward the health food store, noticing several people wearing masks, cutting off their own oxygen supply and choosing to breathe in carbon dioxide instead, which will eventually cause brain and lung damage. Some of these people are simultaneously blabbing away on their cell phones while holding them up to their heads. I marvel at the stupidity of humanity being willing to participate in its own destruction. Cell phone radiation is well known to cause brain tumors.*[216] [217] [218] .[219] [220] [221] [222] *Why in the world would anyone want o hold one to their head?*

*I quickly do my shopping and then get myself out of that store and out of the city as rapidly as possible. I am relieved when I return to the small town in which I am staying and I think I will not go back to the city again. The sun is shining and I am happy to be staying in a house that has old appliances, all of them fully functional and built to last, unlike the 'smart' appliances of today that are complicated to use and degrade quickly due to radiation-induced rapid-aging. Although I've had A LOT of cleaning to do to get this place in order, I am feeling blessed to be living in a space with very, very low levels of EMF radiation. Wired Internet service is available at this location and was hooked up just a few days ago. All is well. How long this will last is anyone's guess but let's hope this obscure little town remains off the radar for a good long time. If the dark forces get wind of the fact that there is a radiation-safe area in the U.S. or elsewhere, they will rapidly swoop in with their 'clean' and 'green' wireless death grid technologies in order to ensure that not a single life remains. Hence, I will keep my location private and hope that more and more of those who read my story will take action to protect themselves and their loved ones from the most grievous and serious threat we have ever faced.*

*I am wishing everyone a very blessed and Happy New Year. May all good things come to you.*

Fortunately for Jeanice she finally has found a decent refuge and I really hope it works out for her long term. It is so difficult for any of us faced with this absurdly impossible situation to feel secure once we have found a refuge, as we are all too aware of how easily it can be ruined. We crave putting down roots. And in a world where so many are made to incessantly migrate due to the sweeping and often sudden changes to job markets perpetuated by the same forces that cause WiFi refugees to flee, most of us long for the chance to grow roots and plant ourselves in a natural, peaceful, life-sustaining setting, preferably with friends, family and community nearby. Jeanice has gotten out of the collapsing mine, or the boiling pot, but it is a lonely existence having to start anew so often and attempt to find or build community in a new location. But Jeanice is determined to do just that—spend more time in the natural world, build community and grow food, while continuing to educate as always, which she is hoping to do more of in-person starting this year as a personal resolution.

Jeanice decided to share her story after being inspired by a fellow WiFi refugee and overseas friend Paul Gregory, who shared his story about just how difficult day-to-day living can be for someone with EHS—especially at the holidays. I happened to also contact Jeanice at the same time asking if she wanted to tell her story for this book—so the serendipitous timing led to its inclusion here. This same friend also agreed to have his story included in this book. This is how Paul spent Christmas in Birmingham, England, 2021 as a WiFi refugee:

## *Paul's Story*

### ***Christmas Diary of a So-called "Electro-hyper-sensitive" Person***

Paul Gregory, 27.12.21

*Today is Christmas Day 2021. I am not writing this account for self-aggrandisement, but to sing of my blessings and raise awareness about all the people across the world who are in a far worse position than I am. They number in the millions. It is not just other people's 4G and 5G phones, masts, Wi-Fi everywhere, but also the LED lights in homes, shops and Christmas decorative lights, which emit pernicious radiofrequency radiation (RFR). This ubiquitous energy tortures me on SMART motorways with the new weaponized street lighting. It prevents me from going to public places and the homes of friends and family, and it has invaded my home. It causes huge heartbeats and heart arrhythmia. A cardiologist says my heart is perfectly normal. My skin itches, muscles ache, ears ring with tinnitus, but I think I am just hearing this radiation. My mood and motivation ride the highs and lows of these emissions. I am only getting the same symptoms that enemy troops get when targeted with psychotronic and electromagnetic weapons.*

*By the end of this account, you will know that I am speaking honestly about my situation. I enjoy good physical and mental health and am driven to question everything, to seek the truth as best I can, in order to assuage my ignorance that has been a lifetime in the making. I am waking up to a reality that I never dreamt existed.*

*My problem is not within me but is caused by the environment. RFR is five trillion times higher than the naturally occurring level. My situation will soon be your situation. We are being micro-*

*waved alive. It is the same radiation that a microwave oven uses to cook food and no one in the media of is discussing the issue.*

*I wake on this Christmas morn from a refreshing sleep in my BlocSilver home-made sleeping bag in my little bedroom, which I have made as mobile-phone-radiation-proof as I can. All the other bedrooms are too radiation-polluted for me to use. It is 7.15 a.m. and I take off my BlocSilver anti-5G hood, which I wear in bed to protect my indispensable brain.*

*Once out of bed, I slip on the BlocSilver full-length protective underwear over my pyjamas (because I would otherwise start feeling ill from the radiation straightaway) and go downstairs to greet my wife who is making a pot of hibiscus and ginger tea for us to drink in bed together.*

*She listens to BBC Radio 4 till we get to the first round of the daily dose of Covid propaganda. She agrees to turn it off to stop my arguing back on every lie that is uttered.*

*After about 20 minutes we go downstairs to have our breakfast. Mine consists of different organic nuts and berries, flaxseed, wheat germ and porridge, nicely soaked in organic grass-fed raw milk.*

*My wife sits at the dining table, and I sit with her for a few minutes before she and I notice me drifting off into a half-waking catatonic state caused by the radiofrequency radiation. She quickly rouses me, and I trot off to the lounge to sit on the floor between the door and the sofa. It is the only place downstairs where I can watch television without experiencing the effects of the new streetlights outside, which beam radiofrequency radiation (RFR) into my house day and night courtesy of the local authority.*

*Breakfast completed, I have come round and now feel normal, in fact, quite good.*

*Christmas Day is unfolding. Towards 11 a.m. I drink a pint of filtered structured water and have a sandwich of organic whole meal sourdough bread and organic butter made from grass-fed raw milk, followed by a square of organic 99% Vivani chocolate, complemented by an organic date, which I have discovered reverses the low mood that can be triggered by the effects of RFR on our brains. Research has discovered that it causes a decrease in blood flow to the brain along with a reduction in blood glucose, serotonin, dopamine and acetylcholine neurotransmitters. I am so blessed to have come across this information because it explains all my symptoms exactly. My body is under attack from the radiation-polluted environment.*

*We have two guests for Christmas dinner. The first knocks at the front door and I quickly exit the downstairs toilet/office/RF sanctuary where I have been reading in order to let him in. I have to secrete myself behind the door to avoid a blast of RFR from the streetlight. Our first guest understands and comes in. He is double-jabbed, 'boosted' and flu-vaccinated and lives with all the Wi-Fi gadgets you could think of, all carefully coordinated by Alexa. I talk to him in an animated fashion for about five minutes, getting him a drink before my half-waking catatonia starts. My wife swiftly reminds me that it is the radiation he is emitting that is doing me in. Back in the toilet I recover completely in 15 minutes, just in time to emerge to greet our next guest. She is totally unvaccinated and well informed about what is going on, but has to live in the 5G emissions that bathe our town of Sutton Coldfield.*

*I get her a drink and retreat to the toilet as the same cycle repeats. After 10 to 15 minutes, I am once again myself and can open the toilet door and the lounge door opposite and take part in the conversation from a distance. I am doing really well and gaining confidence, so I join our guests in the lounge. It is so good to have a 'crowd' of non-socially distanced friends talking animatedly, conversing like we used to do in the era Before Covid (BC). Ever optimistic, I forget to protect myself and soon my wife realises that I have stopped talking. She suggests that I retreat to the toilet in order to be well enough to carve the Christmas chicken. From my sanctuary, I overhear our first guest surmising that my plight is probably due to the huge RF signals generated as people use their Wi-Fi devices more and more through the day.*

*Half an hour later I feel better and feel up to carving the chicken. Dinner is ready, and my wife anxiously inquires as to whether I will be okay sitting at the table in the dining room without getting a 'hit.' I tell her I'll risk it, but I'm ready to retreat if it gets too much.*

*The drinks oil the wheels of cordial interaction. Our first guest pours me a big glass of organic red wine, which I consume with pleasure. We help ourselves to vegetables whilst I keep an eye on the RF level. It is okay; I can't feel anything. This guest wonders if it is because people have put their Wi-Fi devices aside in order to eat their Christmas dinner.*

*The meal is fantastic. I am able to stay at the table for two whole hours. The conversation is exhilarating. If the lockdowns have taught me anything, it is that human interaction, whether at work or with family or friends, is the most precious activity there is. This pandemic fraud can be understood with 10 minutes' research that first confirms that the PCR test does not identify any virus or variant (the Government says: 'RT-PCR detects presence of viral genetic material in a sample but is not able to distinguish whether infectious virus is present') and second the so-called vaccine is of a type never used on humans before. Do I want to risk taking it? No! Why would I want to take a vaccine for a virus that has not been identified? And if the so-called Covid symptoms are not caused by a virus then it must be something else. It is the Wi-Fi and 5G because the symptoms of radiation poisoning are identical to those of Covid. Electro-sensitive people are testimony to that. It is that easy.*

*Both our guests understand my situation and do not dwell on it. My wife had invited a lady who would otherwise have been on her own at Christmas as she was bereaved two years ago. She has been badly traumatized by the Covid propaganda. However, she phoned earlier to say she preferred a quiet day at home. This was sad because the campaign of fear and terror waged by the government has robbed her of her confidence to do almost anything. When reflecting on such criminality I usually tell myself a joke. My latest is a new meaning for the phrase 'a hung parliament.' When the people—or 'the masses' as the elected prefer to think of us— find out what has really been going on over the last century, they may wind up regarding 'hanging' as the more merciful of the options available to them. But I digress.*

*Between glasses of wine, I nearly forget to protect myself. Our first guest is telling us how the usual weekly night with his mates at the pub has been cancelled because one of the group has a daughter in her thirties, a doctor married to an anaesthetist, and the husband is having difficulty coping with looking after their children whilst his wife deals with the 'Covid' crisis at the hospital. This means that the grandparents are on standby to go and help out their daughter and son-in-law with the children and this drinking pal did not want to catch Covid from his friend. We were all curious as to why a young anaesthetist might need help with looking after the children. It turned out that at the age of 26, about 10 years ago, he had had a successful operation for an aortic aneurism, but two or three months ago had become seriously ill with infected heart valves, necessitating a life-saving operation.*

*He was still recuperating and facing the possibility of never being able to work again as a doctor. I am eager to ask for more information, such as, has he been double-jabbed and boosted, when our second guest intervenes to remind me of our prearranged pact not to raise the issue of Covid or vaccinations in order to keep the peace.*

*I feign listening to the conversation whilst I try to join the dots. I bet this anaesthetist has been vaccinated and it has caused pericarditis or myocarditis in the same way that so many international footballers, athletes and teenagers have suffered. We are told that the putative cause was an infection caught from the mouth of a patient he was attending to in the course of his work during an operation. 'A likely story,' I thought, 'They blame a bug without even considering all the toxins in the vaccines. How very convenient!'*

*The conversation is back on a safe track and our meal ends as the time draws near for the Queen's annual Christmas speech to the nation. Everyone gets comfortable in the lounge whilst I resume my usual seat beside the sofa. Too much RFR, so I put on my hood, but I am still in too much pain so I quietly abandon the Queen and resume reading in the toilet, once again comfortable.*

*My situation frequently leads to disappointment but I try not to dwell on loss. I may plan to watch a television programme with my wife, but if the radiation is too high to cope with, even with my RF-blocking suit and hood, I have to give up and retire to the toilet. I make return visits to cafés or countryside that were previously RF-safe only to find that now they have an RF signal I cannot cope with, and my plans have to be altered. The signal-free environment has become smaller and smaller. It is loss, part of depression. I don't want to be forced into this mood, so as soon as possible my wife and I work out a partial solution; we try to beat it or get round it. So right now I switch my attention to the story I had been reading all morning. It is a short story called 'The Machine Stops,' written by E.M. Forster in 1909.*

*It is about a future with AI satisfying all the needs of the world population as they live underground, alone in hexagonal little rooms, rarely going out, but communicating with friends and family around the world via video screens. It feels a bit like my situation, but the difference is that they are so grateful to the Machine. Constant propaganda has taught the masses to like this life till the day that the Machine stops. It is a brilliant story and so relevant to our contemporary world. I just heard on the radio that food delivery is getting so good that people do not need to leave the house or even cook now that a pizza (which is highly toxic, of course) can be delivered 24/7.*

*I hear our guests leaving. I reappear to say goodbye. They have really enjoyed the afternoon, as have I. We congratulate my wife on a superb Christmas dinner.*

*I am not ready for rest so I start reading my next book 'Tragedy and Hope 101. The Illusion of Justice, Freedom, and Democracy' by Joseph Plummer, a précis of Professor Carroll Quigley's encyclopaedic history of the activities of the elite (Tragedy and Hope: A History of the World in Our Time), who wanted a trusted historian to document their take-over of the world. But he reneged on his agreement with them to keep the information private and had it published. He was lucky they left him alive.*

*After finishing the first chapter, I decide I have had enough excitement for one day and want to have a lie down. I try between the sofa and the lounge door, but there is too much RFR so I put on my hood, but still it is too much and I have to go to my last resort for a daytime rest. I move my laptop out of the toilet and lie down in there. It is about 4 feet square, so I pride myself on my athleticism at 75 being able to lie out full-length with one foot on the cock-stop and the other on the toilet lid. I had the forethought to take off my sweater as it gets hot in there. I keep my hood on just in case and take in my Qi Shield RFR-protection de-*

*vice. I used to find it effective, but much less so now. Relaxing, I realise I can still feel RFR, so I lower my hood to cover my face such that I am now totally covered. It works. I can relax and slip into meditation. I listen for what God is saying. He is silent but what comes to mind are all my blessings, a caring wife and family, a home, being free from pain most of the time, warm Internet friendships with people I have never met, and a growing number of awake friends in my area. These are riches beyond compare. I know of people who have had to leave their houses and live rough or in their car because they cannot cope with the radiation coming into their house. I know people who are ill, probably because of ignorance of the polluted, toxic environment they live in.*

*At 8 p.m. I find that the RF signal has gone down so much that with my hood on, I can watch television from beside the sofa.*

*I have had a good day this Christmas Day. I will continue to work on staying well and wait patiently for better days to come. After all, Solomon is said to have wisely observed that all things will pass.*[223]

I relate all too well to Paul's struggle in trying to socialize with the tech-addicted as an electro-sensitive. Many, even after turning off devices for us, still carry artificial EMF charges on their bodies, also called "body voltage," an actual measurable field.[xlii] And as people become increasingly contaminated with heavy metals (contained in vaccines, especially those with highly conductive graphene oxide—more on this presently—found in the latest generation of "vaccines", and in our polluted environments) they also act as antennas, not unlike iPhones, which relay or "ping" off of other iPhones in the vicinity. Ultimately, we may even turn fully into human batteries[xliii] made to power the wireless grid—our own

---

[xlii] You can find body voltage meters at: lessemf.com

[xliii] Already there are patents for wearable tech in the form of jewelry that can be powered by our body's own electrical energy. https://gajitz.com/creepy-parasitic-jewelry-is-powered-by-the-bodys-energy/

invisible prison, while also living as avatars in a virtual simulation of reality. This idea presented in the Hollywood *Matrix* movies is not as farfetched today as it seemed in the 1990s.

And I agree with Paul—*BC* (Before COVID) was another world, even though already challenging for canaries (and others, for a myriad of reasons), this seeming New World Order that has descended upon us makes me long for BC. Now in *AD* (Artificial-Intelligent Design?), I feel a sense of nostalgia for my "kid fears," compared with today, and feel a deep mourning over the sudden sweeping loss of culture and humanity the world over.

I also agree that the E.M. Forster story, "The Machine Stops," is highly relevant to our world today which seems to be moving quickly from the realm of fiction to non-fiction as people's dependencies on toxic, portable machines are fast encroaching into every aspect of our lives, including taking center stage at important social gatherings, even holidays like Christmas, long before set apart as a day to spend meaningful, quality time with close friends and family.

But this hijacking of our lives by "the machine," is intentional and meant primarily as a means, not only for insecure governments to keep citizens under constant surveillance, but as a way to overtake their minds and emotions all together.

## *Battle for the Minds of Men*

Mind-control experiments from the 1940s-1970s (and doubtless continuing today under new names) conducted by the US intelligence and military agencies, disclosed in public Senate hearings[xliv] in the 1970s, revealed that thousands of unwitting US citizens had been experimented upon using a combination of pharmaceutical[xlv] and other drugs, electroshock, hypnosis, isolation, sensory deprivation and radiofrequencies, by the CIA and NSA.[224]

---

[xliv] Conducted by Senator Frank Church, and thus called the "Church Senate hearings."

[xlv] With some of them commonly prescribed to our children, such as Ritalin.

In 1964 Dr. Jose Delgado demonstrated before a stunned audience how mind control could be implemented via radio waves and remote-control devices. Delgado was able to stop a charging bull in its tracks by way of a small remote-control box, after having first implanted an electrode in the bull's brain. He proved that an animal's brain could be altered so dramatically by radio waves as to turn a raging bull into a docile creature in an instant—at the flick of a switch. Similar tests were conducted on lab monkeys where the monkeys, wearing sensor-equipped helmets, whose behavior could be changed from docile to aggressive (the reverse of the bull demonstration)—also at the flick of a remote-controlled switch.[225]

Delgado was a fierce proponent of creating what he called the "psycho-civilized" society by physically manipulating and controlling members of society and making them into "model" "psycho-civilized" citizens—the same way he controlled the bull, using radio waves and brain implants. All, of course, for our "own good" and to make us, "happier, less destructive, and better balanced than the present man."[226]

And the CIA and NSA experiments into mind control, while conducted under the guise of countering psychotropic weapons being developed by the Soviets during the Cold War, also held the promise of controlling another "enemy"—that of the free-thinking citizen. By subduing and controlling its citizens, the US government, now under the thumb of the military-industrial complex would more easily extend and expand its empire.

Hence the creation of the cyborg is in the direct interest of governments wishing to control and monitor is citizenry. Use of radiofrequency-powered technologies represents one means to this nefarious end. Other means, also tested during CIA mind-control experiments (that went under the code names of MK Ultra, and Project Bluebird among others), included drugging, isolation and hypnosis. In today's world, hypnosis is used on a regular basis by mainstream-media journalists, public-health directors, and world politicians—who are all well-versed in the art of mind control through sophisticated use of suggestion, hypnotic language, sound and colors, and through fear mongering.

Isolation—a technique that proved to be one of the most effective at breaking down the human psyche so that it could be reshaped as desired by its controllers, has been used to now break humanity's spirit under the excuse of "protecting us" from one another. Thusly cut off from necessary human contact, suicide rates alongside mental illness have reached staggering new heights, and much of humanity has been broken to the point of accepting all sorts of draconian conditions and incredible impositions on their freedoms, as the "new normal." And, unsurprisingly, while being forcibly cut off from one another, people have turned increasingly to devices and robots for company.

And drugging, already a serious problem in our societies today—one driving all kinds of social malaise as opiate addiction in particular steadily rises, is also being used to further the goals of world leaders in creating the "psycho-civilized" society described half a century earlier by Delgado. The real push behind vaccine mandates comes not from a benevolent desire to save humanity from disease, but instead, from a more sinister wish to better control a people proving to be more resilient to being fully "psycho-civilized" than our self-proclaimed "masters" had anticipated.

## *Vaccine "Immunity" and Graphene Oxide*

The so-called vaccines, which do not even claim to offer immunity of any kind to recipients, as earlier (pre-Covid-19-pandemic) vaccines alleged, but merely promise a reduction in symptoms should one be smitten by the wily phantom "virus"—the scapegoat for the inevitable results of turning up the heat under the "frog pot" with the addition of 5G and the Internet of Things—a "net" set to drastically tighten its grip on humanity in the near future. And when this happens, another and even "deadlier" virus will surely be blamed.

All vaccines, not just for "COVID," only offer "immunity" to the pharmaceutical companies and shareholders who are *immune* to lawsuits ever since the US government, in all its "wisdom," set up the Vaccine Injury Compensation courts in 1986 after passage of the National Childhood Vaccine Injury Act of the same year. By so doing, vaccine manufactures became fully "immune" to liability as the government (who also gets a nice cut of all drug profits in the form of subscriber taxes since 1992[xlvi])

---

[xlvi] US Congress approved the Prescription Users Fee Act—PDUFA in 1992; an added tax on pharmaceutical drugs of which the FDA receives a large cut, thereby helping to fast-

began shouldering the burden for them. This decision came about after pharmaceutical companies in the early 1980s were facing bankruptcy from the overwhelming number of lawsuits they faced as a result of the massive levels of injury and death caused by their vaccine drugs. As it now stands, by forcing vaccine-injury victims to seek recourse in a special government court, not only are vaccine companies never penalized for their mistakes, they are also not particularly incentivized to make their drugs safe, and those injured by vaccines face the incredibly daunting task of trying to navigate an extremely bureaucratic system in seeking recourse—one designed to deter even the most determined complainant from pursuing justice. (One also must report the vaccine injury or death within a *year* of vaccination in order to file a complaint in the government vaccine-injury courts, even though many injuries from poisoning show up years later in the form of cancers and other slow-manifesting disease.)

The latest drugs being passed off as prophylactic medicines (vaccines) come in the form of the newest wave of pharmaceuticals, genetically modifying nanotechnology drugs—ones never before used on the public prior to 2020.[xlvii]

They also have been shown to contain high amounts of graphene oxide[xlviii]—a substance considered to demonstrate extremely conductive properties—meaning that it is a type of metal highly conductive of EMFs. This substance is incredibly toxic and it seems is serving to make those injected[xlix] with it also more electrically conductive themselves. This

---

track drug approval and line the pockets of the FDA to the tune of hundreds of millions since the act's initiation.

[xlvii] For a more in-depth analysis on this topic, see my co-authored book, *Welcome to the Masquerade—Prelude to the Coming Reset.* (Available on Amazon.com)

[xlviii] Mario informed me that the molecular number of graphene oxide just so happens to be 666.

[xlix] This includes those vaccinated with flu shots from at least 2018 onwards, as these too have been found to contain graphene oxide nanoparticles by the Spanish research team La Quinta Columna. Their very important research can be found at– www.laquintacolumna.net

could explain why so many COVID-vaccinated people have reported having become magnetized so that metals stick to either just the injection site or other parts of their bodies (as the graphene oxide moves to other areas). Thousands of videos demonstrating these magnetic[227] people have been posted all over the Internet—too many to dismiss as a mere hoax. And I personally met a recently jabbed woman in the summer of 2021 who had set off airport metal detectors after receiving the injections and subsequently had to undergo a physical search. She related how the airport personnel attending her had waved a metal detector wand repeatedly over the site of her injection in her left arm in attempts to root out the hidden metal.

Graphene oxide[228] on its own, an already highly poisonous substance capable of generating enormous levels of oxidative stress and cytokine storms when glutathione levels drop severely due to its presence, is also specifically activated by 5G frequencies, causing lung inflammation (bilateral pneumonia), stroke, heart arrhythmia, blood clots and cardiac arrest (among other things). And as a nanoparticle less than 35 nanometers in size, it can cross over the blood-brain barrier and break neural synapses to literally map our brains, collect and transmit memories, emotions, and even thoughts and feelings. The presence of graphene-oxide nanoparticles in the body can be stimulated wirelessly through 4G-5G networks and send biological data to an external AI-controlled server or "cloud."[229] As incredible as this may sound all the evidence points to the very real potential for a nefarious covert agenda to turn the human race into cyborg slaves.

## *Brain Links—Cyborg Stage One/Human 2.0*

Add to all of the above, the fact that several leading Tech companies are racing one another to develop brain implants—two-way transmitting microchips that will link us to a literal hive mind, that of "the cloud", and the agenda to turn humans into cyborg slaves starts to sound a little less farfetched. Consumers are being sold on the implants with promises of dramatically increased "intelligence," the ability to "download" skills like languages (as illustrated in the Matrix movies) and to access the "infinite wisdom" of the Internet at any time or anywhere, and not be "outdone" by AI as billionaire Musk has warned could happen. The other promise is that we will be endowed with telepathic powers so that we can text or email anyone at a distance with our minds while having (or pretending to have) a conversation with the person in front of us.

What we are NOT told is that the communication with our brain link to AI will not be one-way and of course, could always be hacked. Also, we are not told *who* will control what information will be accessible from the Internet in this manner. If we can only access Google-"fact-checker"[l] (ever *checking* to make sure no actual *facts* reach the public) approved information, then all that a brain chip will enable is further brainwashing via directly downloaded propaganda, so that instead of expanding and extending our mental capacities—it will simply further dumb-down and enslave us.

Tech and pharmaceutical giants are increasingly merging into one beast, (as for example, Google[li] is now owned under the Alphabet umbrella co., a parent corporation that also owns pharmaceutical companies, and tech companies are increasingly involved in "tele-health" and "biotech" ventures), so trusting the same companies who have been poisoning, killing and robbing us of our intelligence and humanity for decades and more, to suddenly give us back our stolen health and mental prowess, is likely not a good idea, and can have no good outcome.

## *Digital Addiction and Avatars*

The cyborg is truly stalking us and it is about to spill over from sleeping into waking reality. Digital addiction is so destructive today that it is now understood to be a problem on par with heroine or meth

---

[l] So-called "fact-checking" companies are self-accredited and funded by the likes of the Bill and Melinda Gates Foundation and other entities with direct financial interests in Big Tech and Big Pharm companies, which is exactly why fact-checking companies are actively censoring topics related to vaccine injury and harm from 4-5G networks. For more on this topic see Chapter Four of my co-authored book, *Welcome to the Masquerade*, on Amazon.com

[li] Google now also owns YouTube and, in this position, actively censors any information that may threaten its profit margin. Case in point, while preparing this book for publication in June 2022, a podcast I was interviewed for was taken down by YouTube due to alleged "medical misinformation" presented. The show as on the topic of EHS, which is a confirmed, officially recognized disability and we certainly did not present any so-called "medical misinformation" on the show. But since we pointed to the fact of EMFs, including 5G causing biological harm, the show was removed.

addiction and it so pervasive as to require its own detoxification institutions for people whose lives have been totally destroyed from their attentions having been so pulled into ever more convincing (in the imitation of reality) virtual gaming platforms and other simulated "worlds" that they stop eating, sleeping, going to work, leaving home or speaking with other humans.

In one alarming and tragic case, a South Korean couple with a new baby, repeatedly neglected their newborn by forgetting to feed and change diapers (or when feeding giving it rotten milk and spanking when it cried), in favor of a virtual "baby" on a *Sims*-type game (*Prius Online,* similar to *Second Life*) which they kept perfectly well nourished and cared for—with the neglect of their actual three-month-old baby reaching the point at which it died of starvation and the couple faced homicide charges and lengthy prison sentences.[230]

But sadly, this is not an isolated incident. Babies dying due to neglect at the hands of gaming-addicted parents is now such a problem, in military families in particular where gaming addiction abounds, as to merit its own label—death due to "electronic distraction"—by employees of the Department of Defense tasked with investigating these infant deaths within the military.

*Glow Kids* author Nicholas Kardaras, Ph.D., digital-addiction researcher and clinical therapist, received an email from the clinical director of the Air Force Family Program, a division of the Air Force Medical Operations Agency asking him to speak at an annual gathering of mental health providers. The reason for this request was revealed in a series of emails in 2016, relevant excerpts of which are presented below:

*"We have seen increased issues with gaming addictions in the [military] parents of young children and we have seen 5 cases where infants died as a result of physical abuse or neglect related to parents' constant gaming... I chair the Air Force Domestic Violence and Child Maltreatment Fatality Review Board. That is why I am aware of these gaming addiction-related fatalities. All of DoD is now, beginning this year, to track child deaths that have been identified for 'electronic distractions'... If you Google 'Stinky Airmen come from stinky houses' you might see an interview I did where we identified airmen with personal hygiene issues who are red flag[ged] for gaming addictions as they don't take care of the house, themselves, the kids or even the pets when they are gaming. They don't even stop to go to the bathroom, they drink power drinks then*

*they urinate in the bottles and they are lined up under the TV they are gaming on."*[231]

Through his years of helping the digitally addicted, Kardaras related one of the more alarming moments he experienced in his clinical work during which in 2007 a 16-year-old *World of Warcraft* addict entered his office in a highly agitated, dazed and confused state. As the teen was clearly disoriented, Kardaras asked him repeatedly if he knew where he was. The boy finally managed to stammer, *"Are... are... we still in the game?"*[232]

Staggeringly, this child did not know what world he inhabited, that of the tangible physical reality or that of the virtual gaming world. And he is not alone in his confusion. This state of psychosis is now common enough to have earned the clinical label of "Game Transfer Phenomena" (GTP), the effects of which (generally occurring after days of continuous gaming combined with sleep deprivation) are on par with similar drug-induced experiences usually attributed to hallucinogenic drugs like LSD, angel dust or mescaline—such is the power of digital gaming addiction. So potent, in fact, are its physical effects that specially tailored games are designed as drug substitutes for burn patients who otherwise, due to their high requirements for pain-relief medication, risk opiate addiction if given only narcotics for their pain. Apparently, the pain-relief effects experienced by patients gaming instead of taking morphine *rival* those provided by morphine and similar. It is no wonder then that Kardaras and others are now calling digitally addictive platforms and devices "digital heroine."

Not confined to the United States alone, gaming addiction plagues masses of hooked players, especially in Asian countries like Japan and Taiwan where social acceptance of digital addiction and gaming started earlier than in the West, where hordes of young teens sit at public gaming stations wearing diapers so that they do not have to get up from their games at "key" moments.[233]

And as previously mentioned Japan is anticipating a time in the coming twenty to thirty years in which *half* of all its citizens will "live as avatars." No longer will people *have* to go outside or leave their homes again—ever. They can "experience" the world though multiple robots living their lives for them. This society, also called "Society 5.0", is one the Japanese government is actively pursuing as an initiative under the Moonshot Research and Development program. This new society is meant to "free" people from, "physical, mental, temporal, and spatial constraints by 2050."[234]

In other words, "free" them from *living* altogether. This "freedom," can be achieved, we are told, by "combining AI, biotech, and an ultra-high speed network [read—5G], a single person could control up to ten avatars at once to perform a single task from the comfort of their own home and regardless of their own physical limitations."[235] This proposal might appeal to the couch potatoes of the world, whose increasing physical limitations are often a direct result of overdependence on technologies in the first place (or the result of poisoning by toxic pharmaceuticals, some of which cause permanent paralysis), but certainly should not be a model adopted by an entire nation (certain to pave the way for other competing nations to follow suit) as the ideal way for its citizens to "live."

An example given as to just how living like an avatar might allegedly benefit an individual is illustrated with the following excerpt from "Japanese Government proposes Cyborgs and Robotic Avatars for All by 2050" published in *Japantoday.com* May 13, 2020:

*"For example, let's say you wake up one morning and want to see what it's like to work on a fishing boat in the high seas. Well, you can just jack into a team of avatars stationed on such a boat and try to get a nice haul of skipjack tuna. And once you realize it's every bit as soul-crushing as Hemingway described, you can move on to something else.*

*"This fishing boat can also be run both by people and autonomously from time zones all over the world 24 hours a day without stopping or exploiting people. You wouldn't even need to put pants on, let alone travel to the worksite, sparing wear and tear on your wardrobe and the environment in the process."*[236]

People, not yet living as cyborgs or avatars, but coming very close to it, are already so caught up in this other virtual "reality" that not only do they neglect one another, their children, babies, pets and their own physical needs, they fully fail to take note of the dwindling natural world around them. So that when thousands of trees[237] are cut down in cities all over the world, to make way for thousands more 5G antennae, because the trees, we are told, "block" the signals of millimeter waves, which is why (we are also told) they need to go—a bit ironic for establishing the "smart grid" that is supposed to help us "protect" the environment.

Geraint Davies, British MP for Swansea West (who clearly agrees with my sentiments on the matter) spoke before parliament, accusing government officials of environmental negligence by stating (July 2019), "Are you aware of the concerns that 5G can't penetrate trees? As a result, we are looking at the destruction of thousands of trees. This destruction has already started around Swansea. How can we possibly be serious about our ambitions of zero carbon if we are destroying the trees and also have this huge carbon footprint [from 5G infrastructure]? It doesn't add up, and it's clearly environmentally ridiculous."[238]

However, it turns out the trees may not even block the signal, but instead they may serve as too much of a glaringly obvious statement as to how harmful these frequencies really are, as many videos[239] have shown the effects on trees that have remained standing near to new 4G/5G antenna installations (the 4G is still coupled with 5G, but now brought down to street level to be even more directly destructive), where the side closest to the antenna has turned brown and died, while the farther side is still green and alive (for now).

Following our last exodus shortly after arrival to the west coast, I bought a surfboard and thick 5mm wetsuit in order to pursue a long-time personal dream of learning to surf. While initially extremely challenging, it has been a labor of love and immensely rewarding, while being both pleasurable and healing. For me there is nothing that compares with being in the ocean waves—it is a meditative, indescribable joy.

Salt water, and water in general, is said to absorb EMFs and the mineral in the salt (like the hot tubs so prized by Ruth and Gary) and negative ions they create serve to mitigate biological damage caused by EMF exposures. Hence, a lot of canaries find being near water, especially large bodies of it and far from cell towers, quite healing.

EMF activist and educator Sofia Smallstorm, related a story to Mike Winner on an Alfacast podcast in spring 2021, about a friend of hers. This friend is a surfer and he takes his older sons out surfing regularly. But the youngest son, 15 years younger than the others, prefers to play a video game of surfing over going out in actual ocean waves. When the father tried to entice him into joining them, this youngest gaming-son protested that he did not want to, "get all wet and sandy," but wanted to stay home and play his surfing game instead. According to Sofia, "the father said, 'Son you are not surfing, you are moving a joystick.' But the son didn't care. He wanted to stay home... He did not want to get wet. He didn't want to get sand between his toes and have a wave womp him. Surfing the Game, was painless and if you get womped in the game, nothing happens to you—you laugh. And that kind of experience, that is the experience of the future."[240]

It is because the younger generations raised on technology (who do not even know a time before the Internet or even before it became a portable, palpable, integral part of their existence) lack appreciation and direct experience of the natural world, and because they have been trained to expect access to not only cellphones, but also the Internet, everywhere they go, that our natural world, now including even the vast oceans, are under threat of total extinction.

Life as we know it is poised to change beyond recognition, as all of it, with no exceptions, will be bathed in non-native, unnatural frequencies—so that truly the world will only be suitable for non-biological, robotic "life forms," and remaining humans, so physically degraded that walking, perhaps even getting up from a chair, and most certainly going out of doors and pursuing a sport of any kind, will prove an impossible feat. These future generations, if they are able to survive at all, will only do so via help from machines and will only experience the outside world through robots—trusting too that what the robots show them and what sensory feelings are relayed to them are accurate and true representations of the great wide world "out there."

## *Wayfaring and Losing Our Way*

Already our traditional wayfaring abilities, thanks to the world's current dependency and overreliance upon GPS navigational systems, have all but disappeared, and with this disappearance, so too has our sense of place in the physical world disintegrated. Even the most remote-living indigenous people, like the Inuit tribes, for whom finding their way across the barren ice-covered land without even maps for guidance, has been embodied as an ancient hunting skill—a pride of their traditional culture—one that is now being lost and set aside in favor of reliance upon GPS systems mounted to snow mobiles that have replaced dog sleds. This drastic and recent change is not only robbing the people of their ancient cultural heritage, it has led to an increase in fatal accidents as (especially the young) hunters lose their own sense of place and ability to navigate when (invariably) electronic devices malfunction.

Ship captains and crews, no longer versed in celestial navigation, also lose their way when GPS tools fail and they ignore their own senses that alert them to danger, and as a result, veer far off course as they grant machines superior knowledge and ability, over their own. But losing this sense of our surroundings, not only can lead to steering us off course in our travels, it can also lead to a sense of physical displacement, as our brains are no longer required to create "grids cells," sets of neurons used to create maps in our minds. Because...

***"...who we are is tangled up in where we are. We can't extract the self from its surroundings, at least not without leaving something important behind."***[241]

An inherent problem in handing over wayfaring autonomy to GPS systems is that they are, as Nicholas Carr explains in *The Glass Cage*, "not designed to deepen our involvement with our surroundings. They're designed to relieve us of the need for such involvement. By taking control of the mechanics of navigation and reducing our own role to following routine commands... [GPS systems] end up isolating us from the environment."[242]

By never having to "confront the possibility of getting lost," is also, according to Carr, "to live in a state of perpetual dislocation. If you never have to worry about not knowing where you are, then you never have to know where you are. It is also to live in a state of dependency, a ward of your phone and its apps."[243]

Carr further expounds on the problem of relying on GPS and losing our wayfaring capacities with...

*"A GPS device, by allowing us to get from point A to point B with the least possible effort and nuisance, can make our lives easier, perhaps imbuing us, as David Brooks suggests, with a numb sort of joy and satisfaction of apprehending the world around us—and of making that world a part of us. Tim Ingold, an anthropologist at the University of Aberdeen in Scotland, draws a distinction between two very different modes of travel: wayfaring and transport. Wayfaring, he explains, is 'our*

*most fundamental way of being in the world.' Immersed in the landscape, attuned to its textures and features, the wayfarer enjoys 'an experience of movement in which action and perception are intimately coupled.' Wayfaring becomes 'an ongoing process of growth and development, or self-renewal.' Transport, on the other hand, is 'essentially destination oriented.' It's not so much a process of discovery 'along a way of life,' as a mere 'carrying* across, *from location to location, of people and goods in such a way as to leave their basic natures unaffected.' In transport, the traveler doesn't actually move in any meaningful way. 'Rather, he is moved, becoming a passenger in his own body.'"*[244]

*"While we may no longer have much of a cultural stake in the conservation of our navigational prowess, we still have a personal stake in it. We are, after all, creatures of the earth. We're not abstract dots proceeding along thin blue lines on computer screens. We're real beings in real bodies in real places. Getting to know a place takes effort, but it ends in fulfillment and in knowledge. It provides a sense of personal accomplishment and autonomy, and it also provides a sense of belonging, a feeling of being at home in a place rather than passing through it. Whether practiced by a caribou hunter on an ice floe or a bargain hunter on an urban street, wayfinding opens a path from alienation to attachment."*[245]

## *Waves Upon the Ocean—the Internet of Underwater Things*

One of the more horrifying proposals, already being tested, is to turn our world's oceans into the "Internet of Underwater Things" (IoUT) by way of introducing millions upon millions of sensors to the ocean floors and swarms of underwater robots within the waters as well as denizens of autonomous vehicles upon the waters' surfaces. It is indeed the stuff of nightmares and, if fully implemented, will, through projects like SEANet, funded by the US National Science Foundation, "enable broadband wireless communication from any point on or in the oceans to anywhere else on the planet or in space." [246]

The underwater sensors/antennas or "nodes" (as they are being called) will be placed at different depths and employ not just radio waves, but acoustic (AKA sonar) waves, LED lights, magnetic induction and lasers.

The sonar/acoustic waves being tested can reach up to 202 decibels (139 decibel equivalent in the air) well above human pain thresholds and in frequency ranges directly affecting dolphins' hearing,[247] and so will also directly impact both dolphins and whales (as navy sonar already does) who rely on their own sonar waves for navigation, communication and hunting. The subsequent noise pollution inherent to the Internet of Underwater Things proposal will doubtlessly destroy already sick and dying oceans and all life forms supported by them—which means us, too.

Do we want the next generations, our children and grandchildren to live in a world in which the fish, dolphins, turtles, seals, and whales of the seas are replaced by teaming robots and autonomous vehicles? Or perhaps they will not even notice, as they will be too busy playing surfing, snorkeling, SCUBA diving, or fishing virtual-reality video games? Or they themselves will be turning into robots, so will not be as offended at the idea of robot "companion" fishes and sea lions.

The oceans of the world, as unhealthy as they are currently, have still served as safe havens and places of healing for electro-sensitives and so many others. It is one place that seemed too vast and wild to tame with technology and so, at least I, held out hope that it would still be a place of refuge for me personally. So, when I heard about these latest plans to turn beaches and oceans into more SMART places—ruined by artificial *waves* as they are set to overtake the natural beautiful rolling waves of salt water, my heart sank and a bit of my spirit drifted away, as if ready to abandon ship—if this the ocean upon which my soul-ship sails is really to be altered so profoundly.

If this proposal to destroy our oceans in addition to our ionosphere goes through, along with the very ignorant and short-sighted plan to make our oceans and beaches "smarter" in the name of actually "saving" them, then no longer will I or others (particularly the more sensitive canaries among us) benefit from the few places we have found for healing and refuge. No longer will I enjoy my newfound passion of surfing. "Surfing" will only happen "online" while "surfing" webpages and playing video games mimicking actual surfing.

And the waves of the oceans will be polluted with other invisible waves—ones that will most definitely have a horrifying impact on all of the life supported within the oceans, and we will witness destruction on a massive historically unprecedented scale—the fallout from which will not be blamed by placing it where it is due—on the polluting industries responsible for the drastic alterations, but will be blamed instead on ever-evasive "climate change" and will be blamed on you and I for daring to breathe or move about on this planet, and for daring to travel and "consume resources" or to ride upon the ocean waves on our surf boards or in our boats.

Our world is being intentionally reshaped. The new environment created could turn out to prove only supportive of "non-biological" life, so that the only people who will be able and allowed to participate and live in this new-world-order will by necessity be drastically altered as well—so that only cyborgs will find a place in it.

The rest of us will be further pushed out to live on the edges until the edges cease to exist all together.

As Kathryn Stuaffer, from the *WiFi Refugees* documentary stated...

*"It's just crazy, why should we have to live like this? It's supposed to be 'first do no harm' right? If something hurts one person, we shouldn't be doing it. We are going down the wrong path. We need to rethink this."*

The two paths spoken of by Robert Frost in his poem are before us.

*Which one will you take?*

# Afterword

# By Sam

As I sit in our current living situation, off of the electric grid in a beautiful California forest, I am so grateful to whatever forces in the universe have allowed this to happen. Nevertheless, as I stew in the uncomfortable realm of memories, as I review the years spent as Shannon's partner helping her through her ordeal as an EHS canary in a coalmine, I have to admit that our current situation was also the result of an enormous amount of endurance, patience, and will power from both of us.

During the troubled years with Shannon, I spent a lot of time shooting video for a documentary that I conceived to be the one that would put me on the map and set my film career in full motion. I thought how brilliant that this unique situation should arise (Shannon's electro-sensitivity) just as I finished a graduate degree in documentary filmmaking! (That's called looking on the bright side.) Although I felt bad for Shannon and wanted to help her in any way that I could, I was also set on giving another shot to a career in film. I wasn't able to fully grasp the severity of the circumstances for months to come and thus I thought that I could juggle everything. I think there was a moment when I actually thought that all I had to do was live my journey with Shannon, do my best to support her and record it all! That was going to be the film in a nutshell. Do I even need to say that it wasn't as easy as it sounds? There is a reason most filmmakers need crews to help them. We both also, perhaps naively, thought that Shannon would make her recovery relatively quickly and that I could also record it and the film would serve to document the whole process, with the end of the film equating to our happily-ever-after ending in real life... but real life is not like the one lived out on screens, and nothing went exactly according to plan.

I remember Shannon's first breakdown at her apartment in Takoma Park [Washington DC neighborhood]. It was the first emotional collapse out of many that I would witness—a cacophony of tears shaped by fears and frustrations as she melted in my arms. Her despair about not knowing how to proceed to exist in the world with her condition was about as crushing to me as it was for her. I thought that part of being a good partner meant that I had to hide my feelings of despair the best I could. The only thing that I knew for sure then was that I had to be my partner's strength for as long as that required. I had no idea how to deal with a lot of what was confronting me and I think the camera was my crutch. I was comfortable behind it.

Looking back, the whole filmmaking aspect of our relationship—making a film about something that you can't see happening—is very tricky. I would have needed to capture the contrast between Shannon feeling fine and then something EMF-related affecting her. Even then I don't think it would have had the impact that was needed. Instead, I ended up having a bunch of moments that were recorded after the fact because I could never get the camera out fast enough. I had to be available and react to Shannon personally while also being the filmmaker. The near impossibility of this didn't hit me for at least another two years when I eventually decided to drop the film all together and start being more present in my life and in my partner's life.

So, there we were in her apartment and she cried in my arms and told me how she could not remember what it felt like to have energy or enthusiasm. I could not believe how low she was feeling. The critical point in our story arc was when she said, "I'm not sure how much longer I can live here." I knew what she meant. She meant the metropolitan area. She meant the suburbs. She meant anywhere within 20 miles of a city and preferably 50 miles from any cell tower. That last part turned out to be a fantasy, there is no place left on the planet 50 miles away from a cell tower. Then I got the tripod out, put the camera on as fast as I could and sat down on the couch with Shannon. (All you have to do is act natural.)

There were so many challenges and problems to solve the year we first moved that I don't know where to start and I obviously can't describe everything. My goal in writing this is to hopefully provide a unique perspective on this topic as a partner to someone with EHS. There are moments that stick out without having to try very hard at remembering.

The weeklong drive from Maryland to Arizona was a breeze for me, but it was very taxing for Shannon. We had to plan our accommodations very carefully based on locations furthest from antennae, hosts willing to turn off WiFi routers, and avoiding chemical fragrances, noises—basically anything that could prevent Shannon from getting a decent night's sleep.

Sleep was the key, as it still is. As we journeyed, we stopped about every two hours so that Shannon could find some place to ground herself and rest. This usually meant lying down on the little grassy spot at truck stops where all the dogs do their business, but it was worth it. There I was standing over her aiming my video camera at her, capturing it all like a pro (or an aimless idiot, I am not sure which).

After driving across Kansas for eight hours in separate vehicles we had to make the last push to my relatives' house near Denver where we planned to rest for a couple of days. The weather had taken a nasty turn and a snowstorm hit us just after entering Colorado. At first it was just a few flurries for about half an hour. So, we stopped and gathered ourselves to make the last push. I was leading the way and trying to keep a slow enough pace for us to stay together. (Out of necessity we had had to make the cross-country drive in two separate vehicles, which further compounded the stress on Shannon who at the time was heavily impacted by the EMFs from the car's engine and so having to drive on her own did not have the luxury of resting as a passenger.)

It was dark by then and we merged onto the bustling highway for the home stretch but it seemed like every car on the road besides our own vehicles was driving with no regard to the change in weather. It was difficult to see even with the wipers on the highest setting, so of course I was driving slowly. I was nervous and trying to hold it together and reading directions while keeping an eye on Shannon's car behind me. There was no way for me to tell how she was doing. After what seemed like hours of this while managing to stay together through all of the lane changes, exits, and turns, and as if guided on the wings of an angel, we arrived at my aunt and uncle's house about two hours behind schedule.

I was so relieved and I just couldn't believe the day had ended with that kind of intensity. I got out and walked over to see Shannon. She opened the car door and I was faced with a storm of tears and unintelligible sentences. When I had first met Shannon, she had a strong enduring quality so it was very disturbing to see her completely fried—pushed totally beyond her limits. It was easy to tell that her body was working from its final reserves. I remember thinking then that I've never seen her that upset and fatigued before and I think that is still true. I felt like we had escaped something terrible—slipped through the fingers of the mighty hand of devastation.

From there the situation got stretched and distorted a little and from my perspective resembled something from a dark comedy. My relatives welcomed us. They were obviously sleepy-eyed and ready to hit the sack—not at all ready for what was entering their house. Shannon was still in shambles. I remember her sitting on the couch describing how dreadful the conditions were and how strung-out she was when she abruptly directed her attention to me and said, "Sam, maybe you should be filming this." Soon enough there I was standing with a camera in my hand feeling like an idiot trying to capture a moment that had already passed but sticking through it just to do it. My aunt and uncle were standing there behind one of the couches in their pajamas as Shannon described the events of the previous two hours.

When there were no more tears the energy in the room quickly depleted, the camera shut off, and we crashed safely in bed. We spent the whole next day resting and visiting, which felt about two-hundred-percent better than the night of our arrival. The downs were usually followed by ups and we left there rejuvenated and in high spirits.

After settling in at our first home in Arizona, I will never forget having to walk over to our "delightful" alcoholic, meth-head neighbor's property to explain our concerns about him burning trash. (His fire pit was about twenty yards from our house.) With a cigarette hanging out of the corner of his mouth he was tinkering with something under the hood of a truck as he watched me enter his gate and walk down his long driveway. There was no smile on his face. I introduced myself and shook his greasy hand. I could tell that he wasn't interested in having a conversation with me and I knew this was going to be an awful experience. I told him where we were from and why we moved out there, trying to lead into the trash topic as easily as possible. A sort of frightened looking woman, possibly his wife, poked her head out of the trailer door and introduced herself softly. She started to ask me something but he gave her a look that shut her up quickly and she scurried back inside.

He then said to me, "So what do you want?" I tried to make it clear that Shannon had a potentially life-threatening allergy to chemicals from trash fires and he said, "Look around man, you don't know what its like! How am I going to move all this stuff? I don't have a trailer." His yard resembled a mini landfill and right near the fence that separated our yards was a ten-foot-wide burn pit with a couch next to it. I asked him if there was a way to work it out so that he could burn on days when we were not around (the logistics of which would admittedly be difficult). He replied in an agitated tone, "Yeah, sure, okay, maybe that can work."

The rest of that interaction is a fog but I knew he wanted me to leave and I really wanted to leave so I think I said thank you and left as quickly and casually as possible. Of course, as Shannon related already, this neighbor never did accommodate our request and instead hurled insults at us from the other side of our fence whenever he saw us outside our house—prodding us into yet another relocation.

Shannon was always a very social person, so moving away from our home city to the desert wilderness presented a huge challenge for both of us from a social aspect. One reason we chose an area outside of Tucson for our relocation was because Shannon had a couple of old friends from college living in that city, whom she hoped to be able to visit with regularly. After the move, we naively thought that Shannon would recover enough in a short period of time to be able to make regular trips into Tucson for social outings. But soon enough we found that she could not tolerate very much time in town or the long car rides.

She invited her friends to visit her at our new home but having busy lives of their own they were not able to easily make the 180-minute/120-mile round-trip drive for social visits, and ironically for Shannon, they admitted to being nervous about driving through areas outside of cellphone range. In the five years we spent in Arizona, only one of Shannon's two friends was able to find time to visit on one occasion. And only one of Shannon's family members and one of her friends from back east ventured out during those years. In my case I either managed to travel back to D.C. or family and/or friends came out to see me about once a year.

After escaping the "neighbor from hell" and finding a home in Cascabel, the prospect of socializing again after a very isolated first six months, was rather enticing. We found Cascabelians to be very welcoming neighbors who held regular social events at the local community center and at neighbors' homes. But even though there was little-to-no radiofrequency interference throughout that valley, it turned out that the community center itself offered "free WiFi" to community members, meaning that the WiFi was always turned on. After explaining to our new neighbors our reasons for our move, we asked if the WiFi could be turned off for Shannon's visits to the community center. At first, we were given the green light to do so, as long as we made sure nobody present at the center during our time there was in need of the WiFi. But after a couple of occasions when we forgot to turn the WiFi back on upon leaving the center or others who promised to do so for us forgot, causing neighbors visiting the center in order to use the WiFi to think that it was broken, prompting calls to the local Internet provider to come out to service the modem, this privilege was quickly taken away and Shannon was crushed as it meant renewed social banishment. Visiting neighbors' homes was equally problematic as their own WiFi was generally always switched to on and sometimes neighbors or their guests wore clothing washed in conventional, scented detergents or fabric softeners or wore perfumes, none of which Shannon was able to tolerate.

So, her social world quickly shrank down to me and our two cats, for the most part. It wasn't uncommon for her to not leave the boundary of our 150-acre property for weeks at a time, and on occasion, months.

Lots of simple things became challenges after Shannon became hypersensitized to EMFs. Her hearing became very sensitive so I had to be very mindful of what noises I was making—a mindfulness which took me a long time to develop. Any metallic scraping noise seemed to be far more than just an annoyance. If I were cooking with a steel pan and a metal spatula and accidentally scraped it a certain way I would hear an "Ahh, scraping!" sometimes coming from another room, as the high-pitched frequencies created by metal scraping against metal translated into physical torture for Shannon.

Living with the power turned off to most of the house required using flashlights at night and Shannon had a physical aversion to the blue light created by LEDs, so we made filters out of warm colored gels to place over the lights and also invested in tungsten flashlights. Going to the grocery store was mainly my responsibility for a few years as Shannon could not tolerate any time spent in towns or cities due to the high concentration of EMFs and chemicals. It was an all-day affair, factoring in the ninety-minute commute into Tucson, driving around town to do all the errands, ninety-minute trip back, and the half-mile walk to and from the car to the house, while carrying heavy bags through horrendous heat in the summer and bitter cold in winter, and often through ankle, thigh or waist-deep muddy waters of our seasonal river.

In 2018 Shannon experienced dental problems that necessitated surgery. But with her intolerance to EMFs and chemicals it took a lot of doing to find a holistic dentist who could accommodate her, and in the end, she simply had to endure the torture of a dental office in Phoenix that was heavily laden with strong-smelling chemicals. But first we tried a dentist in Mexico who performed the first extraction. Due to the six-hour drive each way, and need for multiple visits and not being able to stay in hotels, we camped on a beach for several nights in a row while

Shannon was made to recover from surgery in those rough conditions with no amenities.

The visits to Phoenix also often required overnight stays camping outside of the city. Her second extraction coupled with bone-graft surgery there, was an ordeal I will not easily forget. Originally, I was going to film the surgery as part of the documentary, but since I also needed some cavities filled and our time and financial resources were limited, we scheduled our dental visits for the same time in the same office. And since Shannon was scheduled for multiple subsequent surgeries, we put off the filming on that day.

While I was on my dental chair down the hall I heard a blood-curdling scream that caused me to jump from my chair and run down to her room to see if she was okay. Her dentist explained that she had reacted to an injection and Shannon, who could not speak with her mouth propped open as it was, nodded her head to assure me she was okay while tears still streamed down her eyes from the horrific pain she had just experienced due to the injection possibly hitting a nerve. I was to find out later that this injection followed many earlier ones for the numbing shots that were not working as intended. It is a common occurrence for chemically sensitive people to react unexpectedly to drugs, and as a general rule because of Shannon's bad experiences with pharmaceuticals she normally avoided them. And if an environment is already saturated with chemicals this can interfere with the effect of a drug given in such an environment to someone chemically sensitive, as Shannon's research confirmed after the fact.

After this terrible ordeal we had to camp outside of the city in order to make the follow-up appointment in the morning and as Shannon could not handle too many hours of driving in one day. She was in terrible pain all that night and hardly slept, her mouth was a swollen mess and she could not take regular pain meds. Ultimately the bone graft didn't take and she had to endure another similar surgery for its removal, this time with another dentist whose office was more tolerable, a fact which seemed to aid in the numbing drugs working this time around, but a little too well, as Shannon reported seeing double for the first several minutes after the injection—something the new dentist said was definitely unexpected.

Those dental excursions were a nightmare for us both and ended with the removal of Shannon's two front teeth with no suitable replacement teeth as she reacted badly to all of the materials used in dentures and could not do another bone graft for implants. In the end, she made her own denture out of moldable plastic, and has vowed to never step foot inside another dentist's office again. If living as she has had to, has done anything, it has been to make her and myself a lot more resourceful, which sometimes has its benefits.

It was not always bad, though. Life has ups and downs and it was no different in our case. We always would find some way of enjoying ourselves and each other's company. I think it is odd and kind of funny how, after Shannon started to get much healthier, how much I really enjoyed simply going shopping with her. Being out again with her, away from home, was a very uplifting experience for me.

As a supportive partner to someone living with EHS and/or MCS, one may end up feeling like one's own life is sacrificed too much in the process, which is probably why so many relationships put under this kind of stress don't last and why so many canaries end up feeling abandoned. In my case I only learned after years of sacrificing pieces of my life and being devoted to my relationship that a lot the old habits and aspirations that dropped away as a result were aspects of my life that needed to change anyway.

I have to admit that the turmoil of all the trying experiences with Shannon were nothing compared with the stress of managing my other personal relationships in the face of this new knowledge. When we started to really examine the facts about the biological implications of harmful EMFs my understanding of the severity of this on a worldwide scale became an insurmountable hurdle for me. I didn't know how to integrate this information. At some point I knew that I had to prepare myself for how to talk about this with friends and members of my family. Not just so they would understand Shannon's condition better, but so that they would consider the effects of EMFs on their own lives.

These were not comfortable conversations for me. In fact, they always felt instead like confrontations and never went the way I had hoped. My tendency is to want to convince people to see things my way and when they don't I get upset. It was never about wanting to just be accepted for my and Shannon's way of life. It was about truly wanting to help the people that I love to be healthier. When the best that I could achieve was my parents agreeing to keep the WiFi off while I visited them I had decided that my words were a waste of energy and the best thing I can do is live by example. That was not such a terrible adjustment to make, but I've found that it needs to be met with a high degree of confidence and humility, which is a tricky balance to strike.

By following the road that was most demanding of me I ended up in some very unique and special places meeting very interesting people who have all added something valuable to my inner life, while also pushing me into new levels of creativity necessitated by our rugged, remote living situations.

And I also found that I too was "electro-sensitive" to a certain point. After years of being forced to live in cleaner electrical environments I concluded that many of my own reoccurring health problems had been a result of high EMF environments, especially from close proximity to WiFi routers. In the new "EMF sanitized" environments, sleep problems improved. I got sick about 95% less than I used to. I also had a heart-beat irregularity for years and when I decided to have this examined by a doctor, years before I met Shannon, he professed that the condition was "normal" for me. But that condition magically disappeared while living in Arizona. I think the biggest change I have noticed in myself is an increased cognitive function. Gradually my brain fog cleared and I remembered what it was like to feel normal again.

What first seemed to be a curse cast upon each of us, turned into a gift, leading to a deepening of our connection with each other and our natural world and a healthier saner lifestyle—one currently sadly out of reach to most people in today's societies.

# APPENDIX A

# Solutions

Because the medical mafia (aka, American Medical Association) rules just about every aspect of our lives in these "uncertain times," I must officially state that I am not (thankfully) a medical doctor and as such my personal views and experience do not count as "medical advice." But as a non-medical-doctor "layperson" I have a lot more freedom to criticize the medical profession as a whole without running the risk of losing a medical license.

Seriously, I do think it should give us a great deal of pause when it becomes a federal offense to speak out against allopathic medical practices—this should not be the case in the "free" world and certainly medical professionals should not be totally controlled, as they are currently, by the financial interests of pharmaceutical companies. It is these companies who legally define "standard of care," and if medical doctors dare defy these "standards," they risk not only censorship but also total loss of their livelihoods. Medical doctors today are prohibited from advising patients to quit prescribed drugs (even if doing so could save their lives) and from recommending patients take medicinal herbs and vitamin or mineral supplements in lieu of these often-lethal drugs.

But as I stated, since I am *not* a medical doctor, my hands are not thusly tied. However, I am legally bound (by this same medical mafia) to point out that any health advice I give is simply for "educational", or better still, "entertainment" purposes.

I will admit here that I did try the medical-doctor route to help with my "EHS," but ultimately found that the treatments harmed more than helped while also costing me a small fortune. Prior to EHS, I had suffered various ailments in my life, as does just about everybody, many of which in my case became chronic after multi-dosing on vaccines for university and international travel in the early and mid-1990s, and I had oftentimes sought conventional medical help for my pains, but again, found the treatments, particularly in the form of toxic pharmaceutical drugs, to harm more than help and to further complicate my health while also adding to my overall toxic-body burden.

Being thusly repeatedly disappointed by the allopathic-medical approach to health, I have spent decades studying and using so-called "alternative" medicines (but really ancient and traditional) in the form of botanical herbs, dietary changes, detoxification practices and mineral and vitamin supplements. And I now refuse to see allopathic medical doctors at all, for any reason. Granted, they have their place, primarily in treating traumatic injuries, so that if I had a car accident or broke several bones, a hospital might be a good place to get "patched-up," however even then, I personally feel it is, more often than not, safer to avoid hospitals as they are actually *the* leading cause of death in the United States—simply from the doctors and medical staff following Standard of Care (maybe what should be called Standard of *Scare*) medicine. This "death by doctor" or "iatrogenic effect" kills roughly 800,000 people in the US annually,[248] yet the bulk of those deaths occurring during the alleged current pandemic are now being conveniently labeled "death by Covid-19." Actually, they do not even claim that a gunshot wound or fourth-stage cancer death is *caused* by Covid-19, they simply say that the patient died *with* Covid-19 if they tested positive, regardless of actual cause of death, so that the deaths can still be added to "Covid-19" death statistics. Additionally, deaths from *any* cause, even for patients testing negative or not tested at all, are also regularly being listed as a "COVID-death."[249]

When one appreciates that it financially benefits each state in the union to declare as many "Covid-related" deaths as possible due to increased allocated federal funding to the states claiming the most pandemic victims (and awarding individual hospitals up to $13,000 for each death listed as COVID-related even if the "primary cause" of death is clearly from something else),[250] and when one understands that it specifically

benefits hospitals when patients leave in body bags rather than on their own two feet (or in a wheelchair), since the hospital makes far more money, generally speaking, over our deaths (paid out by insurance companies, estates or relatives of the deceased) than our treatments,[251] then it becomes easy to understand that our actual health and wellbeing is not in the best interests of these institutions who profit from our collective demise, chronic ill health and eventual deaths. In short, we are worth something to hospitals and doctors while sick but worth more dead than alive—and worth nothing at all if healthy.

Incidentally after a number of years "unofficially" carrying around the EHS and MCS medical labels, I did find a bona fide medical doctor (MD) to write me an official diagnosis stating that I suffer from both. I did not seek out this label for my own satisfaction or to prove anything to friends and family[lii], but I had hoped it would help me to stop my utility company from placing noxious microwave-emitting SMART meters on my home and enable us to stay in it longer while still living in Arizona. As I mentioned in chapter four, I took my "certificate" to HUD, the fair housing department, and pleaded my case that by forcing this kind of equipment on my home it constituted an offense in full defiance of the American Disabilities Act and directly discriminated against me, someone with a legitimate disability. But ultimately the HUD representative who had started out very hopeful of winning my case soon came back to me saying that regretfully there was nothing at all they could do to help. *Nothing.*

In 2020 I quickly obtained a medical exemption from vaccination, noting the political climate and fervor moving towards force vaccinating the entire world. At the time of this writing, at the very beginning of 2022, we are seeing historically unprecedented moves for governments around the world to prohibit participation of its citizens in all aspects of society without so-called "full vaccination" with drugs that are still mostly unapproved by the FDA (not that this label really means anything when considering the FDA's approval track record of allowing sales of a long list of lethal drugs), and that use experimental gene-editing nanotech—the kind of drugs which before now (under the Emergency Use Authorization) were not allowed to be used for *any* purpose due to concerns over extreme toxicity and dangers to recipients—including deaths[252] (many of which we have sadly witnessed in the months following widespread "vaccinations"[253]). I personally know of both deaths and disablement caused directly by these new drugs. My former neighbor and friend Amy,

---

[lii] However, I grimly noted how much more support I received in general from that camp when I was "receiving care" from medical doctors and dentists—particularly when I needed to request donations in order to continue treatments since my state medical insurance was not willing to cover these expenses. This demonstrated to me how much more comfortable people are about health matters being handed over to the "authorities" than turning to holistic/alternative medicines and practitioners and/or engaging in self-care. There is a general attitude held by the public that by not seeking out "professionals" one is being irresponsible and negligent with one's health, however my experience dictates the opposite to be true.

whose story I related in this book, announced to me the tragic death of her close friend's 22-year-old son, who died of a brain aneurysm shortly after being made to take the "vaccine" for re-entry into his university this past autumn.

And Mario's own 89-year-old mother died after being pressured by hospital staff in Canada to submit to the Moderna Covid-19 vaccine. She had been living in a cooperative building for the elderly, one featuring banks of SMART meters and several antennas installed on its rooftop. Having been health conscious her entire life, having always eaten good wholesome foods and abstained from pharmaceuticals, she had begun to decline after living in that building, especially after the lockdowns started and she was forced to live in isolation. "She was alone. She couldn't have visitors and she wanted to die," is what Mario recently related to me about her situation. But finally, Mario's sister went to see her in defiance of the imposed restrictions. When she did, she found their mother in such a terrible state she had her taken immediately to hospital, which is where the vaccine was "really pushed on her until she relented." Soon after receiving the shot his mother developed blood clots and was treated for those. For a time, she was better and able to go home but soon found herself in hospital again after her legs swelled up "bigger and bigger." Soon after this she died. While Mario's family does not want to make the connection and instead views the cascade of serious health crises in the wake of the injection as coincidence, Mario knows that it is not.

Even though mandates are increasingly being imposed for vaccine requirements needed for shopping, traveling, working and going to school, major insurance companies have decided to not cover deaths caused by the new vaccines since the drugs are still considered experimental, as such the resultant deaths are being listed as suicide since the recipient "volunteered" (even though in most cases this "volunteer" status falls under intensive coercion) to submit to the experimental drug.[254]

If this trend to force-drug the world's populations continues I will have to keep renewing my medical exemption and hope that it is recognized as many states have been trying (and often succeeding) to disallow medical as well as philosophical and religious exemptions for vaccination.[liii] It is bad enough that any of us are made to work within this tyrannical system and place ourselves before medical doctors and medical boards for scrutiny year after year and prove that we are sick enough to be placed in the "high-risk" category for not tolerating their toxic alleged "medicines"—it is humiliating, degrading and a violation of basic human rights to body sovereignty, but to not be given the option for exemption at all, of course, is much worse.

We do not live in a free society if we can be force medicated, for any reason. And we do not live in a democracy or republic when a small group of individuals can dictate to the whole what is good for the group and what is not, and who can also decide "risk-benefit"[liv] costs to drugging all members of society and decide acceptable amounts of "collateral damage." Whoever feels comfortable playing god and making these assessments and decrees should offer themselves or their children up for sacrifice in the collateral-damage group.

And it has already been hard enough for the canaries of the world, already extremely marginalized due to constantly being pushed out of participating in a society that increasingly requires use of cellphones for engaging in commerce and the job market and requiring "upgrades" to wireless-enabled computers and household appliances that no longer come with a WiFi *off* button, but now cellphones are actually being forced on greater numbers of people for use as "contact-tracing" devices, which must be kept always-on and ever-near.

And now, I have been recently hearing about the trend for insurance companies to only cover motorhomes/RVs if they are equipped with always-transmitting trackers, at least in the UK so far, as related to me recently by author of this book's foreword, Claire Edwards. If this trend is

---

[liii] See my chapters on vaccination (nine and ten) in my co-authored book, *Welcome to the Masquerade; Prelude to the Coming Reset,* for more info on the history of vaccination, vaccination politics, ingredients, and potential for biological harm.

[liv] See the excellent book, *The Lethal Dose,* by Dr. Jennifer Daniels to get an insider's view of how these assessments are made and how doctors routinely prescribe lethal doses of pharmaceuticals to patients.

allowed to continue, in all probability to it will extend to getting insurance for *any* vehicle, and then perhaps any home as well.

On this current trajectory the future does not look very promising, not only for the canaries, but so too for anyone. What do we have left when we come to find we have lost all rights to body sovereignty and (especially *unmonitored*) freedom of movement?

But we must not lose hope and we must still seek out solutions, which is what this section[lv] is all about! (Please note: I have no commercial affiliations with any of the products recommended in this section.)

## *1- Don't Panic!*

First and foremost, do not panic! I know this is all a lot to digest. Truly understanding what is causing us harm can be an overwhelming and frightening experience, even if, ultimately the knowledge will benefit us. Take things one step and one day at a time.

Stress itself is harmful as it, just like the EMF exposures we are trying to escape, puts us into a "flight-or-fight" response, triggering our adrenal glands to produce cortisol in an attempt to protect our bodies from harm. The problem with this scenario is that the chronic release of adrenalin will overtax our adrenals and lead to adrenal fatigue and eventual collapse. Most of us today, as a result of being under constant stress from EMF exposures and also societal stressors, are experiencing adrenal fatigue, and thus, similar to a car running out of gas, while living our lives on fumes, we are ever nearing collapse. But since we are used to this chronically fatigued and stressed state it often surprises us when we do finally collapse. And that collapse can take many forms. For me it came as a literal collapse in bed where I spent several days unable to get up after my last overseas trip. But for others it can come as a

---

[lv] A small portion of the listed solutions included here I have extracted directly whole or in part from an article I wrote for ElectricSense.com dated April 11, 2020 titled, "The Microwave Injured Masses - Canary Warnings about 5G," which can be found at: https://www.electricsense.com/5g-covid-19/

frightening diagnosis, like cancer or a brain tumor as Liz and Olga experienced respectively.

So, the first thing we need to learn is how to reduce stress in our lives, by avoiding and reducing EMF exposures, and also by learning stress-reduction techniques such as meditation practices (this can take the form of moving meditation like Qi Gong or Tai Chi as well—as I am someone not easily able to sit still for regular meditation I gravitate towards Tai Chi and other flow arts like hoop dancing), and taking time out of one's day to be still or listen to relaxing music. Taking a hot bath, or a walk in nature can also make a big difference. But these practices need to be performed regularly in order for us to reap the most benefits.

And for those who are grappling with a new diagnosis of EHS, please understand you are not alone, and there are things you can do to help yourself heal and move forward. As Olle Johansson said, most EHS sufferers first find themselves at rock bottom, but manage to climb up out of the hole and reach a stable, more tolerable level. Generally, they do not go back to pre-EHS days but they do level out to a manageable state and still find ways to not just stay alive, but also to thrive.

## *2- Take a (mainstream) Media Fast... Permanently!*

*"As many of you know it turns out this whole pandemic of Covid-19 is a fabricated web of deception. It is a tool of manipulation and control. Picture for a moment that there were no news reports, no masks, no social distancing, lockdowns or store closings. Would we see that anything was different from all the past years of our collective experience? Have we seen people dying in the streets? Have we all lost people unexpectedly who were in good health? Are we burying bodies in our backyards? Or is there only the illusion of a pandemic?"*

—Dr. Andrew Kaufmann, 2020 *Wake Up Call* Documentary

The real virus is that of the one infecting our minds due to the incessant barrage of propaganda aimed directly at us thanks to industry-owned mainstream media. This propaganda definitely comes in the form of "news" but it also takes on a stealthier guise as "entertainment." We are "programmed" via "programming," as we are told what to believe, how to feel about a wide-range of topics, and how to behave. Media is being used as a powerful tool of conditioning and it attacks us at every turn now that is it no longer confined to newsprint, radio or even television. Nowadays it follows us everywhere as we are chronically assaulted via multiple online-connected portable electronic devices.

Rather than relying on our own senses, our own observation, experiences and our own (once-common, now endangered) "common sense," we give all of our authority away to outside "authority" sources to dictate to us our own life experience. As Dr. Kauffmann so profoundly observes, "Are we burying bodies in our backyard? Or is there only the illusion of a pandemic?" What does our own experience and observation tell us about our world? Do we pay attention when we feel unwell when using computers, cellphones and WiFi, or are we feeling sicker and sicker when we engage with these signals and fields? Are we inhabiting our bodies consciously enough to make definitive links on our own, between certain exposures, whether they be to toxic food, chemicals, EMFs, news or people, and our states of health, or are we exclusively relying on outside authority sources?

## 3- *Take ACTION!*

*"I said, 'somebody should do something about that.' Then I realized, I am somebody."*

—Lily Tomlin (actress)

Get involved in the fight to halt and roll back the implementation of all so-called "smart" technologies. It can be hard to rise up out of complacency and do something. Those of who are considered to be "activists" did not necessarily consciously choose this path. We would, for the most part, have loved to have stayed seated on the sidelines holding a bag of popcorn, cheering on others, or simply ignoring the struggles altogether and distracting ourselves with mind-numbing "entertainment."

And today like no other time in history there is a literal endless assortment of entertainment in a wide variety of forms from which to choose and it is possible to indulge in these distractions from just about anywhere at any time. I do not think that this is by accident, coming at a time when there are more serious and immediate threats to the wellbeing of the human race and the planet than ever before, and happening in a very condensed timeline. The number of threats is so overwhelming as to easily induce a state of apathy as we shut down due to feeling powerless and insignificant. It is most definitely a David-and-Goliath fight, but remember, in the end, David beat the giant.

And we can do the same, but we cannot do it alone, we have to remember there is strength in numbers. And even the smallest action can make a difference when added to the many small actions of many others. (See the *Resources* section at the end of the book for links to help get you started.)

## *4- Avoid unproven EMF remediation gadgets*

Like most people grappling with EHS, especially in the early confusing days, I wasted a lot of money—hundreds, if not thousands of dollars, on EMF-protection devices. There are many such products on the market that claim to reduce, eliminate or protect from EMF radiation. Some of them employ crystals or totems to be worn on the body, or use symbols drawn on stickers to place on wireless-radiating devices. Others claim to produce healing frequencies (like the Schumann resonance) to combat and override the harmful ones.

I admit to having tried crystals and even having put shungite rocks in water and drunk the shungite-infused liquid and I have used what we affectionately dubbed a "schumannator"; Schumann-frequency-emitting device, and did not find that any of the above helped. If they help at all it will be on a subtle energy level that I did not find strong enough to make a notable difference. If pendants and charms or anything else requires belief for the products to function, I advise not to waste your money.

## *5- How to Wear a Tinfoil Hat—All About Shielding EMFs*

Some shielding products may help to reduce EMF exposures, but these also have their limitations. A homemade faraday cage or faraday bed net and metallic-cloth-based shielded clothing can help if one is able to measure the reduced field with an EMF meter, but this approach is tricky as use of conductive metals also serves to amplify low-frequency fields as one reflects away the high frequencies. Low frequencies also cause chronic health complaints and so it can be like jumping from the frying pan into the fire. Aside from picking up and amplifying low frequencies, reflecting high frequencies can also result in creating unwanted "hot spots" in various parts of one's home or environment around the shielded areas. When using shielded clothing it is nearly impossible to create a full faraday-cage effect and so the radio waves may simply enter into gaps in the clothing and create these same hot spots within.

Although I did make use of metal fabric shielded clothing and even made a hat with it (yes this is definitely like wearing a tinfoil hat, except that the metal fabrics are more comfortable and breathable), I did not find relief to be significant or long lasting. I did notice that for a time wearing silver anti-static gloves reduced pain and electrical sensations in my hands when typing on computer keyboards, however there is a concern that the metals penetrate into the skin and increase metal toxicity in the body.

Also, when I have tried to measure EMFs after putting up shielding between myself and a wireless source, I have not been able to show any reduction in EMF exposures to myself. And it is important to note that the newer 4G and 5G frequencies boast of the ability to penetrate metals and other materials. While I have not been able to reliably measure this difference, I have noted that natural fabrics (especially wool and also leather) have been more effective in helping me feel better in the presence of EMFs than using metal-fabric shielding. I have found wool hats particularly helpful in reducing headaches on trips to town, for example.

Gary and Ruth also notice wool and leather aiding in the same way. In their view the wool or leather serve as a "second skin" of sorts and thus adds a small layer of protection. What they really found useful has been using earth or dirt as a barrier. They have each fashioned small blankets filled with dirt by placing the dirt in plastic Ziploc bags taped together with duct tape and contained between two pieces of fabric sewn into a blanket. I too have tried this method and did find it helpful at times—albeit bulky, heavy and a bit messy (over time dust from the dirt seeps out). I was even able to measure a reduction in EMFs with my EMF meter. Ruth and Gary also discovered that sleeping on the earth directly really aids in obtaining restful sleep and discharging EMF build-up in the body. In Ruth's case of sleeping in her camper truck, unable to sleep on the ground, they came up with an ingenious way of "bringing the ground to her" by sandwiching earth between sheets of plywood to make an "earth bed."

## *6- UNPLUG!—Reduce, Limit and Eliminate Exposures to All Harmful Sources of Manmade EMFs*

Easier said than done in today's EMF-saturated world, I know. My life was turned upside down by EMFs, I had to move several times and thousands of miles, give up my career, and leave friends and family behind to live in very isolated remote locations. However, if you have not yet been injured to the point of hypersensitivity as I have, you may take a proactive approach and save yourself such a fate.

Firstly, reduce use of ALL electronics. Even those that are non-wireless emit some type of radiation (in the low-frequency range). For these it is especially helpful to stand away from sources—sit far back from TV and computer screens, do not stand near and lean upon refrigerators or other appliances with motors, and turn off, and if possible, unplug, any electronics not in use. (Unplugging helps, as electromagnetic fields also emanate from cords while plugged in, even if the attached appliances are powered-down.)

***Hardwire your Internet connection**

Everyone can benefit from simple actions to reduce exposures. The best thing to do is to hard-wire your Internet connection. For most people living in towns, suburbs and cities and even rural communities, this should not be too difficult to accomplish. Sadly, the days of WiFi on/off buttons on Internet modems are long gone and it is no longer as simple as pushing a button, however it is also not overly complicated either. The first thing to do is to call your Internet provider and tell them you want the wireless function in your modem disabled. They can do this remotely via simple checking and unchecking of boxes on screens that send commands to modems. You can do it too once you are given instructions as to where those screens can be found within your account management pages. You can also turn off 4G and 5G features included in newer modems, in addition to the WiFi.

Once you have disabled the wireless functions, you can run Ethernet cables from your modem to your computer. If you have multiple people using multiple computers in your home you can purchase an inexpensive Ethernet "splitter" that works as a hub between the modem and the various laptops and computers in your home. Ethernet cables themselves can be purchased in rolls of hundreds of feet if needed. Also, for newer laptops and tablets that do not have Ethernet ports it is easy to buy a USB-to-Ethernet adaptor.

Yes, I know, it is messier to have so many wires around the house! But the trade-off is most definitely worth it. Once you have wired your computer to your modem remember to turn off the WiFi function on your laptop. This setting is usually found under "network settings" or via an icon at the top bar of your screen. If you do not disable it you will still be exposed to wireless radiation even if it is not being used for anything while you are plugged into the modem via Ethernet.

By using Ethernet over wireless connections, in addition to protecting yourself from WiFi frequencies you will experience faster and more secure Internet connectivity.

***Ditch the phone!***

But "my phone is my life," you say! I know, I have heard it before and once I too felt similarly. But your phone is *not* your life, your *life* is your life, and your phone is robbing you of that. However, you do not have to get rid of phones altogether. You can replace your so-called "smart" phone with a "dumb" old-school flip phone if you really need an emergency portable phone. Even then, consider how often you actually use your cellphone for true emergencies. Today almost everything constitutes an "emergency," including a friend with insomnia who has no qualms about subjecting you to the same in his or her "dire" need for moral support in the middle of the night. Remind yourself that you are not the emergency services, and even people working for such services take turns in shifts and are not on call 24/7. If someone really needs emergency help, you are likely not the person who can help them—instead they need to dial 911 for the police, fire station or ambulance.

Replace your cellphone with a landline where possible. Where not possible demand your right to have one installed. The FCC is currently facing a number of lawsuits, one of which relates to misappropriating funds[lvi] set aside for the expansion and maintenance of landline-based

---

[lvi] To learn more about the lawsuit, one can visit Irregulators.org and also purchase Kushnick's book, *The Book of Broken Promises: $400 Billion Broadband Scandal & Free the Net* on Amazon.com.

telephone and Internet connections and using them for the expansion of wireless grids instead. These funds were paid over decades by landline ratepayers and total in the billions.

Had this not happened, wireless grids could not have proliferated as they have done and would not have been profitable. The massive profits have come at the expense of taxpayers, *not* as a natural result of fair business practices. It is your right to demand access to landlines, ones already paid for by you or your family and friends before you. If we do not demand this right and do not fight to keep landlines available, they will be taken away. When was the last time you found a functional public payphone? *Exactly.* But not too long ago there were functional payphones all over the nation and world.

Television shows, movies and even "news" documentaries love to instill a sense of fear and panic in the public when relating stories about the recent past in which cellphones were not readily available and supporting networks limited—portraying a world in which emergency help was much harder to get when needed in the absence of wireless networks.

What they fail to point out is that with the existence of public landlines, in addition to plentifully available payphones in towns, emergency public telephones were also spaced out every mile on public highways (but now sadly replaced with cell towers) and available free of charge for emergency calls. Nor are we told that our world is *not* safer today than it was before the proliferation of cellphones and the encroachment of the wireless grid. Children still go missing, in larger numbers than ever, and women are still raped and murdered, and men are still shot and robbed. When storms cause power outages cellphone networks fail but landlines still function. When disaster strikes, like in the case of the World Trade Center attacks, cellphone networks get overwhelmed with use and again, fail. Landlines offer more reliable emergency help than cellphones and do so without causing physical harm.

If a landline is not available in your home or community, you can still wire your phone calls by using a VOIP (Voice Over Internet Provider) service. A VOIP modem connected via cable to your Internet modem can be used with a traditional landline cord connected to a traditional landline phone.

---

Also you can sign up for an Internet phone service like Skype that enables you to make phone calls and even send and receive text messages over your (hardwire-connected) computer.

Do *not* use DECT cordless (or any type of cordless, but DECT is the worst offender) phones when using landline or VOIP, as this will totally negate the benefits to be had from ditching the cellphone since cordless phones emit similar types of wireless radiation, sometimes even more powerful than from cellphones.

If you still cannot ditch your cellphone for whatever reason, or you feel you need to make these changes in baby-steps, always keep your phone on "airplane mode" when not in active use. This function can easily be found via the airplane icon, which you only need to tap to turn on or off. By activating the airplane-mode feature you have effectively disabled all wireless functions, meaning that your phone is no longer sending or receiving signals. (But you may also need to turn off GPS locating services as well to eliminate ALL wireless signals.) Many assume that once in areas out of cellphone range that the phone automatically is not sending and receiving. Well, it may not be receiving but it will be actively engaged in sending out signals in search of a tower or another phone off of which to "ping." In the case of iPhones in particular, many iPhone users are unaware that their phones are acting as antennas for other phones in the area, meaning they act as substations for the main cell towers. This use of these phones increases the amount of radiation to the user by very dangerous levels, much in excess of already poor standards set for radiation safety limits.

If you are "off the grid" your phone will actually be sending out stronger signals than usual in its search to find the "mother ship," so to speak, and so the levels of radiation you will receive from your phone in your pocket will be higher than usual in low-signal or no-signal areas. Do yourself a favor and put it on airplane mode, not only in places where you cannot use it anyway, but any time at all it is not in use. Contrary to what you may have been conditioned to feel, you do not need to be "on call" at all times in all places and you do not need to immediately answer every text or be alerted to every social-media post and "like." Give yourself a break from being always in demand and "always on." Your body and the canaries around you will also thank you.

In airplane mode you can still use your phone for your alarm, or camera, or video recorder or mp3 player, etc.

Also only use speakerphone when making calls. Using a wired headset only serves to carry the radiation from the phone via the wire directly into your head, the same way it does when you hold the phone to your ear. However, there are "air-tube" headphones that do not carry the voice via a wire but by using a tube, and these can dramatically reduce the EMF radiation to your head.

Best still is to *ditch the phone*. By being a cellphone subscriber, your subscription contractually obligates the telecom company to provide you with "full coverage," which increasingly means expanding these networks into every last reach of wilderness left on the planet. If you want to "be the change you wish to see in the world" as Mahatma Gandhi once said, then you need to take *real action* that will bring about *real change*. Maybe we can one day have a wireless-communications network that is not biologically and environmentally destructive, but we cannot have it if we continue to support the current destructive system.

For now there really is no totally safe way to use unsafe technology, and nobody (friend, family member, or employer) has the right to make you use health-destroying tech, or place it inside, on top of, or near your home. And it is important to consider that with each call or text or web search one is playing a dangerous game of Russian Roulette as it is not possible to guess at what moment cumulative effects reach the tipping point leading to either brain tumors, cancers, heart failure/cardiac arrest, or electro-hypersensitivity.

Now may be a good time for you to reassess what you consider to be a "necessity." I also once thought my cellphone to be necessary but when push came to shove and I had to give it up in order to reclaim my health I found a way to do it. I have not owned a cellphone of any kind for over eight years now and have found ways to manage without, while continuing to work, study and communicate. What seemed devastating and impossible at the time (giving up my phone) turned into a an unexpected liberation when I found myself free from the constant demands these devices place on our time and was able to reclaim the privacy[255] they inherently rob.

But if you still absolutely MUST have a cellphone, and I do understand that there are times and places that might prove exceptional in some cases and in the interim while we fight to topple the telecom giants where no other alternative options are available... then remember to most especially put the phone on airplane mode when you go to bed so you can be assured a better night's sleep. Also turn off your WiFi at night if you have not yet figured out how to hardwire your Internet connection. And do the same for all other "smart" tech in your home.

This leads us to the next essential item on my list...

## *7- Get Restorative Sleep*

Getting restorative sleep has played a huge role in my own recovery from microwave injury.

Reducing and eliminating EMFs from your bedroom or other sleeping area, can go a long way to aid in correcting poor sleeping patterns. Today's modern world is rife with insomniacs and it is no wonder when you read the studies on how EMFs and blue-light exposures disrupt circadian rhythms.

All electrical screens emit blue light and this light frequency matches closest with noonday sunlight, resulting in our bodies being constantly confused by this chronic all-day and all-night exposure to frequencies we should only be getting at midday. This action alone causes serious disruption to signals and circadian rhythms in our bodies. It is not, contrary to popular belief, just at night and in darkness that our bodies produce melatonin. This process happens 24/7 and depends directly upon exposures to specific light frequencies at specific times of day.

It is possible to reduce this blue-light interference by use of blue-light filters for computers and tablets, iPods and phones. Programs or apps like *Iris* or *Flux* can provide these filters. Also blue-light blocking glasses like those found at Blublox.com can also help and will also reduce eyestrain while using screens.

Likewise, most of today's light bulbs emit blue light. Sadly, the move toward replacing (and even banning) traditional tungsten light bulbs with compact fluorescent, LED or halogen bulbs, based on claims of energy savings and reduced greenhouse gases in the environment, has led to an overall unforgivable decline in human health. These blue- and green-light emitting bulbs are responsible for severe damage to health because these shortwave frequencies are extremely unnatural and harmful.[256]

They also do not save energy at all—especially when one considers the big picture and how the mercury content contained in the fluorescent bulbs is harvested and disposed of—a substance that can cause serious long-term injury from exposure to leaked vapors from cracked or broken lights, and the common improper disposal of the mercury-containing bulbs contaminates our environment leading to serious long-lasting impacts. Also, in my own tests comparing alleged "energy-saving" bulbs with traditional tungsten counterparts, I have found the latter to be longer lasting and requiring less overall energy consumption. Many of these new bulbs are now also "smart"—enabling them to be turned off remotely and to remotely report energy usage to utility companies—meaning they emit additional harmful radiofrequencies. (And they often contain hidden surveillance cameras as well.[257]) Exposure to blue light, including these new types of bulbs has also been linked to an increase in skin, brain, prostrate, and breast cancers and tumors.[258]

Invisible wireless frequencies also emit a type of (invisible) light so that when one goes to bed at night and leaves these devices on it is akin to sleeping with the lights on and just as bad for sleep for this reason alone. But the irregular pulsing of the frequencies also interferes with sleep.

Without proper sleep cycles we cannot produce adequate melatonin which is a master hormone responsible for regulating other hormones and important biological functions throughout the body. Maintaining proper levels of melatonin also helps repair DNA damage caused by environmental toxins including EMF exposures.

However, I would advise against taking melatonin supplements as it is very difficult to properly dose and not overdose (something that can also be very harmful). Additionally, taking any synthetic hormone[lvii] can result

---

[lvii] This holds true for iodine supplementation as well. Hypothyroidism is also at epidemic proportions, I believe directly as a result of EMF exposures but also from heavy-metal-containing vaccines as well as other toxic exposures in the environment. As such, many alternative health practitioners may advise iodine supplementation, especially in high doses. I personally was given such advice years before become hyper-sensitized to EMFs, and followed through for a number of years only to find myself in a worsened state as I became dependent on increased dosages that led to goiter. I was eventually able to carefully wean myself from the supplements. Today I take a pinch of kelp powder

in dependency on this pill as our bodies stop producing the necessary hormone on their own. But I will say that if in dire need the best way to ingest melatonin would be by way of a natural source, like you can get by drinking **tart cherry juice**, which contains natural melatonin. This is *tart* cherry specifically, not black or any other cherry. And best to avoid any brands containing added sugars as this can counteract the benefits provided in the tart cherry juice. A little goes a long way, even in this natural form, so go easy with dosing.

Again, the best thing you can do in aiding your own body in producing melatonin itself is get your body back in touch with natural circadian cycles. Turning off all EMF sources at night including plugged-in clock radios next to beds, turning off lights, WiFi, and all other wireless devices and as many electronic appliances as possible is a good start. Some find added relief from turning off breakers to bedrooms at night as well.[lviii] But reducing overall EMF exposures during the day is also key to reestablishing these rhythms. When using lighting at night and early in the day, be sure it is as warm-toned as possible–if necessary, place yellow, orange or red gels over blue-colored light sources. This helps mimic natural light, which coming from the sun is in the infrared range the first few hours after sunrise and the last few hours of the day before sunset. If possible, turn off all nights for one to two hours before going to sleep. And going to bed at the same time every night[lix] is also paramount to reestablishing these necessary rhythms and providing one with deep, sound, restorative sleep.

---

regularly, especially in winter and I feel this is helpful in providing enough iodine along with other nutrients while not overdoing it or developing a dangerous dependency. If you do choose to use seaweeds for supplementation, pay attention to your body, if symptoms of swelling in the thyroid area arise after ingesting seaweeds, reduce or stop the dosing.

[lviii] For more tips on how to create a low EMF bedroom see-
https://www.electricsense.com/where-is-the-place-you-absolutely-must-start-if-you-want-to-protect-yourself-from-electromagnetic-radiation/

[lix] Getting to bed at the same time each night is more easily accomplished if one also rises at the same time. Equally important may be going to bed before 11 pm. I find that getting to bed well before 10 pm, gives me the best night's sleep.

## *8- Increase Exposures to Natural Sources of EMFs*

It is crucial to recognize that not all EMFs are created equally. In other words, some sources are harmful and others are beneficial. Natural sources of electromagnetic fields such as those emanating from the earth under our feet and down from our sun overhead, nurture and sustain all life forms.

All too often, financially motivated industries seek to obfuscate facts by comparing natural EMFs with manmade sources simply because they may share a similar frequency range on the light spectrum.[lx] But UV from sunlight and UV from artificial light have very different biological effects because one is native to our environment and the other is not. We were created to coexist with one and not the other.

So get as much sunlight as you can, without the use of toxic sunscreens. To achieve this without causing sunburn you will need to expose yourself in increments while emphasizing early and late-in-the-day exposures when the sun's rays emit primarily infrared (non-burning and very healing) radiation. Dr. Jack Kruse[259] has a lot of useful information on how to take the most advantage of this amazing life-giving source of energy.

You can also practice grounding[260] or "earthing" by walking barefoot on the beach or in other natural low-EMF environments. Lying in the earth in nature can also be very beneficial. The idea is to get as much exposure as possible to the healing frequency which naturally emanates from our earth, also called the Schumann Resonance. But be careful not to "ground" in densely saturated manmade EMF environments, such as in cities, since there may be an unnatural harmful ground current present, which can in some cases override the natural one.

---

[lx] I have had dental assistants tell me that getting doses of x-rays or CT scans is the same as getting ten minutes of exposure to natural sunlight. This is a gross misrepresentation, and extremely irresponsible for any medical staff to relate this type of misinformation to patients. Ten minutes in noonday sunlight is beneficial, no exposure to x-rays or CT scans is healthful—these exposures only harm us.

## 9- *Clean up that Dirty Electricity*

In addition to reducing and eliminating exposures to sources of wireless radiation, cleaning up "dirty electricity" can result in huge benefits to health and work to prevent cancers and other terrible, often lethal, chronic illness—and help heal microwave injuries. Dirty electricity refers to interruption of electrical current flow in the form of high-frequency voltage transients or harmonics (radiofrequency and electro-magnetic interference) on house wiring, usually caused by EMF overload on the electric power grid, but also from dimmer switches and appliances equipped with transformers[lxi]. Wiring in homes and the power lines that carry electricity to this wiring themselves were originally designed to carry a 50-HZ (in Europe) or 60-HZ (in the US) capacity of electromagnetic fields, ones which, have been chronically overloaded (showing as transients from 2 to 100 kilohertz) in the decades since, due to the massive increases in personal use of transformer-enabled electronic appliances, including electric utility meters—particularly of the "smart" or radiofrequency-transmitting variety, but also from modern farm irrigation systems—a real problem in rural areas; places to which canaries often flee to escape EMFs in cities.

Telecoms are also now using power-line infrastructure upon which to place additional antennae for expansion of their wireless communications' grids (including 5G) and antennae for "smart grids" used to collect electricity-use data from "smart" radiofrequency communicating meters, also piggybacking along power-line poles. Power lines themselves are also used by electric companies to send information from digital ("non-smart") meters via the power lines—these signals along with broadband Internet signals carried over power lines, are further adding to the increase in over-powering of these lines.

The resultant harmonics from this overload can actually take the form of radiofrequencies riding on a building's electrical wiring. With home,

---

[lxi] Computers, TVs, CFL blubs, and stereo equipment—all use transformers to convert 60HZ AC to lower voltages. Transformers change smooth-flowing alternating current to shorter bursts in order to save energy, however the chronic stopping and starting results in feedback on building wiring creating harmonics, transients or "dirty electricity." Smooth 60HZ sine waves have minimal biological effects—it is when the waves are disrupted by these bursts, or *pulsed* (as is done by wireless communicating devices) that serious health effects occur.

school or office wiring now emitting EMFs at dangerous levels, the wiring in the building serves to create a kind of EMF cage in which an individual inside is trapped.[lxii]

The best way to measure these fields is to invest in a Graham-Stetzer (named after its inventors Drs. Martin Graham and David Stetzer) meter designed to plug into electrical outlets and provide measurements that display the average rate of change (dV/dT) of transients on wiring in Graham-Stetzer (G/S) units. Measurements over 50 indicate a dirty electricity problem. But for optimal health, especially when electrically sensitive, it is best if one can get the readings down to near 20.

Some readings I have seen, especially on wiring hooked up to off-grid solar-power sources (that generally, sadly, use "noisy" inverters when converting DC power to AC), have read in the thousands of G/S units. In these environments, even in places without any cell-tower or power-line interferences, I have immediately felt the impact of this high level of DE (Dirty Electricity) in the form of intense head pains. But chronic exposures to DE can cause a myriad of adverse health effects including cancers, depression, allergies, Lou Gehrig's disease, infertility, attention deficit disorders, obesity, and chronic fatigue, to name a few. The problem of dirty electricity is a serious and prevalent one and should be considered as important as reducing exposures to cellphones, WiFi and cell towers when working to reduce all sources of harmful EMFs.

Many schools can have very high levels of DE as well. In these same schools it has been common to see "cancer clusters," where large percentages of teachers and/or students develop cancers while working at or attending the school. In one such case at Vista Del Monte Elementary School in Palm Springs, California, dirty electricity researcher Sam Milham, author of the excellent book, *Dirty Electricity,* was able to bring down the readings from over 5,000 G/S units to less than 50 units and witness a marked improvement in the levels of wellbeing of staff and students and immediate improvement in behavioral issues among

---

[lxii] The excess EMFs can also extend to carry along metal plumbing in buildings, effectively electrifying the water that travels through it and further amplify the EMF cage effect in a building. EMFs can also ride along these metal pipes via EMF-contaminated ground as, at least in the US, the ground is used as a return back to power stations for excess electricity carried to homes. In most other countries a neutral wire is used for the return. (See Sam Milham's *Dirty Electricity* book for more information.)

students normally hyperactive and disruptive who became quite calm and quiet after the remediation was employed.

This remediation came by way of plugging in Graham Stetzer DE filters into outlets. These filters can be purchased from various online vendors[261] for around $35 each. Generally speaking, it takes from 10 to 20 filters to reverse DE issues in standard houses. The filters work by countering the spikes or harmonics in wiring by producing its own balancing frequency. Some very sensitive individuals feel that the filters themselves cause symptoms and do little to help them. My own experience is that I have found them to be helpful, however I ultimately needed the help of a professional EMF-remediation expert[lxiii] in the proper placement of the filters to bring about the desired effect.

Another filtration system rumored to be more effective than use of the plug-in filters is a whole-building filtration system installed at the entrance of power (breaker box/electrical panel) into the home or building. These "sine-wave tamer" systems called SineTamer typically run in the $2,000 range but are said to be well worth the investment as they can bring down transient activity to near 1.5% of the peak spike levels and filter radiofrequency and electromagnetic interference from 1kHz to 10MHz. Olga Sheean, whose story I related in this book, testifies to the SineTamer[lxiv] working in her own home, providing her significant reduction in symptoms.

In the case of the school Sam Milham helped, the school board voted to uninstall the filters, refuse Milham's help, and ignore his findings (presented to the board via PowerPoint presentation)—even though the installation had been offered as a free-of-charge donation and had only shown benefit to school students and staff.[262] It seemed the school board was nervous about accepting the gift as any admission implied or overt to having willingly or unwittingly previously exposed students and staff to harmful EMFs (especially that resulted in cancers) could constitute grounds for legal nightmares resulting in costly lawsuits. Sadly, this decision to protect the school's financial interests at the cost of student and staff's health is not uncommon and provides an example of the rule

---

[lxiii] These experts are often called "building biologists." We personally got help from Total EMF Solutions EMF Consultant, Paul Harding of sleepbiology.com, you can read his personal EHS story at his website, plus contact him for solutions.

[lxiv] More info on SineTamer at www.cratuscanada.com

over the exception. Unless we reverse this attitude and trend, we will continue to pay a high price—often one costing the lives of students and their educators.

## *10- Find or Create Radio Quiet Communities and Other Low-EMF Locations*

This is easier said than done. The first thing we did when seeking a safe haven was to scour the Internet in search of already-established Low-EMF refuges. Upon first glance there seemed to be a number of possibilities, but upon further inspection my partner and I found that none of these communities were actually up and running—but instead were still newly in the making. Of course, by the time you read these things might have changed and it is always worth looking into. There was a plan some years ago for a refuge in northern Arizona, in the White Mountains, but the last we heard the founding members were running up against barriers to obtaining permits for tiny homes and resident camping. Oftentimes these great ideas do not play out as hoped primarily due to insurmountable legal obstacles. The other big issue that arises, of course, is safeguarding the property from future development—namely wireless-grid infrastructure or high-tension power lines or even wind turbine or solar farms and Doppler weather stations—all or any of which can instantly pollute a large area with EMFs, making a once pristine area uninhabitable for canaries seeking refuge. At this moment in time there are no laws to help us protect areas from EMF intrusion by government, industry or military. This is something that needs to be urgently addressed and remedied so that we can protect the few remaining radio-quiet zones from future contamination to ensure safe housing for affected individuals. Some efforts toward establishing radio-quiet zones for this purpose are underway in some parts of Europe, in France and possibly Portugal. But I am not personally versed on the details. If an international move is possible for you, it may be something worth investigating further.

It is my sincere hope that more options will be available soon. In the meantime, many of us are finding pockets here and there that are working for us—for the moment. And I have discovered the best way to find these places is to search for off-grid listings on Craigslist or Real-estate sites or Intentional communities or work-trade opportunities through organizations like WWOOF (Willing Workers on Organic Farms) or

WorkAway. Often inquiring with Airbnb listings that are off-grid and do not offer WiFi may lead to possible long-term rental options, even if not listed as such.

But be aware, the description of "off grid" does not guarantee freedom from wireless radiation as the same properties may be near cell towers while off of the electric grid, but as many are located in remote locations there is a better chance of finding options in this way. It can also help to check cellphone coverage maps and look for the areas not yet covered in "plague-red" markings. This can be a way to home-in on possibly suitable areas. However, finding a place away from power plants, power lines, cell towers, WiFi, and wind farms (that create massive dirty electricity and health-robbing infrasound up to 15 miles away[263]), is most definitely a tall order in today's world, so all we can do is find the best options available. Any reduction in EMF exposures will help. And many WiFi refugees do well to start by going on regular camping trips in wilderness areas. Sadly, group campsites generally no longer provide low-EMF refuges as so many are now equipped with WiFi or cell towers for the "convenience" of campers.

Change will only happen if we demand it. So even if you are not personally in need of a refuge, help others who are, by complaining to local officials, state park staff, etc., and asking that these places be kept free from this intrusion—for the sake of sensitive people and also sensitive wildlife.

## *11- Deal with the Fungus Among Us*

Fungal infections, unlike those allegedly caused by viruses, can be scientifically verified and are a big problem for many of us in todays' world due to the prevalence of monoculture and pesticide-based agricultural practices that deplete our soils of important beneficial microorganisms, which serve to keep things in balance.

Due to the lack of these organisms in our foods (even in organically grown foods) our guts are also lacking in the same organisms, in turn, creating a terrain supportive of yeasts and fungi (which have an important role to play in a *balanced* state) not kept properly in-check from overgrowth.

Natural herbal antifungals like **Oregano, Black Seed Oil, Pau D'arco,** and **Garlic**, can be very helpful in combatting fungal overgrowth as can minimizing EMF exposures, since artificial EMFs have been demonstrated by Dr. Klinghardt to encourage growth of molds and fungi up to 600 times more than they would in a low-EMF environment.[264]

**Sulfur** supplements (typically sold as *MSM*) are also very powerful fungal reduction aides. I personally add a scoop of the powder to my water anytime I feel I am beginning to notice excessive mucous in my sinuses, throat or lungs. The sulfur quickly reduces this overproduction. I have also recommended it to friends who have felt overpowered by stubborn flu symptoms that do not show signs of easing after several days or weeks. In the cases where friends heeded my advice they reported "miraculous" recoveries.

Garlic, onions, broccoli, cabbage and cauliflower are each high in sulfur and recommended as part of a regular diet for those suffering from chronic yeast/fungal overgrowths. However, it is best to use caution with garlic as it is a powerful medicine and can cause stomach irritation and excessive thirst as well as flatulence, so should be used sparingly. Also sulfur supplements may need to be taken in small doses as sulfur also helps to detox heavy metals from the body, but if one detoxes too rapidly it can cause too much discomfort. Best to go slowly, especially to begin with, except in the case of acute flu symptoms when higher doses (up to one full ounce divided in four parts taken over a 48-hour period) seem to be more effective.

Interestingly some types of beneficial fungi can also be used to combat other harmful fungi and mitigate damage from harmful EMF exposures. Since fungi of all kinds grow more rapidly in the presence of EMFs, Dr. Klinghardt and colleges have begun researching the potential to use medicinal mushrooms in countering the effects of radiation damage from EMF exposures. Being in the fungi family, **mushrooms** also have been observed as not only surviving in high levels of radiation, but also actually thriving in it. This points to an inherent protective mechanism in the mushrooms, one that is hoped to have benefits by way of ingestion for humans. Last I read, Klinghardt did not have specific mushrooms yet to recommend, but I have personally heard of **Chaga** as one strong contender. And I have used Turkey Tail (as it grows in abundance where I live) with some noted benefits, similar to those experienced via sulfur supplementation, in reducing excessive mucous and strengthening lungs.

Raw-foods advocate and author of *We Want to Live*, Aajonus Vonderplanitz, experimented extensively with counteracting the negative health impacts of vaccines with ingestion of **moldy berry juice**, presents other research into how fungus or molds can actually heal. Below is the excerpt in its near entirety on the subject from Vonderplanitz's book:

> *I experimented with vaccines in animals (rats, cats and rabbits) and found that the mutations compounded with each succeeding inoculated generation. In the second generation, with all three types of animals, mutations started as slight deformations of an ear, or eye, or jowl; or a shortened limb; or scoliosis. Glandular malfunctions were prominent. In some animals, temperaments became unruly. In the sixth and seventh generations' mutations were severe: loss of glands, organs, limbs, features, motor and neural functions; brutal suicidal and homicidal tendencies; and impotence resulting in extinction.*
>
> *In my experience, of all the pharmaceuticals that are accepted as miracles, vaccines are the most dangerous because the side effects are most often subtle, or attributed to another problem, or ignored. These side effects become obvious after the third and fourth generations. That is when the mutations start to become exaggerated and pitiful. With humans this will probably be apparent by the year 2015...*
>
> *While I was experimenting with raw moldy foods as possible remedies for particular health problems, Owanza, who was assisting me, discovered that eating moldy berries affected the mutant antibodies caused by vaccines. She placed raspberries, blackberries, and strawberries in separate glass containers with lids that were not tights. She allowed them to sit at room temperature until a healthy mold had formed. Then she placed them, still in glass containers with loose lids in refrigeration for ten weeks. After these ten weeks, Owanza strained the moldy berries through a porous sieve, separately. The mold-juices were ready for my subjects.*
>
> *Blood analyses were done on eleven people with long histories of vaccinations, with specific focus on amino acids. All subjects had minor complaints, ranging from headaches to joint pains. For one time only, each person drank ½ cup of one kind of mold-juice—of the particular berry that normally appealed to*

*their taste. Their diets were not altered. Blood analyses were done 30 days after drinking mold-juice. Analyses showed an average of 30% increase in stable amino acids over previous analyses. There was only one side effect as a result of drinking the mold-juice—everybody became lethargic for a period of 30 days. After that period, everyone's level of energy and health increased sharply above what it had been before drinking the mold-juice. Health complaints either diminished or disappeared.*

*Between 13 and 18 months after they drank the mold-juice, three of the people planned to get vaccinations for travel reasons. Blood analyses were done before inoculations and again a month after inoculations. The post-inoculation analyses showed a 30% drop in stable amino acids.*

*Twenty-four months after taking the mold-juice all eleven had blood analyses. In eight of them the stable proteins in their blood remained high. In the three who had recent vaccinations, stable proteins were still down and the health complaints that they had prior to drinking mold-juice returned.*[265]

Intrigued by this account I personally harvested an 8 oz. jar full of wild blackberries and left them out in late summer 2021, to grow a fuzzy blanket of mold just as described. I never did put the jar in refrigeration afterward since living off grid means we have no refrigerator—but not long after the mold developed, the weather turned cooler so that effectively my moldy juice was stored in near-refrigeration temperatures for about three months. Hesitant to drink the juice at first, fearing both the taste and my body's response (and also unsure of how well I could handle a full month of increased lethargy), I put off the juice consumption an additional few weeks until I got my courage up just after the new year and drank my allocated strained half cup of moldy juice. I am happy to report that I have not experienced any ill effects at all apart from possible increased muscle aches—but have instead felt a dramatic increase in energy right away. I most definitely did not experience stomach upset and the taste was surprisingly quite pleasant.

## *12- Detox Heavy Metals*

This is an important one. When comparing notes most all of us who deal with EHS have found that we also have high levels of heavy metals in our bodies. Besides upsetting many important intricate functions in the human body and causing brain inflammation and damage to other organs, heavy metals are conductive, and as such interact with EMFs, serving to amply signals. Many of us with EHS feel like walking antennas and the heavy-metal poisoning could offer one explanation for this feeling, or it could be contributing to the problem by damaging nerves.

There are inexpensive ways to measure these levels, the most popular being the use of **hair mineral analysis,** however many of the labs wash the hair samples in chemicals that strip a lot of the minerals so that the results cannot be depended upon. There are a few labs that do not do this or wash in a very mild soap. Dr. Lawrence Wilson is a known alternative medical doctor who works with a number of other practitioners who use this type of lab for the hair-analysis tests. The practitioners are also trained by Dr. Wilson to properly interpret the results, and help tailor diet and supplementation programs to bring down these levels. I personally worked with one of Dr. Wilson's practitioners for about a year or two and had positive results in reducing the heavy-metal burden in my body.

The most effective tool I was provided with during this process was instruction in the use of **infrared saunas** for detoxification through the largest organ–the skin. Out of all of the detox methods I have tried this proved most potent while also being the least invasive and least expensive.

But I must warn the reader that there is an important distinction to be made between *far* and *near* infrared saunas with the former (far infrared, the type of dry sauna usually found at day spas or gyms) produce high levels of EMFs and so should really be avoided by those with EHS or anyone seeking to lower their EMF exposures. However, there may be some exceptions, always check the EMF output levels for any sauna before using or purchasing.

I built a **near-infrared** sauna at home for less than $100. All you need do is purchase three to four 250-watt red-coated bulbs (these can be found at farm feed shops and hardware stores, they are sold as heat lamps for raising baby chicks), along with reflector lamps and either mount them on 2x4 lumber pieces screwed onto a wall, or obtain light stands (I still had some from my professional photography days so I used the latter). Place the lights in a triangular or square pattern at a level and spacing to cover your torso. The lights should be mounted or placed in a small room, closet or shower stall (I used a shower) where one can benefit from the most heat in the enclosed space while also allowing for enough fresh air flow to enter from a cracked door or window. I used a swivel barstool so that I could turn each side of my body to the lights (placed 1 to 3 feet away) for a few minutes at a time. Basically, if you set it up correctly you will give off the appearance of a rotisserie chicken spinning around next to red heat lamps.

Using this sauna setup in colder months in our house without central heating took much longer to break into a sweat. But even without initiating sweating, the infrared light exposure has amazing benefits as it penetrates deep into tissues with healing frequencies and counters our excessive blue-light exposures. Doing a sauna a few hours before bed can also aid in restful sleep.

Start with short exposures, ten to fifteen minutes, and build up to an hour at most and not more than one to two hours daily. Otherwise, the detox effects will be too powerful and sudden, which can cause a "healing crisis" in the body. Best to go slowly with any kind of detoxification process as to not overburden the body.

While sweating in the sauna, be sure to wipe away excessive sweat with a towel (so that toxins do not reabsorb into the skin) and be sure to properly hydrate before and after with high mineral water. Adding a pinch of unprocessed sea salt or clay to the water can help mineralize the water. And be sure to shower soon after the sauna to remove toxins released onto the skin.

**Dry skin brushing** can also aid in this process of removing toxins from the skin (before or after the sauna). It also helps circulate the lymphatic system—another way to aid in removal of waste from the body and improve overall blood circulation. (Plus, it feels really good!)

Taking a little **charcoal supplementation, clay and/or kelp** after sauna therapy (and in general) can help remove heavy metals from the digestive system by absorbing them for removal via the bowels, kidneys and urinary tract.

**Silica supplementation** is also really beneficial in safely removing aluminum in particular, from the body. You can buy silica or purchase (food-grade) Diatomaceous Earth (DE) powder (ground-up fossilized shells which contain high amounts of silica) and add a large spoonful to a half-gallon or gallon-sized jar, then fill the jar with water and leave to sit until the cloudy water clears (at least partially, this generally takes 24-48 hours). Sip this water throughout the day. Adding a splash of lemon juice or apple cider vinegar to your glass will aid silica absorption. The DE powder can be reused several times before replacing. This is the cheapest and probably most effective way to get silica into your diet, as it is a hard mineral to absorb and is more easily assimilated after it is absorbed into water molecules first.

**Chlorella (and other blue-green algae like spirulina) cilantro, parsley, and seaweeds** all help to remove heavy metals but many of us have found that chlorella and cilantro in particular bring on too much of a healing crisis and can be harmful in that cilantro, rather than binding the metals to it, works to mobilize them, meaning that moving mercury or aluminum from bodily tissues where they may have been doing less harm, could be moved to other places where more harm may occur before it can be safely removed by one of the algae or seaweeds. I cannot really explain why I also had trouble with the chlorella except to say that it may be less easily digestible than other algae, or could have had something to do with the way it was processed.

I would not personally advise anyone to use any of the pharmaceutical drugs intended for heavy-metal chelation, as I have heard literal horror stories about the damage they have done to the recipients. These, I feel, are too powerful and have the potential to do more harm than good, even under the administration of a qualified physician.

**Safely removing mercury amalgam fillings** can also help reduce one's overall heavy-metal burden. But be sure to find a qualified dentist who understands how to remove the filling(s) without further exposing you to vapors from the mercury in the process. Using a proper dam so you also do not ingest it and supplying you with proper respiratory protection is paramount. Otherwise removing the fillings can do more harm than good if not done safely.

**Avoid further heavy-metal poisoning by avoiding heavy-metal containing drugs**. This means ALL vaccines, as they ALL contain some type of heavy metal. While mercury was finally removed from most childhood vaccines in the past decade, aluminum (equally and possibly more dangerous than mercury) was not. And mercury is still included in flu shots. Also, the new experimental gene-editing vaccines being used for Covid-19 contain large amounts of graphene oxide—a metal considered to be a "super conductor," and extremely toxic.

Metals and other toxic ingredients in vaccines injected intramuscularly reach into both the blood stream and the lymphatic system. Up until as recently as 2016, medical professionals and scientists assumed that the white-blood-cell-carrying lymphatic system did not reach into the brain but stopped at the brain stem. However, in 2016 this was proven to be false and showed that when stressed and going into "fight-flight" the lymphatic system will carry white blood cells, used to carry away toxins from the blood stream, into the brain. If children or adults are stressed when they receive these heavy-metal-containing injections the lymphatic system can carry small amounts of aluminum, mercury or graphene oxide into the brain. In the case of aluminum, it only takes a very small amount to cause *encephalitis*, or swelling of the brain and subsequent brain damage—what we now call "autism." The reason vaccine-induced autism affects a larger percentage of boys than girls is that boys more readily go into fight-flight while under threat, whereas girls show a greater propensity to "tend and befriend" attackers as a coping mechanism, meaning that their own lymphatic system will not be triggered at that crucial moment and thus not carry the metals to their brains. However, some girls of course, will prove the exception. And when the brain is not directly affected, the metals get stored in granules within bodily tissues and sometimes stay there protected from harming the rest of the body—but often open up for various reasons at later dates and suddenly wreak havoc many years down the road after receiving a vaccination.[lxv]

---

[lxv] Medical researcher and author Forest Maready explains this process in great detail in his excellent book, *The Autism Vaccine.*

## 13- Reduce Inflammation

EMFs are linked to triggering an inflammatory response in our bodies. Reducing EMFs in your environment will naturally reduce inflammation in your body.

But since some of the effects of EMF exposures can be long lasting and chronic and because it may be impossible to eliminate or reduce all sources of EMFs in your environment, there are some other ways you can help ease inflammation.

These are some things that have helped me immensely—

-Anti-inflammatory herbs like **turmeric, ginger**, and **kava kava**[lxvi]

-Regular **Epsom salt baths** (these baths also have a grounding effect and help discharge built up body voltage)

-**Magnesium supplementation,** a lifesaver for most electro-sensitives, although my own experience is mixed as I have found it disruptive to regular bowel movements, often surprisingly causing constipation rather than the opposite—an effect for which it is more commonly reputed. Using magnesium trans-dermally in the form of Epsom salt bath—a form of magnesium, or as skin spray-on like the ones sold by Ancient Minerals, may serve to help avoid the bowel movement disruptions.

-**Essential fatty acids** rich in omega 3 & 6's, like **fish, hemp** or **olive oil.**

-**Natural essential oil rubs** containing mint and eucalyptus and also winter mint, clove and/or cinnamon applied to the forehead can quickly reduce tension headaches and applied elsewhere can reduce inflammatory pain in other parts of the body. Be sure to avoid brands

---

[lxvi] *Be aware* that turmeric can be constipating, ginger can cause acid reflux/heart burn, and kava kava—an herb so nice they named it twice—can cause excessive thirst and dehydration, but the rumors (started by pharmaceutical companies) that it can cause liver damage are incorrect and have been since disproven.

containing chemically based perfume oils and petroleum oil bases and be sure to avoid contact with mucous membranes as the contact can cause serious irritation and inflammation.

## *14- Mitigate EMF Damage*

When I spoke with Dr. George Carlo[lxvii] in 2015, he explained to me the importance of the dynamic between the rate of compensation by the body and the rate of EMF exposures to it, when it comes to healing from EMF exposures. His recommendations involved a list of supplements coupled with dietary and lifestyle advice to help increase my body's own rate of compensation vs. the rate of exposure to harmful EMFs.

From his point of view, it is crucial to first open up the cells in the body before aiding them in detoxifying–this can help to increase our "adaptive capacity", according to Carlo.

I was told the program he outlined for me should be followed from six months to two years depending on the severity of damage done. For my part, I faithfully followed his program for a full two years but did not find it as helpful as I had hoped. I believe the main issue was related to taking so many synthetic supplements, many of which contained artificial flavoring or coloring, that in my case of also dealing with MCS in addition to EHS, was ultimately counterproductive as I had to cope with the added burden of chemicals from the supplements.

However, I did find his advice to follow a daily yoga, or other light exercise program that involved stretching, to be helpful. And following Carlo's advice to incorporate creative artistic pursuits along with foreign language lessons into my daily or weekly routine I likewise found benefi-

---

[lxvii] Carlo led a clinical cancer research study into the biological effects of cellphone radiation for a wireless lobbyist group (CTIA) headed by Tom Wheeler (who later became chair of the FCC, tasked with regulating the wireless industry–so no conflicts of interest there!) who hoped he would find in favor of the industry and not report adverse health effects. However Carlo did find cellphone radiation to cause cancers/tumors in lab rats and advised CTIA to include warning labels on products. As a result of these findings, CTIA tossed out the $23 million dollar study and Carlo along with it. See Carlo's book for more info: *Cell Phones; Invisible Hazards in the Wireless Age.*

cial. These activities are meant to help activate the parasympathetic nervous system, the one associated with higher levels of relaxation and lower levels of stress.

But his advice to regularly lift weights resulted in repeated injury in my case, even though I had taken care to start very with very light weights (not more than 5 lbs.). The damage already done to my joints from EMFs prior to that time may explain why weight lifting for me was not the best plan, however it may benefit others.

Possibly his supplement routine could also be beneficial for those less sensitive to artificial and synthetic ingredients. At the moment I do not recall what they were exactly except that B-vitamins, anti-oxidants, a few amino acids, and multivitamins were part of the plan.

Many EHS sufferers do swear by **antioxidants** as playing a crucial part in their own healing journey. This is probably because exposures to microwaves cause oxidative stress and damage, and antioxidants are used up in the repair process. Also since EMFs damage metabolisms, it can be extra difficult to obtain necessary nutrients from foods (especially considering how compromised our food sources are already, being genetically modified and heavily doused with glyphosates), and so adding nutritional supplements to diets may be wise, however the same damage to metabolisms may also make it extra challenging to absorb nutrients from supplements as well, especially if they too are synthetic, laced with chemicals or are genetically modified.

I feel the best course of action then is to choose supplements wisely and also get organic food-based sources wherever possible. Best is if you can get herbal sources in bulk and unmodified, and better yet is to grow your own herbs. The only *real* super-food according to Dr. Jennifer Daniels, author of *The Lethal Dose*, is the one you pick either from your own garden or from the wild, yourself.

Specific nutrients that can help to repair damage from EMFs are listed by Dr. Goldberg in his book, *Would You Put Your Head in a Microwave Oven?,* and include the following:

**Melatonin,** for its antioxidant, anticancer properties and ability to prevent breaks in DNA in brain cells and prevent kidney damage from cellphones. Goldberg recommends not taking more than .5 mg doses, but as I mentioned before, I recommend not taking a synthetic form of any hormone, but instead, to make lifestyle changes that can aid one's body in producing the hormone itself–in this case the changes necessary I have listed out in the *Restorative Sleep* and *Increasing Natural EMF Exposures* sections of this Appendix. Again, if you need to take a form of melatonin, try **Tart Cherry Juice**, a natural form, over a synthetic pill.

**Zinc,** because of its ability to protect eyes from oxidative damage and preserve antioxidants in the blood, while offsetting cell-membrane damage. Dr. Goldberg does not recommend a specific dose. I have personally used zinc often over the years but ultimately found that it was causing me nausea and learned that taking zinc can cause an imbalance in copper but taking copper can also be toxic. The challenge with taking *any* vitamin or mineral supplement is going to be in maintaining the proper balance of overall nutrients. For this reason, I consider it best to obtain these nutrients from food sources. Zinc is present in abundance in meats, eggs, pumpkin and flax seeds and almond and Brazil nuts and copper is available in many nuts as well.

**S.O.D./Super Oxide Dimutase,** for its ability to protect DNA from damage caused by microwaves and for producing hydrogen peroxide in the body. I have not had any personal experience using this supplement but I can say that using **ozone** therapeutically has helped me quite a lot. I bought a **cold corona**[lxviii] ozone machine from Synergy co. in the US for about $500. You need to obtain an oxygen tank for it to work, which can be done without a medical prescription by purchasing oxygen tanks from welding supply companies. The quality of oxygen is exactly the same grade, regardless of purpose for use. You will need to get the correct adaptor for regulating the flow of oxygen (called a *regulator*) into the ozone machine. I used ozone therapeutically by both insufflation in my ears and rectum at very low levels. The amount breathed by placing in the ears is minimal and understood to be safe. The effect of ozone therapy on the body is demonstrated by increased levels of oxygen, as ozone is a form of oxygen (O3) and produces peroxides in the body.

**Gingko Bilboa,** is on the list as a powerful antioxidant that helps with blood circulation and prevents oxidative damage in kidneys, eyes

---

[lxviii] It is essential that cold corona and not hot corona be used for ozone insufflation therapy as using hot corona can be very harmful if used in this manner.

and the brain. You may have heard of its ability to aid memory function, this is likely due to the protective effect on the brain. I have not yet experimented much with this herb, but plan to. My running joke is that I have Gingko in the cupboard but keep *forgetting* to take it. (If you are having trouble comprehending the joke maybe you need some too!)

**Bilberry extract,** is said to specifically help with vision by protecting the eye from oxidative damage and minimizing retinal fatigue. I have never tried this but it is another on my list to try in the near future.

**Caffeic Acid,** allegedly has similar protective properties as melatonin and can prevent oxidative and DNA damage from microwaves. It can be found in many foods, such as apple cider vinegar (ACV) and bee propolis. I have noticed many benefits from adding a splash of ACV and eating spoonfuls of raw unheated honey containing propolis, so there may be something to this claim.

**Catalase,** is an enzyme that allegedly protects mitochondria from microwave damage. I have no experience with this one.

**Coq10,** is recommended as another super antioxidant, this one specifically found in the respiratory chain, and said to maintain cellular metabolism and protect from oxidative damage to the cell membrane. I have only taken this sporadically as it is quite expensive, but have generally heard it to be a worthwhile supplement helpful in tissue repair and important for protecting eyes and the heart.

**DHA** together with **EPA**, are essential fatty acids found primarily in fish oils and work well as anti-inflammatories and also protect delicate organs like the heart and eyes. I have had good success with fish oils generally although they can go rancid easily and need to be taken a little at a time to avoid stomach upset and acid reflux. Fish oils also can provide a good amount of natural vitamin A and D—the latter of which is especially important in the northern climates in winter when vitamin D from sunlight is in short supply. Vitamin D is also plentiful in wild mushrooms, which also grow in winter in the same climates. (Make certain you are *one hundred percent* sure about wild mushroom identification if you plan to gather mushrooms yourself as some species have poisonous look-a-likes that can be deadly if accidentally ingested.)

**Acetyl-carnitine,** is listed as important for energy production in the cells, working by transporting fats to the mitochondria. I have used this supplement on occasion and felt it helped combat muscle fatigue and soreness to some extent. It might especially be good for anyone suffering with fibromyalgia.

**Lycopene,** is last on Dr. Goldberg's list and touted as another powerful antioxidant that has anti-cancer effects, but not one I have used personally. I believe it is found naturally occurring in tomatoes and have noted benefits eating a lot of fresh tomatoes from my own garden. Store-bought tomatoes on the other hand (that may be sprayed with citric acid concentration as preservative), cause me to develop sores in my mouth and upset my stomach.

Not on the Dr.'s list but also worth mentioning, are **glutathione, glycine, GABA, and NAC.** Glutathione is an especially important supplement for anyone who has been exposed to graphene oxide in vaccines or other medicines, as the graphene oxide dramatically reduces glutathione levels. The supplement is best taken in liposomal form. Also doing coffee enemas (search for instructions online) can help boost glutathione levels in the body.

## *15-(How to) Avoid antibiotics and other toxic "medicines" (Safe alternatives to common ailments)*

Antibiotics are routinely and (in my opinion) recklessly prescribed for many ailments to the point at which most of us have experienced its harmful long-lasting effects. The word *antibiotic* appropriately and literally translates to "anti-life," and any use (not just overuse) can lead to a long list of chronic impairments in basic bodily functions. The same holds true for many other modern-day allopathic pharmaceutical drugs.

The two largest-grossing industries today (Big Tech and Big Pharma) both profit from our collective demise and serve also as primary cause for this same demise. Dependency on health-destroying technologies creates a dependency on pharmaceuticals and pharmaceuticals create dependencies on harmful technologies (pace makers, implants, nano-tech, etc.) and also on more pharmaceuticals. Given that these two monstrous industries support and feed one another it is no surprise that many tech companies are also buying pharmaceutical companies and developing "bio-tech" medicines and medical devices. So, it stands to reason that if we can break our dependency on either tech devices or pharma drugs we can break our dependency on the other. Whatever we can do to stop feeding our oppressors will help free us from their tyrannical hold, and most especially help us to regain our health.

Here are some natural remedies for a wide variety of common health complaints I have found helpful in replacing these drugs and easing my former dependence upon them. (Listed in alphabetical order.)

### *Allergies (Seasonal)*

Allergy symptoms during seasons of high pollen count can be extremely aggravating. For most of my adult life my heart sings at the beginnings of spring when flowers are blooming and tree buds blossoming until my hay fever dampens my joyous heart song and replaces it with the incessant beat of post-nasal drip, watery eyes, fatigue, headache and sneezing.

While the basic go-to holistic remedies for seasonal allergies include **nettle-leaf** teas or powder-filled pills and **Quercetin** supplements, Ruth and Gary taught me a more effective and immediate cure. Gather samples of all (non-poisonous) **flowering plants** and **infuse** in hot/just boiled water for several minutes, then drink this tea repeatedly until relief is achieved. Eating **raw local honey** can also help.

I also discovered homeopathic eye drops for my very aggravating watery eyes (sometimes accompanied by thick discharge), which turned out to result from actually being too dry. The brand of drops that worked wonders for me is *Similasan.*

(See *Flu/Colds* for more information about managing symptoms of sinusitis.)

### *Anxiety/Depression*

This is an epidemic-level ailment among our populations (thanks in large part to the proliferation of EMFs and gaming and Internet/media addictions) that drives us to reach for magic pills. Many who have been subscribed antidepressant and anti-anxiety medications can testify to the sometimes-terrible side effects that can also continue long term, as well as cause drug dependency. Of course, one of the side effects is an increase in the symptoms the drugs are meant to treat—and this can mean suicidal ideation and sometimes, tragically, execution. Hormonal disruptions are also reported, including infertility issues, loss of libido, and drastic mood swings. Also, side effects of increased anxiety or aggression, or turning into an emotionless zombie are also reported.

Here are a few herbs and supplements that have helped me, and others: (Use as directed.) **Saint John's Wort herb, SAMe, GABA, Glycine, Kava Kava**

Also, **cold-water therapy** has been reported to ease these symptoms. This can take the form of cold showers and/or baths (even with ice in the bath), ice packs held on torsos, faces dunked into ice water and plunges into natural bodies of cold water like the North Pacific Ocean. Both Dr. Jack Kruse and Wim Hoff (aka the "Iceman") offer instructions for using cold-water therapeutically. I do *not* however think that Hoff's training using hyperventilation is a good idea or necessary for cold-water submersion, although of course it is a natural reflex to rapidly intake one's breath when first entering cold water but this does not mean it is a good idea and after one gradually adjusts to cold-water exposures one should no longer find it necessary to alter one's breathing in any drastic way and should instead be able to breath calmly and steadily. I personally began my own cold-water therapy nearly two years ago upon moving to the west coast and am now able to swim in the ocean (temps varying between 45 and 55 degrees Fahrenheit year round) in only a bikini for up to twenty minutes. I know of others who can sit in baths filled with ice water set at 35 degrees for half hour intervals.

Cold-water therapy offers a wide range of benefits beyond mood elevation and stabilization, but please *proceed slowly* and with caution to avoid hypothermia or even (in the case of ice therapy) frost bite!

**Changing your diet** (esp. **eliminating sugars, alcohol, and stimulants like caffeine**, and possibly increasing healthy fats and proteins and reducing carbs) and also **adding beneficial bacteria/probiotics** to the diet can also help reduce symptoms of anxiety and depression since the health our gut also affects our moods since there are as many neurotransmitters located in our intestines as in our brains.

### *Asthma*

I personally suffer with this condition and once was given a steroid inhaler by a doctor to use when it was triggered from polluted air while living in New York City. Fortunately, I did not use inhalers for a prolonged period, since use of steroids in any form is extremely harmful to health and disrupts hormonal balance leading to a cascade of other health problems.

I mainly practice **avoidance** to keep my condition in check. This means avoiding smoke inhalation, especially from trash burning or cigarettes, and avoiding inhalation of powders from cayenne or other strong peppers (meaning I cannot be near anyone in the process of sprinkling it on his food). Exposure to certain chemicals, especially ammonia, is another trigger. And breathing too deeply through my mouth while exercising or laughing can also trigger an attack. I have found the ***Buteyko*** method (developed by a Russian doctor by the same name for asthmatics) of breathing very helpful in controlling my breath intake and preventing attacks as well.

EMF exposures are also directly linked to increases in incidence of asthma and other lung ailments including lung cancers. People living in close proximity to power lines in addition to living near cell towers and using cellphones and WiFi, have been statistically shown to suffer higher rates of asthma.[266] So lowering EMF exposures can also help reduce asthma and other lung disorders.

There are also a number of herbs that can help reduce chest congestion and ease inflamed bronchial tubes. **Shizandra berry** (also helpful for adrenals), **Hawthorn** (also supportive of the heart), **Elecampane, Eastern Skunk Cabbage, Lobelia, Black Cohosh**, **Rowan tree berries**, and **Turkey Tail mushroom** are all excellent herbs indicated for asthma. **Natural ephedra**, also called **"Mormon tea,"** a desert plant, can aid someone in the midst of an attack, and so too can **Lobelia** (help end an attack). Steam inhalation of **chaparral** from boiling the plant in water and standing over the steam can also provide relief.

### *Back pain*

Unfortunately, I have suffered with more than my fair share of back problems. I find it all too easy to "throw out" my back. **Prevention** of back problems mainly involves taking care when lifting heavy items.

"Heavy" for one person may not be "heavy" for another. Do not judge yourself. Know your limits. And when lifting take care not to use your back, but instead bend at the knee and bear the weight in your legs. If your legs are not conditioned, lifting anything remotely heavy will present an added challenge and danger. Regular leg-strengthening exercises can help reduce the possibility of injury when lifting. I prefer **Tai Chi** exercises to this end.

If you do injury your back, the first thing to do is, of course—**Rest!**

In the first hours and day or two after injury, **ice** applied to the pained area can be helpful. After this time heat should be applied and possibly alternated with ice. Using **infrared red-lamp heat therapy** near the site of injury can serve to rapidly speed healing. Take care not to sit too close to avoid accidental burning. Also keep the lamp away from any fabrics, papers or other flammable materials.

If you are able to get into a bath, a **warm bath with Epsom salts** added can also reduce inflammation and speed healing.

Natural **sore muscle rubs** applied to the area of injury can also work to relieve pain and reduce swelling. My preferred rub is called, *Narayan Extra Strength Sore Muscle Balm.* It is non-petroleum, beeswax based with menthol and white camphor.

For sore, but not injured backs, **massage** and **acupressure** (self-applied by use of a **"Backnobber"** if unable to get help from another person) can also help a great deal.

### *Bladder, Kidney, and Urinary Tract Infection and Pain*

Antibiotics are commonly prescribed for urinary tract infection (UTI), and bladder and kidney infections. All too often the drug first alleviates the horrific pain associated with these conditions, but soon after the short reprieve, the drug serves to intensify the condition and terrible pain. At this juncture doctors tend to give patients another "stronger" dose of a different antibiotic, which can again lead to this vicious unending cycle, particularly when the UTI then leads to bladder and then kidney infection, sometimes resulting in emergency hospitalization and further administration of toxic drugs.

I personally have been down this road too many times. Fortunately, I eventually found safe and effective remedies for recurring UTIs and bladder pain that have kept me free from needing any pharmaceutical drugs for more than 15 years.

For some time, unsweetened **cranberry juice** did the trick, if the signs of UTI were noticed early enough. I also found that **D-Mannose** powder worked in a similar way but more powerfully and immediately. When those two options were not available, I also successfully used saltwater flushes and lemon juice to "nip" the UTI "in the bud."

To do a **saltwater flush** simply obtain a half teaspoon of sea salt and dissolve in a glass of water and drink down in one go. Then chase the saltwater drink with a glass or two of regular, unsalted water. Relief, as the urinary tract is flushed out by the salt, should come quickly.

For the **lemon-juice** remedy, squeeze juice from one half, or up to two whole, lemons into a glass and drink undiluted. Next, chase this mixture with a glass or two of fresh water. The remedy works on the same principle of the saltwater flush—to clear out the interference in the UT pathway.

Another amazing remedy—the one I have found most potent and also strong enough to ease and cure bladder and kidney infections, is the use of **Manzanita-leaf tea**. This knowledge is derived from Native American medical wisdom and is the best remedy I have come across to date. If you live in an area where Manzanita trees grow (much of the southwest and northwest parts of the US) then you can harvest leaves yourself. Alternatively look for sources online. (I harvest my own.) You need only a few leaves per cup of tea. Boil the leaves for several minutes, simmering up to twenty minutes first. Then drink a cup of the tea at least three times per day for three days in a row for best results.

I have cured extremely acute and painful UTIs with several of these methods over the years. In one case I was urinating blood and the lemon juice alongside a hot bath rid me of the pain completely and quite quickly.

For women prone to these events, it can help to address **vaginal yeast infections** (sometimes these are asymptomatic and go unnoticed by the sufferer—until it turns into a full-blown UTI or bladder infection) as a preventative measure. Using **yogurt** with live cultures or **coconut oil** in a

**suppository** or doing an **apple cider vinegar douche** (with the ACV diluted in water) can help restore vaginal flora and keep yeast overgrowth in check. Likewise, **tea tree oil** dabbed in one's underwear can also help. (Wearing loose-fitted natural fiber underpants can also help reduce the incidence of infection by providing adequate airflow.) Also, avoidance of soaps applied directly to the groin area is helpful, as exposures to soap ingredients (even natural ones) to this area can cause infections.

### *Bruising*

It is said that increasing **vitamin C** intake can reduce bruising, but vitamin C can also act as a blood thinner and easy bruising often happens due to problems with blood-clotting, so that idea seems counterintuitive to me and personally when I have tried increasing vitamin C to help curb my easy bruising, it has not seemed to help.

However, I *have* had success with **vitamin K** (also found naturally occurring in **leafy green vegetables** and **butter**) supplementation. Also, I tend to avoid high-impact sports as a preventative. Obviously, blood-thinning medications should be avoided for people who bruise easily.

**Arnica** is also a homeopathic supplement that is a potent remedy for healing bruises. When I worked in seal rescue centers (rescuing and treating injured and orphaned wild seal pups) arnica was so well prized in Scotland and Ireland it was used regularly and quite successfully for treating seals with serious bruising due to traumatic injury. Arnica can be taken in pill form or applied topically in commercial rubs.

### *Burns*

**Aloe Vera** is an amazing remedy for healing burned skin—even working in severe cases. One of our two cats, in his first year, before he was quite fully grown, when first learning about wood stoves and the fact that they can get very hot, jumped up onto the top of the stove while in use before we could stop him. He immediately jumped off again but the damage had been done—all of the pads on his paws had been burned off completely, exposing raw pink flesh underneath. I immediately treated his burns with fresh Aloe Vera and he, fortunately, made a very rapid recovery with no obvious scarring.

Using **butter** or other **saturated fats** on burns can help soothe the skin as well. (And do not let anyone tell you using butter is an "old wives' tale" because butter will "cook" skin, butter on its own cannot cook anything, it can only be used for cooking if applying a powerful heat source to it. Butter does not do the cooking, heat does. While burned skin may feel hot, it is not hot enough to cook.)

**Do NOT apply ice** directly to burns, especially second degree or worse (with blistering and missing skin). Ice may feel good at first but will worsen the pain after removal as the extreme cold can further damage delicate burned skin. You can, instead, apply strips of wet cloth or gauze first dipped into cool or icy water directly onto the skin in order to experience relief in the first hours after burning.

**Honey**, applied to the burns (especially after cooling with wet cloth) can also help reduce pain and speed healing. Raw unprocessed honey is best.

### *Cancer*

The American Medical Association has a long track record of suppressing natural cancer cures and discrediting, vilifying, censoring, threatening, removing licenses from, and personally and financially ruining any medical or non-medical person advocating, selling or freely offering non-invasive, non-poisonous cancer therapies. So again, I must state that I am an "ignorant nobody" simply listing the following for educational and/or entertainment purposes and I have absolutely no affiliation, financial or otherwise, with anyone offering natural cancer treatments.

With that disclaimer out of the way, I can say that I have heard of successful treatment and cures for a variety of cancers using **ozone therapy**. The ozone needs to be applied via insufflation (in ears, rectum or vagina) and by way of a cold corona (not hot) medical-grade ozone machine[lxix] set to very low output levels and using oxygen tanks connected to the ozone machine.

---

[lxix] Synergyozone.com is a good resource. Sold for non-medical (water purifying) purposes only, of course.

**Chaparral** herbal preparations and **CBD oil** (from cannabis), drastic **dietary changes** (namely cutting out all sugars and sometimes also increasing saturated fats) are other natural remedies rumored to send cancers into remission.

Some have also reported success using the **Gerson method** (involving coffee enemas and carrot juicing among other things).

You can research any of the above online to find out more. Do not forget to also reduce/eliminate EMF exposures as these (especially from cellphones and WiFi, etc.) are definitively linked to the manifestation of many cancers.

### *Cataracts*

Several years ago, I read on a health forum (Curezone.com) a man's personal account of how he had at age 35 years old developed cataracts after six months working for a telecom company installing and working on cellphone towers. He was convinced his cataracts were the direct result of exposure to the intense levels of microwaves he received in his new line of work.

This is not surprising and can help explain the rising levels of cataracts and other previously rare eye conditions alongside the ubiquitous increase in exposures to microwaves. Microwave radiation exposure standards were set based only on thermal heating effects and only on these effects upon the brain from a 2nd generation (2G) cellphone held nearly an inch away from the plastic mannequin head[lxx] and for only six minutes duration.

Heating effects upon the eyes or thyroid (a gland situated closely beneath the skin) were not tested, nor were exposures from cell towers, multiple antennas, multiple cellphones and other wireless devices or wireless routers. Nor were *non-thermal* biological effects accounted for

---

[lxx] Referring to the cellphone SAR/S.A.M. tests done in 1993 mentioned in footnote 2, chapter two.

at all. And millimeter-wave radiation (currently being introduced to the world at an alarming rate for 5G systems) has not been tested for safety *at all*,[lxxi] and has thus far only been used in military weapons applications as a directed-energy heat or "pain ray"[lxxii] crowd control tool. But, we are told "not to worry" since the energy *only* reaches into the first layers of our epidermis. No consideration is made for the effect on the eyes (at the level of our skin) or thyroid or ears. And no consideration is given to the fact that our sweat ducts act as helical antennae to carry the radiation deeper into our bodies.[lxxiii]

With the un-abatement to wireless system rollouts and subsequent continued increase in exposures to 3G, 4G and 5G frequencies of radiation, increases in a myriad of illnesses and cancers, along with incidence of cataracts are sure to increase.

Also, on Curezone.com I read of a man's successful treatment of cataracts using **castor oil**. He placed drops of castor oil into his eyes nightly for eight days and they went away, after which he did not require surgery. Best is to avoid development of cataracts by avoiding microwave and millimeter-wave exposures as much as possible.

### *Cuts/Scrapes*

For deep cuts, get **EMT Gel wound care,** which is sold for livestock and domestic animals, but works wonders on humans too. It is a collagen gel and works like an invisible Band-Aid to close and protect wounds as they heal. Do not use other liquid-Band-Aid-type products, they do not

---

[lxxi] Wireless industry reps admitted that there had been zero funds allocated by the wireless industry on study health impacts/safety of 5G exposures when senator Blumenthal pointedly asked the question as to how much money the industry had committed to support independent research in a 2019 Senate Commerce Committee meeting. Blumenthal said, "So basically we are flying blind here on health and safety." https://mdsafetech.org/2019/02/13/no-research-on-5g-safety-senator-blumenthal-question-answered/

[lxxii] See ADS (Active Denial System) https://www.scienceabc.com/innovation/the-active-denial-system-test-what-is-it-what-does-it-do.html

[lxxiii] See my chapter on 5G (chapter 7) in my co-authored book, *Welcome to the Masquerade* for more details about this.

work as well or at all. I have seen cuts that normally would need stitches heal rapidly with this stuff. Truly amazing.

For scrapes just rinse or wash and let nature do the rest. Really no need to "disinfect" with alcohol or hydrogen peroxide. If you feel the need to do this, only do it once. Repeated use will slow healing, as bacteria are needed in the healing process. In fact, there have been studies showing that wounds *not* cleaned with soap/disinfectant vs. ones that have been, fair better and heal faster.

### *Constipation/IBS*

This can be a real problem for many people in today's world, especially as chronic use of antibiotics and poor diet serve to disrupt microbial intestinal balance. I have personally struggled with this and sometimes with the opposite problem. When someone fluctuates from constipation to diarrhea, this is called IBS, or Irritable Bowel Syndrome, by the medical establishment. Interestingly sometimes antidepressants are prescribed for dealing with IBS since we really do "feel with our gut," as there are as many neurotransmitters in the gut as in the brain. But I think that reestablishing gut flora is the better approach to both alleviating constipation or diarrhea and also balancing erratic moods.

After many years of limited success with many different products and dietary changes, I have finally discovered that a very particular type **yogurt** from a local dairy has reestablished regular bowel movements for me. This yogurt is from **A2** type dairy. I learned last year from a man living exclusively off of raw cow milk for nearly a year (and staying strong and healthy on the diet) that different ancestral backgrounds require different types of milk as our ancestors herded different kinds of cattle depending on geographical locations. He found that milk from type A2 Jersey cows worked best for him and I stumbled up on the same, especially benefitting from A2 dairy. Since we both have Irish ancestry this may be no coincidence (Irish herds are still predominantly A2).

But it also appears that A2 is healthier than A1 (A1 is currently the predominant type found in the US) for everyone. A book called *Devil in the Milk*, explores this difference in great detail for those interested in learning more. To sum up, A1 milk contains beta-casein that can generate the casomorphin peptide BCM-7, what is known as "the devil in the milk" and it is this molecule that is said to be responsible for playing a role in causing diabetes (via attachment to insulin-producing beta cells in the pancreas), and neurological disorders (by passing through the blood-brain barrier) and also to cause inflammation, digestive problems and so-called "auto-immune" disorders—while A2 milk does not. A2 was also allegedly the only type of milk humans consumed for thousands of years until a genetic mutation resulted in A1 dairy herds.

Prior to my own discovery of A2 dairy (from cows), even my own home-made yogurt from goats I milked myself did not have this profound effect on me.

Another helpful probiotic drink (if you can get access to the right type of dairy) is making **kefir** probiotic brews. You can also make **water kefir** without dairy, which requires a sugar, water and dried fruit mix as feed. (You can purchase both kefir and water-kefir "grains" and brewing instructions from sellers at Etsy.com.)

There are many herbs that can help relieve constipation, although many of them are very powerful laxatives and should be used with caution; this includes **Senna Leaf** and **Aloe Vera.** A milder safer laxative is **bay leaf tea**. I have only used fresh bay leaves so far and just one or two are needed per cup of tea. But if using dried herb you may need more.

Also, **catnip tea** can work as a mild laxative and diuretic, but beware, since it is also very calming it may cause drowsiness.

In really dire situations taking **castor oil** or **Epsom salts** internally or drinking **sea-salt water** (see salt water flush under UTI, just increase the dose in this case), or doing **water or coffee enemas** can help. But overuse of laxatives can create a dependency and also disrupt regular bowel movements.

For **diarrhea**, especially caused by eating bad foods or drinking contaminated water, ingesting **charcoal** (sold as "activated charcoal" in powder or tablets) works wonders. But also restoring electrolyte balance is important. Get an **electrolyte beverage** or make your own by adding a little **sea salt, honey** and/or **sugar** to it. Eating **bananas** to restore potassium is also a good idea.

### ***Diabetes/Hypoglycemia***

Diabetes and hypoglycemia are generally considered to be two sides of the same coin, as one can flip to the other (generally hypoglycemia can change to diabetes over time). The best solution is to **cut out ALL sources of sugar**, even fruit sugars until the condition is under control. Best is to stay clear of all refined sugars indefinitely. I suffered with severe hypoglycemia for some time until I cut out the sugar. I did this on a couple of occasions "cold turkey" and suffered three days of horrific withdrawal symptoms in the form of sweats, muscle aches and cravings manifesting primarily as dreams about eating chocolate cake. But after the three days, the craving subsided and my blood sugar normalized. Over time I slipped back off of the wagon and had to start the process all over again. These days I stick to fruits or unprocessed powdered stevia when I crave something sweet and sometimes a bit of raw, unheated honey. (Not all "raw" labeled honey is actually *unheated*. Unheated honey is intact with all of its nutritional value and enzymes and does not affect blood sugar the way heated honey does, since the heating process converts it into a simple sugar. Check the label to see if the honey was heated or not.)

Adding adequate **saturated fats and proteins** to the diet is also very important as well as staying clear of refined carbohydrates (white rice, white bread, etc.).

Many studies have shown that diabetic patients who opted to control blood sugar through diet over insulin injections kept their blood sugar/insulin levels in check more successfully and lived longer, healthier lives than their insulin-dependent counterparts.

Diabetics are also known to be low in the mineral **chromium**, so that supplementing in this mineral may help the condition. A natural form of chromium can be obtained by supplementing with **brewer's yeast**.

**Bitter Melon, Cinnamon, Rosemary, Jamun Extract, Clove, Turmeric, Bilberry, Sage, Garlic, Okra, Fenugreek,** and **Ginger** are all herbs indicated for stabilizing blood sugar levels. Research each one further for suggestions on dosage.

### *Earache*

Even very painful earaches can be cured by warming chopped **fresh garlic in a little olive oil** and using an **eyedropper** to drop the oil into the ear—only after it has cooled enough to not damage delicate tissues in the ear. Always test the oil temperature before putting it in the ear. It should be warm or room temperature, not hot. Cold liquids added to the ear canal can aggravate the condition, so make sure it is at least tepid.

You can also add **tea tree oil** to another **carrier oil** (like olive or sesame) and place a few drops of the mixture in the ear canal. Rest on your side so that the oil stays in the ear for at least a few minutes. Putting a cotton ball in the outer ear as a plug can help to hold the oil in the ear longer. Relief should come relatively quickly.

You can also buy commercially available **ear oil mixtures** that have either **garlic, tea tree and/or mullein herb** to cure earache.

### *Eczema/Skin Rashes*

Professor Olle Johansson discovered that time spent in front of TV and computer screens can cause skin irritation and rashes to the point he felt compelled to coin the term "skin dermatitis." So, first notice when your rash acts up, have you been on cellphones or using WiFi or in front of screens for an extended period?

If you have already cleaned up your EMF-act, and still experience rashes, first address dry skin issues by applying **natural moisturizing creams, lotions or oils**. Avoid any products with added fragrances. Even those labeled as "natural" fragrance may be chemically derived. Also avoid products with added preservatives, dyes or alcohol. Best is to use **coconut, olive or sesame oil.** These oils, which normally work quite well for me in average climates did not quite suffice in the desert. While in Arizona I found **raw unprocessed mango nut butter** to be the miracle cure I needed. Many also find **Shea nut butter** helpful, but for my sensitive skin it was too heavy and clogged pores.

For **itchy skin**, again avoid products containing any type of fragrance or other chemical additives and also eliminate these chemicals from your other personal-care products, especially from laundry detergents and other soaps. To alleviate the itch, add **Thuja (cedar) essential oil** to your preferred skin lotion or oil and apply. **Tea tree oil, or thyme oil,** used thusly can also help. (Thyme oil is especially helpful when applied to shampoos in treating dandruff.) **Aloe Vera gel** is another soothing topical that can relieve itchy skin (best is if you can get it fresh, directly from an aloe plant). **Raw honey**, also **mixed with ACV** diluted in water is excellent, if a bit sticky. Also if not in rash form, **dry skin brushing** can very effectively remove dead skin, if this is the cause of the itch.

Adding **probiotic foods** to the diet can also reduce and eliminate persistent skin rashes, especially in the case of eczema, which is really a fungal skin condition. (**Yogurts, kefir, and fermented vegetables** like sauerkraut work as probiotic food sources.) Soaking the skin in **Epsom salt warm water baths** can work wonders in resolving fungal skin issues.

***Eye Infections—Chalazions, Stys, and Pink Eye***

In 2006 one morning to my horror and surprise I woke to find one of my upper eyelids massively swollen and red, only to increase in size and irritation over the following couple of days until I was told by an eye doctor to place **warm-water compresses** (preferably **with salt or baking soda** in it or use warm **tea bags** as compress) over the eye for several minutes at a time and multiple times a day. I was also given an antibiotic ointment to place in the eye. It was a very viscous substance (petroleum based) and served to help the dryness but the medication itself caused me to have chronic eyelid infections for some years after (during which time I kept applying the ointment until I "got a clue" and realized that the medication was causing the problem to repeat itself).

The cause of infected eyelids (this includes stys) is due to eye dryness and blocked tear ducts along the eyelid. You can prevent them by keeping eyes moist and adding lubricating drops (**saline with aloe** or **coconut oil**, and in severe cases **castor oil** work wonders, but only apply oils at bedtime as they blur vision), and keeping hydrated.

If you get the infections, simply do the warm salt water (or baking soda, or tea) compresses (any rag will do, it helps to place another dry towel over the wet one you are using by dipping it into a dish of warm liquid, to retain the heat for longer and to keep the rag from making a mess) and you will open the clogged pores and release the trapped substances and hence clear the infection. There is no need for medicated, toxic, antibiotic creams or internal medications.

The infection is caused by clogged tear ducts, not by a bacterium, even though if analyzed many bacteria are sure to be present, but their presence on the scene does not equate to causation. Trying to kill off bacteria will only serve to harm and not help to heal the irritation. The gentle approach, while sometimes requiring more time and patience is always the best. Once I stopped using the eye "medicines" and stuck to the compresses and prevention methods, I finally stopped getting the infections.

For **pink eye** try **homeopathic drops** by Similasan. Symptoms of pink eye include redness, feeling of grittiness in the eye, thick discharge and/or burning sensation. You can have pink eye without the eye actually getting red or pink or burning or itching. If you have excessive eye watering and discharge, especially along with feeling like dirt is stuck in your eye, using the above-mentioned drops can bring about immediate and long-term relief. They also sell drops for dry-eye relief which if used at first signs of excessive watering (generally actually caused by dry eyes) can prevent development of pink eye.

### *Flu/colds*

*Flu* basically did not exist prior to the introduction of artificial EMFs into our environment. Since I have escaped to low EMF havens, and drastically reduced my own exposures and use of electronics, I have not once suffered with the "flu," and neither has my partner. This is a seven-year period I am referencing. Before I understood how EMFs were linked to poor health, I got "flu" just like everyone else, at least once a year and most often in winter. The reason flu is more common in winter (or was) is because people tend to increase their artificial EMF exposures and decrease their exposures to natural EMFs during this time, since they tend to spend more time indoors in winter and get a lot less sunlight and contact with the earth and natural settings. Blue-light exposures are increased significantly in the winter too and this serves to drastically impair circadian rhythms and lead to increases in ill health. But because people are now spending less time out of doors and more time indoors and on screens year-round, incidences of flu are increasingly showing up during all seasons.

Another contributing factor in winter is due to poor ventilation along with dry indoor environments caused by central heating systems that irritate and inflame sinuses leading to sinus and respiratory complaints. And being inside more also exposes most people to increased amounts of toxic chemicals, especially as people mistakenly attempt to "sanitize" homes and rid them of "germs" by use of dangerous chemical agents. If our attempts at sterilization of our environments actually succeeded, we, along with all other life forms, would cease to exist.

The first thing to do in order to recover from flu and colds is to take a break from screens and wireless radiation and get plenty of rest.

For severe **sinus congestion** that interferes with rest, **boil water** in a pot and **add a few drops of eucalyptus and/or mint essential oils** and **inhale the steam** (better if done with a towel over your head, in which case be sure the stove is turned off and your towel does not catch fire). Also chewing on **fresh bay leaves** will rapidly open sinuses. Using a **Neti Pot** to **irrigate the sinuses with warm salt water** is also helpful. (But the salts can be drying so **add aloe or mango butter or coconut oil** to your nasal cavities afterward to avoid causing nose bleeds later—this holds true for eucalyptus inhalation as well).

Do not take substances to reduce fever. **Fever** is a sign that the body is trying to raise its temperature to induce sweating in order to eliminate toxins. Better to **encourage sweating**. This will naturally reduce fever. While living in small village in Turkey, I was invited to dinner at a neighbor's home at which the neighbor's son-in-law, a medical doctor, was visiting. The doctor was sick when I arrived, sitting on the sofa complaining of flu and fever symptoms. He was in the midst of getting help from his wife by having her bring him as many blankets as possible and instructing her to make a fire in the woodstove. He dressed himself in as many layers of clothing as possible including a winter coat, hat, and scarf, climbed inside a sleeping bag, put several blankets on top of that, and laid down on the sofa near the fire. Not long afterward, he broke an impressive sweat and quickly got up to take a shower to rinse the sweat off. When he rejoined us after having showered and changed, he was a new man, totally cured.

This scene stuck with me so that years later when I felt the same, I tried this approach with great success.

Taking **sulfur** supplementations can also ease flu symptoms.

### *Headache and Muscle Ache*

(See Solution Number 13, *Reduce Inflammation)*

### *Heart Burn/Acid Reflux*

**Avoid eating before lying down,** especially oily food (i.e. sardines). Ingestion of too much ginger can also lead to acid reflux/heartburn. And overall reduction of stomach acids (*not* increases in them as is commonly erroneously believed) can also lead to this condition. **Increasing stomach acids** by adding **apple cider vinegar** (ACV) in small amounts (teaspoon or less) to drinking water regularly as well as adding **nutritional yeasts** to food can help to increase stomach acid, which tends to decrease with age.

**Pressing firmly on acupressure points** along the middle of the ribcage can help "drop" the stomach back into position as a "high" stomach (often resultant of stress) can lead to indigestion, especially manifesting as acid reflux/heartburn. Standing while eating can help in this case as well.

### *Hemorrhoids*

The discomfort from and sometimes very acute pain caused by hemorrhoids can be very quickly alleviated by way of **garlic clove suppository**. Simply peal a clove of garlic and insert (with or without lubrication with coconut or olive oil). The relief is usually nearly instant and the garlic will come out on its own in due course. Alternately purchase **natural hemorrhoid creams/suppositories with witch hazel** added. To prevent hemorrhoids, avoid improper heavy lifting. (see "back pain")

### *Incontinence*

This can be an embarrassing problem, of course, and it happens especially to women as they age and lose control of certain muscles. Men can experience the same problem due to prostate issues.

Fortunately, when I started to have some small issues with this after age 40, I found a simple, yet very effective, fix. It may sound strange but it works. Soaking your "nether regions" in **cold water** for brief periods a few times per day until the problem resolves works like a charm. Doing this is called taking a "**sitz bath**," and there are products on the market that serve the purpose of providing containers (sometimes that can be fitted onto toilets) for the baths. But you can also just buy any small shallow plastic tub and fill that with cold water. Or just splashing cold water in lieu of using a tub can work too.

This remedy is for moderate to mild incontinence. I do not have a suggestion for severe cases, except to say that I do know there is a relationship between dirty electricity and incontinence and that certain pharmaceutical drugs can cause this problem.

### *Indigestion*

**ACV (apple cider vinegar)** added to water can aid digestion. (As mentioned under *Heartburn* above.) Contrary to what we are told by mainstream medicine, it is the *decrease* in stomach acid (which happens naturally with age), not the increase that causes indigestion. Drinking highly acidic apple cider vinegar or **lemon juice** can help to increase stomach acid. So too, can taking **HCl (hydrochloric acid)** supplements, but I have personally found these to be too strong.

Adding **nutritional yeasts** to food also has the effect of increasing stomach acid and aiding digestion.

**Muculent herbs** (that produce a mucous-like substance) like **Slippery Elm Bark, Licorice Root, Cinnamon and Marshmallow Root**, can help by soothing the digestive tract.

Also taking **bitters** (bitter herbs) along with meals aids digestion. This used to be a common practice in our societies until pharmaceutical companies and the AMA gained such a foothold in our world. Bitters were sold as liquid tonics and some health food stores still carry such products. However, most of them now include sugars or other sweeteners, which totally counters the action of the bitter herbs. The point is to stimulate bile flow by tasting the bitters. All one really need do is buy one bitter herb and put it on the tongue (either in powder or tincture form) before or after eating in order to stimulate this flow and aid digestion. The best bitter herbs I have discovered for this purpose are **gentian** and **andrographis**.

### *Intestinal Spasms*

I can safely say that the most severe pain I have ever experienced has been from intestinal cramping. The first time it happened I was taken out of my hotel room on a stretcher by paramedics in the middle of the night and rushed to the emergency room, where instead of being helped, I was made to wait for at least an hour (even though the ER appeared completely empty of any other patients) and sign a lot of papers while in agony and while my teeth were chattering and I was shivering uncontrollably. After begging for multiple blankets (although it was summertime, the spasms had sent me into shock, or so I surmise), and finally regaining my circulation, the pain subsided long before a doctor finally attended me. The doctor simply chalked up my experience to food poisoning and gave me an IV with Benadryl to hydrate me and I am assuming deal with the (already long past) allergic reaction I had experienced (hives on my palms and stomach) along with the cramping and violent diarrhea.

That experience helped me endure the second episode a few years later by teaching me that making myself as warm as possible seemed to end the cramping and that going to a hospital was a pointless and costly waste of time (that first trip had cost me over $600).

So, I learned that **a very hot water bath** (as hot as tolerable) was key to ending the ordeal quickly. I also learned after the second, third and fourth times (from deductive reasoning) what could be triggering the spasms. In my case, dehydration coupled with ingestion of rough foods (nuts and carrots to be specific) and/or too much dietary oil at once (vegetable and nut oils) definitely preceded the events. After I figured out the triggers I was mostly (it happened a couple more times) able to avoid setting off the spasms again.

When I speak of spasms, I am talking about very sudden and acute horrifically painful intestinal contractions.

Getting into a hot bath really was a lifesaver for me (as the pain truly was unbearable) on a few occasions. However once when this happened, while on a camping trip, I did not have access to a bath but discovered that some knee-high **compression socks** I happened to have with me, seemed to do the trick of easing the pain and ending the episode. This was probably due to the circulatory help the socks provided.

My brother is prone to the same problem and had (independently) also discovered hot baths as *the* remedy. However, the last time he had an attack, he fainted from the pain while getting into the bath, slipped, fell and broke his neck. Luckily his neck mended and he is still with us today. But I add the story as a warning to be sure to have help getting into the bath if you choose this as your remedy.

### *Kidney Stones*

My recommendation of drinking **ACV (apple cider vinegar)** added to water (usually one to two teaspoons per cup) saved two friends from having to undergo surgery for removal of kidney stones. The drink worked to break down the stones by making them small enough to pass, unassisted, through the urinary tract.

**Lemon juice** works similarly, but I cannot attest to results from use. Drinking ACV water or lemon water regularly can also prevent stones. Taking **magnesium malate** can also break down the stones.

### *Menstrual Cramps—When "Aunt Flo" Turns Ugly*

In addition to chronic UTIs, I suffered many years with intense debilitating menstrual cramps, to such an extent I became highly dependent on every-increasing doses of ibuprofen and other similar pain-relieving drugs—until eventually they no longer worked at all and I developed an allergic reaction to the drugs. Left having to find my own solutions after modern medicine had failed me (and I am quite certain caused the menstrual irregularities which led to the painful cycles to begin with), I ultimately found several herbs that worked to effectively reduce, and eventually totally alleviate, the terrifying pain—pain so awful several times I doubted I could survive it and it made me want to end my life in those moments when the pain would not cease (at least with the intestinal cramping the pain never lasted as long).

**Motherwort** herb was the first herb that really worked to bring about a change in this harrowing ongoing experience. I took this herb as a tincture and at first in high doses but as it works to balance hormones after regular prolonged use, I found I needed lesser and lesser doses over time. It can (and should) be taken throughout the month to bring about this hormonal balance, but can also be taken to ease acute pain.

**Kava Kava** herb I found to be a bit stronger than **Motherwort** during acute pain crises. I took this in powder form, sometimes in a tea and sometimes directly under my tongue (placing it under the tongue helps avoid choking on the powder). Pills filled with *kava kava* also helped.

Before finding these two miracle herbs I did use **cannabis** successfully for a time. However, I found it worked mainly by moving my point of focus away from the pain elsewhere rather than bringing about physiological changes to help address the imbalance or root cause of the pain, and thus, similar to the pharmaceutical drugs, became a crutch rather than a cure, and also left me with some unpleasant side effects like cloudy headedness, short-term memory loss, and excessive thirst. Of course, smoking the herb also aggravated my already compromised lung condition but ingesting it could be tricky for dosing and a couple of times led to frightening overdose experiences during which I became quite sick.

Fortunately, one huge benefit of having drastically reduced my EMF exposures has been the lessening and eventual disappearance of the terrible cramps. However, I sometimes still experience excess fatigue or muscle aches during "that time of the month," and have found relief by drinking **bay leaf tea** (from fresh locally grown leaves, only one or two leaves per cup is needed) and also from drinking local **raw goat milk**.

Women experiencing **delayed menstruation** can bring on their cycles by drinking a cup of parsley tea made from boiling a bunch of **fresh parsley** in water (only needs a minute or two of boiling). Generally, only one or two cups are needed. (Obviously if one is pregnant and wishes to avoid miscarriage do not drink this tea! Even eating a lot of parsley regularly can have this effect, so should be avoided during pregnancy. And catnip tea can be used to the same end but requires higher dosing and also should be avoided during pregnancy.)

Also sleeping in fully dark rooms can regulate cycles where light pollution is prevalent. If living away from artificial city light interference, making sure that all lights and electronics are turned off at night and sleeping near a curtain-less window allowing the regular phases of moonlight to enter also will go a long way toward regulating menstrual cycles.

### *Menopause/Peri-menopause*

Many of the microwaved-injured masses seeking help, for their debilitating symptoms from allopathic doctors (who tend not to be educated about microwave illness), are told (if they are women over the age of 40) that their symptoms are simply age-related and signify "that time of life." Ruth related in her story how her horrible pains and insomnia were chalked up by doctors and others to being caused by menopause. She found this to be untrue in her case, as have many others with EHS/microwave illness been similarly misdiagnosed.

And as younger populations are becoming increasingly sick from EMR exposures they are now given the newer label of "peri-menopause," which is supposed to mean "pre" menopause—a label that is being given to younger and younger women so that the description has really lost all meaning, particularly when a woman meant to be in the prime of her child-bearing years in her late 20s or early 30s is given this diagnosis.

A friend of my same age group (late forties, early fifties) who like me has neither gone through menopause, nor peri-menopause, swears by melatonin supplements and stands behind the idea that menopause is actually an unnatural occurrence, since no other mammal experiences this. My friend feels that menopause is actually a disease of civilization and, if it occurs at all, should not happen until very old age.

Personally, I think it is happening too early and because of extreme hormonal disruptions from the incredible number of estrogen-mimickers to which we are all now exposed thanks to the thousands of synthetic chemicals found in our personal-care and cleaning products and dumped into our environments thanks to massive use of chemical fertilizers, pesticides and herbicides, so that our food and water sources are contaminated as well.

Like menstrual dysfunction and pain, painful and/or early menopause is a symptom of unnatural hormonal imbalances. Since disrupted circadian rhythms from ubiquitous EMF and blue-light exposures cause plummeting melatonin, low melatonin is most definitely implicated as part of the problem. I would not, however, recommend synthetic melatonin supplements, as it is too easy to overdose and also cause a dependency.

But drinking unsweetened **tart cherry** (containing naturally occurring melatonin) is a good substitute. Better still is to **drastically reduce exposures to EMFs** (especially wireless radiation sources) and blue light by changing LED and fluorescent lighting to tungsten and wearing blue-light blocking glasses and blue-light filtration programs when exposed to screens.

**Black Cohosh herb** and **Wild Yam Root** are potentially helpful herbs in reducing painful menopause symptoms but as these affect hormone balance, it is best to use under the guidance of a qualified alternative-health practitioner.

Also **avoid flu shots and all vaccines**, especially the latest "Covid-19" injections, as they have been directly linked to negatively impacting blood platelets[267] and fertility. Reports of older women who had already gone through menopause[268] but began menstruating again, and young women who were menstruating but then stopped, are becoming commonplace in the vaccinated and in some cases in those who have come into close contact with the vaccinated.[lxxiv] Whether this is due to the graphene oxide content (and interaction with EMFs) or due to the creation of the synthetic spike protein is unclear.

What is clear is that women's reproductive systems are being adversely impacted by these drugs, and in some cases (such as with HPV vaccines and the Covid-19 vaccine manufactured by Johnson and Johnson called *Janssen*), they contain the substance Tween-80, a known sterilizing agent.

---

[lxxiv] This phenomenon is called "shedding" and was included as a warning in Pfizer's own documents about their Covid-19 drug in study Intervention number: PF-07302048, titled, "A Phase 1/2/3, Placebo-Controlled, Randomized, Observer-Blind, Dose-Finding Study to Evaluate the Safety, Tolerability, Immunogenicity, and Efficacy of SARS-COV-2 RNA Vaccine Candidates Against COVID-19 in Healthy Individuals." The document details what they call "occupational exposure" hazards resulting in AE's or SAE's (Adverse Effects or Serious Adverse Effects), which if noted by "investigators" participating in vaccine trials should be reported to the company within 24 hours.

### *Nausea*

**Ginger tea** is very effective at reducing symptoms of nausea, but be aware that taking excess amounts or ginger or ingesting it too often can lead to reduced stomach acid and thus indigestion or heartburn.

Since nausea is often caused by dehydration drinking a **high-electrolyte content liquid** can help ease the nausea. But if the cause is food poisoning, then taking **charcoal tablets** or powder can help along with electrolyte drinks.

And of course, if you are able to vomit this can help too, but be sure to replace lost electrolytes afterward.

### *Nosebleed*

The most important thing in stopping (especially chronic) nosebleeds is prevention. Most nosebleeds happen after nasal passages become too dry during drier times of the year or after moving from a humid to an arid climate. Delicate capillaries inside the nose then break and result in, sometimes severe, bleeding. I have been prone to nosebleeds much of my life but they increased in severity in later years as my ability to form clots decreased (possibly from EMF exposures) and while living in more arid climates. The best preventative measure I have found has been to **apply mango butter or coconut oil to** my **nasal cavities** on a regular basis, particularly during drier seasons. Also ensuing adequate levels of proteins and fats in the diet while reducing sugar and carbs seems to help.

Greater risk of (uncontrollable) nosebleeds (or other hemorrhaging) occurs if one is taking blood-thinning medication or taking supplements or herbs that have the same effect. If taking fish oil, for example, it is best to take it along with butter or other sources of **vitamin K**, as K vitamin has blood-clotting properties. Dietary sources of K in addition to butter include green leafy vegetables, red meats, and eggs.

Also many reports are surfacing of children getting nosebleeds from exposures to powerful WiFi systems at schools, such as happened in Jenny Fry's case mentioned in chapter nine. So reducing exposure to

WiFi, and other sources of wireless radiation can prevent and solve nosebleeding issues in these cases.

If you still get a nosebleed after trying prevention strategies (or not) be sure to **pinch the end of your nose** with steady but not too forceful pressure and **lean your head *forward***, NOT backward. For some strange reason when characters in movies get nosebleeds, they always lean their heads backward. This is a dangerous practice, as you can choke on the blood going down your throat. What you need to do is form a clot and this is why you want to lean forward and hold your nose. It helps to also **insert toilet tissue into your nose** in a wad and you may even want to **tape the outside of your nose with cloth medical tape** instead of holding it together with your hand. This is what EMTs and paramedics do when treating patients with nosebleeds (unfortunately I know from experience).

It may be tempting to check the status of your bleed by taking out the stuffing but this can easily disturb the blood clot too early and start the bleed anew. My worst nosebleeds have lasted 3 hours, probably because I did check after the first hour and restarted the bleed by disturbing the clot. Generally, you can feel if blood is still flowing. Be sure you are somewhere you can rest and stay seated. Do not get up and move about. Especially do not go from sitting to standing and sitting again or crouching and bending. Taking vitamin K during the nosebleed may help to end it sooner. When removing the tissue from your nose afterward, do so with great care. It may help to **apply steam** either from taking a hot shower or boiling water in a pot and leaning over it. The moisture will reduce the chances of the tissue pulling at nasal passage membranes.

### *Pet complaints/Home remedies for pets*

I did not want to leave out our beloved animal companions here. They too can benefit from avoiding pharmaceuticals (as can your bank balance since veterinary bills now rival those of regular doctors). We adopted our two cats as one-month-old kittens who had been weaned too early and as such were very poorly when we brought them home, and I am convinced would not have survived if they had been treated conventionally. As it stands today, at six years old, they are thriving and in excellent health. I have learned a lot of helpful tricks in treating them with natural home remedies and also caring for many other animals (chickens, goats, cats, dogs and even wild seal pups) over the years. Here is a sampling:

### *Ear Mites/Mange—*

Ear mites cause mange so treating ear mites will treat mange. (Most healthy cats or dogs will not get mange even if they have ear mites, but starving animals who are infested with mites do get mange easily and often die because of it. So, making sure animals are well fed is a crucial part of treating mange.) Apply **coconut oil** with a cotton ball/swab inside and outside ears. This will suffocate the mites and clear the infection. Apply when signs are present; like your pet excessively scratching ears and shaking its head and when the inside of ears look especially dirty. **ACV (apple cider vinegar)** diluted in water half/half ratio) applied around the ears and head can also help, especially with alleviating the itching and reducing scratching by pets.

### *Fleas and Ticks—*

Use diluted **apple cider vinegar/ACV** (ratio approx. half vinegar and half water) rinse on fur. Apply with a saturated cloth, focusing on the top of head, ears and back of neck and over the tail and under the legs. Also add a splash (lid full) of ACV to the cat or dog's drinking water. Do both of these things as needed. It works! Even for the worst flea infestations. Our cats have not had ticks (and we live in the woods with plenty of ticks) since treating them for fleas with the vinegar so I expect it repels ticks as well. I have also used **herbal flea/tick collars** with some success but cats in particular do not do well with essential oils and the collars are less effective than the vinegar (which is good for them). Using **flea combs** daily while working to eliminate the problem is also helpful. For fleas in home carpeting try **enzyme powders** designed for breaking down flea exodermises and eggs. Alternatively use ACV for cleaning carpets.

### *Intestinal Worms—*

Worms are primarily a problem with young animals in their first year or two. Use **food-grade diatomaceous earth** added to food in small amounts (sprinkled on food is good) until symptoms clear. Also adding **activated charcoal** supplements to food will help clear diarrhea (for any reason) very quickly. For pinworms I have successfully used **homeopathic remedies** for treating myself and my cats (it is easy for pet owners to get pinworms from pets).

### *Ringworm—*

Same treatment as for humans, use undiluted **thyme oil** applied directly to the rash. Cover with a shirt that fits them (we made one for our infected kitten with a sock when he was tiny) or other wrap to keep them from licking it off.

### *Urinary Infections—*

Add **ACV** to drinking water on a regular basis (only a small amount, approx. half a teaspoon) to both clear up and prevent infections. Symptoms of an infection usually demonstrate as incontinence (peeing outside the litter box/indoors) or pets peeing more frequently than usual.

*Avoid health complications but avoiding ALL vaccines and other pharmaceuticals. And yes, this includes the rabies vaccine. This drug has never been proven effective at preventing rabies in spite of all the hype to the contrary and rabies has never been proved to be contagious and caused by a virus (as the virus alleged responsible has never been discovered or isolated). Rabies, as in the case of many other diseases blamed on contagious microorganisms, is a disease of starvation and neglect. Well-fed and well looked after animals will NOT get rabies. And humans cannot get rabies from animal bites and neither can animals get rabies from bites of other animals. Period.

Also feed your animals high-quality pet foods, avoid anything with artificial coloring, flavoring or other added chemicals. And avoid formulas with corn fillers, especially for cats, who need a low-carb diet. Most commercial animal feed is too low in fats, so giving your animals healthy fats like **coconut oil, butter, raw milk, raw eggs and/or raw liver** is extremely beneficial and your pets will love it! You can also find more natural home remedies for pets at Earthclinic.com.

(Now back to remedies for humans...)

## *Poison Oak/Ivy/Sumac*

A homeopathic remedy called **Scratch-N-Itch** (formerly called **Rusto Lotion** though not actually a lotion but liquid and manufactured by WHP) topical spray provides amazing and fast relief to the nastiest cases of poison oak, ivy, or sumac. In lieu of this try applying a poultice of **clay** and/or herbal **toothpaste** (especially one with mint oil added).

### *Ringworm*

**Thyme oil** applied undiluted to the rash is an incredibly potent remedy for even the most stubborn ringworm rash. Ringworm is a very itchy skin rash that typically demonstrates in a circular red pattern—usually on legs or arms, but can affect any area. In spite of its misleading name, ringworm is not a parasite or worm, but a fungus, commonly found in soils and sometimes in pet's (cat, dog or even livestock animals) fur. Thyme oil helps because it is a potent anti-fungal remedy. It may sting a bit when applied but relief is considerable and fast. (By the best brand you can find, it does not pay to get the cheap stuff as it may not be potent enough. If it does not sting a little when applied it is probably not the best quality.) **Honey** applied to the skin may also help.

Exposure to the fungus that causes ringworm is common but susceptibility to it is related to weakness in the body, generally to due with low thyroid function. Adding a little kelp to the diet or doing other things to help your thyroid (like shielding the thyroid from blue light and microwave radiation) will also help clear up and avoid incidence of ringworm skin infections.

### *Sore Throat*

To relieve sore throat, gargle with **warm salt water**. We did this as children, it is what the school nurse would have students do if they came to her with this complaint and it is what our mother had us do at home—it worked then, and it still works now.

Also, I have found an **herbal throat spray by Zand** to be incredible for providing fast relief. It contains honey, tea tree oil, licorice, sage, Echinacea and some other herbs. **Sage or clove oil** placed into the mouth in drops can also help.

But remember a sore throat, as annoying as it can be, is simply an irritation, it does not mean you have caught a deadly virus or bacteria and are doomed. Often the cause is ongoing postnasal drip from allergies, or irritation from an environmental pollutant or pollen, or from having choked on food or talked too much or too loudly. Be patient, it will resolve itself with time and addressing underlying issues like postnasal drip will speed the healing process and help avoid reoccurrences.

### *Toenail Fungus*

This is a symptom of a deeper issue and may well be linked to EMF exposures but can be managed through application of natural antifungal medicines. The entire twelve years I lived in Washington, DC (in high EMFs) I had a very severe case of toenail fungus on my big toes—one of which could not grow back entirely after falling off completely due to the tenacious fungus. After moving out to low-EMF locations the fungus cleared up and the nail grew back. I no longer have issues with toe nail fungus. While I did, I used **thyme essential oil** applied undiluted to the nail on a regular basis. Make sure to get a high-quality brand as the effectiveness will be compromised if buying a cheap oil. I have also heard that **apple cider vinegar** applied daily will clear the fungus, but did not personally experiment with this for long as I found the thyme oil to work quickly.

### *Toothache*

Aching teeth do not always indicate cavities and often result from misaligned bites. One way to manage bite problems is to purchase a **dental guard** either through a dentist or from reputable sellers. I have found a company, Armor Guard[269], that sells high quality, non-toxic custom-fitted guards for only $40 (compared to $300 and upwards when purchased through a dentist, with no sacrifice in quality).

Wearing the guard nightly on either upper or lower teeth can reduce tooth pain. Also **massaging the TMJ** (temporomandibular joint in the jaw) before bed at night can make a huge difference in alleviating tooth pain caused by clenching and grinding.

Otherwise chewing on **Prickly Ash bark**, used by Native Americans for toothache is a great way to immediately reduce pain (providing lasting effects afterward).

**Clove or sage essential oil** applied directly to the tooth creates a numbing effect. Clove is very potent but also can create a spicy burning sensation in the gums, and especially on the lips, so use with caution. Sage works without the burning and so can safely be used for teething toddlers.

**Kava Kava** herb powder applied directly the tooth can bring amazing and immediate numbing results as well.

If you do have a cavity, know that these can heal without going through the "drill and fill" dental routine. The problem with this latter approach is that good tooth enamel is removed in the process and eventually the continuation of repeatedly filling cavities will lead to loss of the entire tooth.

Many have reported success in healing teeth with **dietary changes**—namely adding **high quality fish oil and/or raw butter and raw milk** seem to help restore teeth to health. I have at times had success with high quality **calcium supplements** as well, but still prefer food sources.

**Prevention** is the best method. At age eight I went to the dentist with a mouth full of pristine white unfilled teeth only to be told by the dentist that I needed a total of eight fillings. I was given them all at once—all mercury metal amalgams. Soon after I was admitted to the ER for the first time in my life due to severe abdominal cramping. I no longer think this is a coincidence and feel it is definitely related to the poisoning my body underwent from the mercury. Years later, I received 3 more fillings and eventually replaced all 11 with porcelain. That was in 1998. I have not had any new cavities since. I attribute the significant reduction of simple sugars in my diet and never touching fluoride again to my success.

Fluoride is an industrial waste byproduct of aluminum manufacturing, can be deadly if swallowed, and is used as a rat poison, plus leads to brain damage and docility in humans. It does NOT protect against tooth decay, but instead results in bone loss, and so also, yes, tooth decay.

**Oil pulling**[lxxv] is another helpful method, especially for healing gums. You simply take a tablespoon of either; olive, sesame, or coconut oil in your

---

[lxxv] Note that oil pulling also reduces plaque buildup, but in some cases plaque may serve to hold teeth in place as has been discovered with ancient Egyptian mummies, so

mouth and swish it around for 10-20 minutes daily. After swishing, spit it out and rinse with water. Oil pulling is a traditional remedy from India and the oil is said to pull the toxins out of the mouth (and actually out of the whole body through the mouth).

**Avoid root canal (and other dental) surgery**. No good can come of leaving decayed dead teeth in the mouth. Sadly in 1991-93 my two front middle teeth died after botched orthodontics a few years earlier, and underwent multiple root canal surgeries, as over time each would fail and ultimately, I experienced pain in the root areas of those teeth from unresolved infections to the point at which I experienced so much bone loss that my teeth became loose and bled.

This latter incidence happened in 2018, so I ended up consulting with seven dentists in Arizona and one in Mexico, in order to come up with a solution. The teeth had to come out, that was certain; they were coming out on their own already (but as mentioned in my foot note, this may have also been precipitated by oil pulling, although I cannot be one hundred percent certain).

However, replacing them was particularly tricky with my chemical "sensitivities." I paid for expensive blood work to help pinpoint what materials I might tolerate but found the test results totally irrelevant as I still reacted to all alleged "safe for me" materials anyway. I also was pushed into getting a bone graft upon extraction so that I might replace the teeth with implants. This "advice" was sold me by a supposed "holistic" practitioner.

I was told I would be given a natural mineral-based implant but on the day of surgery when my mouth was clamped open and I was unable to protest, the dentist decided last minute to insert a synthetic product into my jaw instead. Not surprisingly, my body fully rejected the implant and I had to get the invasive gum-cutting surgery done all over again by another dentist (as the original one refused to admit to any problem with the graft) to remove the fine sand-like grains from my gums. Given the

---

perhaps some plaque is actually needed, especially as we age. Just something to consider, especially as I actually had one of my front teeth come loose after months of oil pulling. It was a discolored root canal tooth, which I still think was best removed, but the oil pulling seems to have precipitated the need for an emergency extraction.

rejection and long recovery time from the surgeries, implants were no longer recommended for me by any dentist. (One dentist I met will not do them at all, since in her experience, she has only witnessed resultant health problems and even some deaths. No matter what the implanted material, it is a foreign object inserted into the body and so there is always a risk of rejection and implants also require very invasive surgeries and often do not last much more than a decade.)

The only other options left for my missing front teeth were to get a dental bridge or partial denture. The bridge in my case would require shaving down the enamel of four other teeth to stubs and placing crowns on the stubs. This was too risky for me as I did not know how I would react to any of the crown material options and also shaving down teeth, like drilling out tooth decay, weakens teeth and can result in tooth death, requiring even further removal of teeth. Potentially losing four more teeth was not a chance I could take. But the dental partial materials (I tried four different ones) all caused adverse reactions in my gums.

Ultimately, I **made my own denture** from plastic nylon beads sold as moldable denture adhesive material.[lxxvi] Over time, after a lot of trial and error I was able to heat the beads in boiling water and mold them into almost natural looking teeth that fit more comfortably in my mouth than any of the professionally made dentures I had fashioned for me by the "experts."

The homemade denture is also much thinner and less bulky so that speech with it is a lot easier as well. The only drawback is that it is not strong enough to use while eating and so must be removed at these times. Also, you need to stain the very white material to a more realistic tooth color by soaking it in a cup of regular tea for a few hours. And they need to be remade every year or so as they do not last as long as conventional counterpart. But the cost is only about $12. I have

---

[lxxvi] The product I use is called *Instant Smile* thermal forming impression beads by Billy-Bob (brand). I tried other similar plastic beads but only these have worked for me so far. https://www.amazon.com/Impression-Material-Included-Billy-In-stant/dp/B01LZVL5A5/ref=sr_1_13?crid=LTIAPRKMMMVU&keywords=insta+smile&qid=1643137253&sprefix=insta+smil%2Caps%2C235&sr=8-13

recommended this approach to several friends—all of whom have been happy with the results.

Another thing the (in my opinion) unscrupulous dentist (Margolis in Phoenix) sold me on was jawbone cavitation surgery. Some members of my EMF Refugees forum had discussed the possibility of clearing up covert jawbone decay through surgery as a means of healing from EHS and MCS. Margolis subjected my mouth (and so also my whole head) to ultrasound radiation in order to determine the presence of these alleged "cavitations." Several areas of my jaw lighted with red on the scans, indicating a potential problem.

But another doctor I consulted after the fact (for the bone-graft-removal surgery) said that reading the scan that way was a matter of interpretation. Red simply pointed to an inhibition of blood flow to those areas, which could indicate a bite problem rather than decay. In fact, it appeared it was impossible to know for sure if there was any real decay at all. But getting a patient to believe this to be true translates into a big moneymaking business for some dentists offering jawbone-scraping services. I personally paid nearly $4,000 for my two surgeries, done alongside the extraction and bone grafts.

I was scheduled to return for four more such surgeries in other parts of my jaw, behind my wisdom teeth. But my experience with this very shady "medical professional" (whom I learned too late had lost his malpractice insurance due to multiple lawsuits against him) and the very long time it took me to recover from the first surgeries (nearly two years) plus learning that jaw cavitations may not be a real threat after all, prompted me to cancel the other appointments. I also tried unsuccessfully to get my money back by way of filing a complaint with the dental board (a complaint that included so much evidence of fraud that it was nearly 100 pages long). My attempt probably failed because Margolis had once been a member on the board and because one more strike against him would have meant loss of his license—something his friends on the board were not willing to allow happen.[lxxvii]

---

[lxxvii] I had emails and even film recordings proving this dentist lied to me and that he falsified my records but I was simply dismissed by the dental board with the excuse of "not enough evidence" without explanation as to what evidence could possibly be missing. I also found out after the fact by requesting his records from the dental board that he had lost his malpractice insurance years

I also learned, too late, that there were other much less-invasive ways to treat jawbone decay if one really wanted to try that. I heard of another doctor in Phoenix who was using **ozone injections** (not requiring surgery) into the jaw as an alternative treatment.

My recommendation however would be to leave it be, or at least only do the ozone injections and to avoid any and all dental surgery when possible. Take your time and thoroughly explore other options before getting sweet-talked into expensive surgeries that make false promises to cure all of one's ailments.

### *Uterine Fibroid Cysts*

Some women have sadly been pressured into hysterectomy surgery based on recurring trouble with these growths. These uterine cysts, sometimes referred to simply as "fibroids," can be very painful and cause heavy bleeding and cramping during menstrual cycles. Development of them has been linked to excess estrogens or phytoestrogens from foods. Eating a lot of soy can lead to this problem and many pharmaceuticals (especially birth-control pills) can upset hormone balance and also contribute to fibroid growth. Chemicals used in personal-care and cleaning products also act as estrogen mimickers.

I have met a number of women who have had success in reducing and eliminating the growths via application of **castor oil packs** to their abdomens. Saturate a cloth (preferably flannel) with castor oil and place

---

earlier due to the vast number of lawsuits filed against him. (In the state of Arizona dentists are not required to carry malpractice insurance in order to practice.) And I discovered there had been several complaints filed with the board against him as well. The board admitted to me that if they approved my complaint, he could lose his license. Personally, I think he should lose it as I do not think he is an ethical medical professional, and my experience along with a much worse experience related to me by another former patient and his record demonstrate this. After this serious ordeal my advice is to thoroughly research any dentist before submitting yourself to any treatment. You can request dentists' files from dental boards. These will be more reliable than online reviews. After his impressive sales pitch on my first visit, I also had left a favorable review!

under wax paper (used in cooking) directly on the skin and then apply heat to the pack by placing a towel over the wax paper (the wax layer is for keeping the oil in the cloth and from spilling out onto clothing, etc.) and placing the heat source (hot water bottle or heated stones, avoid electric heaters as they will expose you to a high electromagnetic field), and leave in place for as long as possible. Half-hour sessions are perfect. Use as needed. Doing the same on sore muscles/backs/necks is also helpful in relieving pain in strained areas.

## *16- Increase Your Body's Structured Water*

The water in our cells does not take on a regular liquid form but a gel-like one instead, which Dr. Gerald Pollack[270] calls the "4th phase of water" or "structured water." The better structured the water in our cells, the better for overall health, wellbeing and longevity.

Things that reduce the structured nature of this water include exposure to artificial EMFs and ingestion of junk foods. When the water in our bodies is not properly structured, reduced blood flow can occur, contributing to heart disease and other serious ailments.

But things that help structure our cellular water include **exposure to sunlight and our earth's natural magnetic field**. Also, **physical contact with loved ones**, including our pets, has been shown to directly influence and create healthy structured water and thus improve health.

And **drinking structured water** (from mountain springs where the water moves in spirals through rocks, or from water moved through appliances[lxxviii] designed to structure water by inducing this same action) and **eating structured foods** (foods higher in fat content are more structured) also aids in structuring our own water.

So, get out in the sun, walk barefoot in nature, drink fresh spring water, get plenty of healthy dietary fats and hug your favorite humans and animals.

---

[lxxviii] See Dancing with Water, www.dancingwithwater.com

## 17- Eat a Balanced, Healthy Diet

Just as our electronic devices are replacing real connections with people, while becoming addictive and giving us dopamine hits with every text and social-media alert and "like," we also look for quick fixes from junk or high-sugar foods, as a way to cope with our chronic fatigue and low-dopamine states.

But contrary to what pharmaceutical companies and other industries lead us to believe, there is no magic bullet to health and wellbeing. Instead, "slow and steady wins the race," and we need to seek real nourishment both in our diets and relationships.

Eating a whole-food based, balanced diet is key to achieving health when suffering any ailment or chronic disease condition. Unfortunately, people in our societies are now plagued with food allergies and intolerances so it can be tricky finding the best diet given all of the limitations. Allergies (of all kinds) were a virtual unknown occurrence before the introduction of mandatory vaccines early in the 20th century. There seems to be a direct correlation between the rise in allergies and vaccinations[lxxix].

This could be due to the long list of toxic ingredients in vaccines, in addition to heavy metal content. Many of the ingredients include egg, peanut, wheat and dairy products, which may explain the rise in allergies to these particular substances as vaccinations are designed with the intent to teach our bodies to reject whatever contents are contained within them (the main one being an alleged virus[lxxx]), so it is no surprise that our bodies would learn to reject other sources of dairy, wheat, eggs or nuts. (Or it could simply be that toxic ingredients in the form of metals, preservatives, etc., could be impairing our digestion of the same

---

[lxxix] See *The Autism Vaccine* by Forest Maready.

[lxxx] But as viruses have not been isolated and proven to exist, what really passes as a "virus" in vaccine ingredients (when included, this is not always the case) is really a cell culture sample of sputum or blood from a sick patient believed to be infected with a particular virus, it is not an isolated virus from those samples. For more on this topic see my chapters on viruses and vaccines in my co-authored book, *Welcome to the Masquerade; Prelude to the Coming Reset*, available on Amazon.com.

through physical damage to bodily tissues.) Another reason for rejecting foods our ancestors ate with relish and no ill effect is the steady contamination of these same foods (and others) from pesticides and herbicides along with genetic modification adulterations.

But we can still purchase organic foods and also grow our own wherever possible. My former late employer (whom I loved dearly, but with whom I did not always see eye to eye) often bragged of surviving all kinds of terrible foods he had had to eat while serving in the Air Force and also at various places of employment over the years. And so (mistakenly I feel) he came to believe that it did not matter what he ate and so he ate whatever junk foods he liked, as he felt he was able to *survive* each meal.

However, I firmly believe that meals should not be viewed as something to survive, but instead as something that nourishes and sustains us. So eating foods that are laden with poisons of all kinds in the form of artificial flavorings, colors, stabilizers, preservatives, and the aforementioned pesticides and herbicides (which also contain heavy metals), does not help us meet the goal of consuming food for its nutrient and life-giving properties.

As we reduce our intake of toxins and find safe ways to detoxify, we will discover that we can eat a wider variety of foods over time, given that they are themselves free of toxins. However, some of us may still be unable to eat certain foods and so, first and foremost we should listen to our own bodies, in favor of following the latest diet trend.

I have personally followed many diet trends in my own pursuit of health, sometimes at the advice of a specific doctor, but ultimately found that each extreme fad diet, while at first seemingly beneficial, led to some other kind of problem down the road. As a result, I now do my best to avoid extremes in diet while ensuring that I get adequate protein and fat alongside carbohydrates.

One issue with which many of us with EHS have to contend is keeping our blood sugar and blood pressure stabilized. As Dr. Sharon Goldberg pointed out, this is a hallmark trait she found in her patients with microwave injuries—the inability to keep blood pressure stable. This is definitely something with which I have struggled over the years and still find challenging. The things that help me with this problem are avoiding simple sugars and eating meals at regular intervals. It can help to eat smaller meals more often and to be sure there is enough healthy (not trans, or rancid) fats at each meal. But I also have to be sure not to eat too much fat as this can trigger painful gallbladder attacks and lead to serious electrolyte imbalances that can cause muscle spasms and cramping.

Also, **dehydration** from microwave exposures can be a big problem alongside the adverse impact of exposures on kidneys. Sometimes just drinking more water does not do the trick in combating chronic or acute dehydration. Adding a little more saturated fat to the diet may help and also avoiding excess salt[lxxxi] in the diet, or other dehydrating foods like excess starches, fiber, or dried foods like crackers, dried fruits or nuts. I have noticed great benefits personally from drinking raw goat or raw cow milk regularly when I can get it. The best solution would be if you could obtain milk from your own goat or cow—or that of a neighbor's. But this of course, is not always a practical possibility. Drinking beaten raw eggs can also really help combat dehydration (if not palatable on its own, you can mix them into a fruit smoothie and/or add raw honey).

Helping the kidneys heal and repair will also help with the problem of dehydration, as the kidneys help regulate our "water bodies." Drinking **marshmallow root,** or **gravel root** tea and adding a splash of **apple cider vinegar** to water can help the kidneys break up kidney stones and function more efficiently. **Black cherry** juice concentration can also ease gout symptoms. Also, it helps to reduce or **avoid diuretic teas** if struggling with dehydration.

---

[lxxxi] There are a lot of different opinions in both allopathic and natural medicine literature about how much salt and what type should be included or excluded from the diet. Certainly, avoiding regular table salt that is highly processed and contains added chemicals is advisable. Some advocate eating as much salt as one craves as long as it is high quality unprocessed sea salt or Himalayan pink salt. I did this for many years only to find it was not doing me the good I had hoped. About a year ago I stopped adding extra salt to my food entirely and have greatly benefitted from doing so. I feel much more hydrated and no longer wake in the night feeling the need to drink a glass of water. Most of my food also does not have added salt (except for cheese as I have been unable to find unsalted cheese) but I take care to eat a high mineral diet with plenty of fresh greens, including wild harvest greens and added pinches of kelp powder at times along with added pinches of clay to my drinking water.

## *18- Take (care of your) Heart*

While writing this section of the book, I received a message from a friend about another friend who ended up in hospital due to severe heart palpitations. While at the hospital he was refused treatment due to his refusal to have a Covid-19 test or be injected with a Covid-19 experimental drug. This stance taken by hospital staff is extremely discriminatory and illegal, but such is the climate of a pandemic-panicked world. He was however given a diagnosis, one of atrial fibrillation.

At one point I too could easily have received this same diagnosis had I gone to a hospital at the height of my own EMF-induced heart arrhythmias. Instead, I discovered the source of/trigger for my arrhythmias (microwave radiation) and got the "heck out of Dodge," as they say, in order to stabilize my heartbeat. My own heart problems have been the most worrisome EMF-induced symptom I have experienced, since prolonged arrhythmias can lead to heart attack and atrial fibrillation (AF) and are linked with increased incidence of blood clots and strokes.

Our bodies function via weak electrical signaling—most especially our hearts. This very important organ is particularly susceptible to interference from much stronger EMF exposures inherent to our many electric and radiofrequency-emitting gadgets.

Heart problems of all kinds have arisen alongside our exposures to EMFs. One case presented in the film *Take Back Your Power*, related the story of a man having experienced chronic arrhythmias after installation of a SMART meter on his home. This discovery was *not* made in time to save him from having to be surgically implanted with a pacemaker. The pacemaker at first offered relief until the radiofrequencies from the SMART meter interfered with the pacemaker causing it to malfunction on several occasions so that this same man subsequently experienced cessation of his heartbeat for life-threatening periods of time. The eventual removal of the meter stopped these interferences.[lxxxii] As

---

[lxxxii] Interference of radiofrequency-emitting devices with electronic implants such as pacemakers and insulin regulators for diabetics and brain microchip implants for Parkinson patients is well documented in scientific and medical literature. Packet inserts for the implants themselves warn patients to avoid using cordless or cellphones and warn to distance from other microwave-emitting devices. One implant can also interfere with the functionality of another, something not always disclosed to patients prescribed

previously related in this book, children are now experiencing cardiac arrest at schools equipped with powerful Wi-Max systems and with cell towers on school grounds. So common has this trend towards heart attack of children in schools become nowadays it has led to the necessity for schools to equip themselves with emergency defibrillators.

My advice to the friend just diagnosed with AF is to reduce all EMF exposures, clean up dirty electricity and eliminate all sources of wireless radiation in his home—NOW!

For those whose hearts have been compromised and especially for those who cannot immediately remedy their EMF exposures there are natural medicines that can help.

**Hawthorn berry** is a traditional herb known for its supportive effect on the heart. Additionally using the supplement **CoQ-10** and Chinese herb **Dan Shen** is recommended for cardiac support. (Do not take Dan Shen if you are already taking blood-thinning medication.)

I also highly recommend anyone with heart problems of any kind to read Dr. Thomas Cowan's book, *Human Heart, Cosmic Heart*[lxxxiii] for incredible insights and further recommendations.

## *19- Support Adrenal Health*

One disastrous effect of chronic exposure to EMFs is overstimulation of the adrenal glands (often called the "master gland"). Maintaining health of these delicate organs is essential to the health of our entire body.

---

multiple implants. There is at least one reported case of a young woman in her 30s dying after her pacemaker failed due to passing through an airport security metal detector.

[lxxxiii] Available on Amazon.com, with more information and alternative purchasing options at www.humanheartcosmicheart.com

You can check the state of your adrenals by **shining a flashlight** in a darkened room at your eyes from the side, while looking in a mirror to check the dilation of your pupils. If your adrenals are healthy then your pupils should contract and remain smaller for up to two minutes. If your pupil dilates within 10-30 seconds, or quivers or stays dilated for 45 seconds before contracting then there is a problem with adrenal function.

As mentioned in the first solution listed, our adrenals are being chronically stimulated by EMF exposures to produce adrenaline and keep us in a state of high anxiety. The constant demand on this gland to release this hormone leads to eventual burn out and even adrenal collapse.

In addition to avoidance and reduction of chronic EMF exposures and getting proper rest, a few herbs and vitamins can help restore adrenal function.

Any herb or food high in **vitamin C** helps the adrenals—this includes (but is not limited to) **kiwi fruit, amla berry**, and **rose hips.** Natural sources of C are best, since synthetic supplements are often derived from toxic substances and make it harder to determine dosing thereby making it be easier to overdo it, since adding a single vitamin tends to lead to other imbalances in the body, whereas getting vitamins in their natural form works in a synergistic balanced way.

**Licorice root**, **ashwaganda** and **andrographis** are other recommended herbs for gently restoring adrenal health.

## *20- Move Your Body*

As we find ourselves living out our lives increasingly online, we seem to be forgetting that we inhabit a physical body in a physical world. As much as I personally love physical activities, especially in the form of water sports and dance, as a writer and artist, all too often I find myself tethered to my desk and computer for hours on end and tend to forget the physical needs of my body—especially the need to move.

Physical movement is not only important for keeping muscles and joints conditioned, aiding us in comfortably navigating our physical world when needed, it is also important in aiding our bodies' capacity for detoxification. Without daily exercise our primary detox pathways cannot function properly without initiating lymphatic flow. Even light exercise will trigger this flow and vastly enhance the detoxification process.

Practicing daily **Tai Chi** is a simple, light, and effective way of achieving this goal. **Qi Gong** is another similar practice, as are some (less strenuous) forms of **Yoga.** **Hula-hoop dancing** is a personal favorite. **Daily walks in nature** represent another great way to keep moving. I also recently took up surfing and find the practice very therapeutic but also quite physically challenging. **Swimming** is another favorite lower impact sport but harder to practice when one is reactive to chemicals used in swimming pools and where the average ocean temperature ranges between 45 and 55 degrees Fahrenheit year round. (As they do for me on the west coast, so I use a thick wet suit that keeps me warm for surfing, however as already mentioned I have been increasing my time spent in the water without a wet suit as well as part of my cold-water-therapy practice.)

Whatever activity you choose, **take care not to overdo it.** While it is important to get cardio exercise as well as engaging in movement to keep the lymphatic system flowing, it is easy to overdo it, especially when first riding the waves of enthusiastic resolution common to starting a new practice. Take things slow and easy. This is key for those of us who have been injured by microwaves and as a result experience weakened, compromised tissues and joints, and adrenal fatigue. We need movement, but we also need rest. Too often the rest aspect of physical health is ignored in favor of pushing physical limits. We do not need to prove anything here or break any Guinness World Records. If we do push our limits we may also find that we become too discouraged to try again after experiencing physical injuries or exhaustion. It is not important how we appear but how we feel. Keep the focus on that.

## *21- Heal with Sound Therapy*

Plants have been known to grow more healthily and quickly if exposed to classical music recordings or people singing to them directly. Crying babies can be quietened by soothing melodies sung to them by caregivers. And music is said to “tame” the wildest and angriest of “beasts.”

The Babanzele Pygmies[lxxxiv] sing so harmoniously with their natural surroundings that their voices take on a rhythm fully in sync with other voices in the forest—those of frog croaks and cricket chirps, all weaving in and out of one another, without one voice dominating over any other, to create a symphony of vibrant life-sustaining frequencies.

Frequencies can harm and also heal. It makes sense that a good way to combat harm from artificial frequencies would be found in the form of natural frequency therapies. This is why increasing our exposures to frequencies from the sun, water and earth can truly help us. But also, sound frequencies produced from musical instruments or even from simple tuning forks can heal as well.

Olga Sheean told me about healing with **tuning forks** and this does seem to be a very interesting emerging area in alternative-energy healing that is worth exploring further. By hitting a tuning fork tuned to a specific frequency and holding it over different parts of your body, you can literally feel a change in vibration in your own being. This practice is based on the idea of resonance—meaning that as beings of frequency we can resonate in harmony with another vibratory frequency in close proximity to our own. It acts as a kind of entrainment, and this happens to us all of the time, both while within natural electromagnetic fields and artificial ones. It is provably healthier for us, of course, to resonate with natural frequencies. So in absence of access to a forest, river, or other natural sources, we can initiate these same vibrations by way of tuning forks or playing a musical instrument. There are many practitioners[lxxxv] now available to help guide you if you want to learn more about working with tuning forks.

Anyone can learn to play a **musical instrument**, whether you feel musically inclined or not. The goal here is to feel the vibrations of the instru-

---

[lxxxiv] See recording "Women Gathering Mushrooms" by Luis Sana, produced by Bernie Krauss found on the album, *Book of Music and Nature: An Anthology of Sounds, Words, Thoughts*, music of the Babanzele Pygmies and sounds of their forest home. https://digitalcollections.wesleyan.edu/object/wupmusic-78

[lxxxv] However, beware of sound healers using frequencies from the alleged "Solfeggio Scale" which appears to be a sham and not even a real scale but uses frequencies, which are very discordant.

ment, not to compose a symphonic masterpiece. Choose an instrument with which you most closely "resonate". I personally gravitate toward string instruments and have played guitar for a number of years now. Olga tells me she has taken up beating upon steel drums. Others prefer wooden flutes, singing bowls or wooden drums. There is no one perfect "correct" instrument–you get to choose.

**Try tuning to 432 Hz.** Current standardized tuning is set to 440. Some call this "the devil's tuning" as crazy as it might sound, it may have been set to this specific number as an intention to weaken and sicken us.[lxxxvi]

However, there is quite a lot of misinformation circling about 440 vs. 432. It is true that there was no standard tuning until the British set the standard in 1939 and then the US adopted it in the 1950s, and it is possible that many musicians were opposed to 440 and in favor of 432, but rumors about Nazi involvement have not been substantiated and claims that 432 was the prior standard (there was no previous international standard) or that it is the frequency of the earth or the ancient instruments were tuned to it is all fictional.

But it does seem that 432 Hz is more harmonious than 440 Hz according to some musicologists and I personally prefer it. But tuning (referring to concert pitch) to another frequency does not work in an exact manner without also changing "temperament" (the interval between each note in a scale) to match and this is not possible without the instrument design allowing for tunings in other temperaments. So, in order to achieve a true 432 Hz scale (in which all notes in the scale adopt the appropriate frequencies to match), temperament (the current standard being now set at "equal" or "even temperament") would have to change to "Pythagorean temperament." Today's standard musical instrument is manufactured to play only with even temperament, so just changing concert pitch to tuning A at 432 Hz, will simply lower the pitch slightly and not achieve effects attributed to the "magical" 432 tuning scale. However, this pitch change creates a more comfortable resonance, in my

---

[lxxxvi] See the book *432 Hertz; The Magic and Mystery of Sound and Music*, by Jonas Maluik, and Roel Hollander's website: www.roelhollander.eu for more information on this fascinating topic.

opinion, and is easier to play and sing to, and as such still has an effect, even if subtle.

Beware of falling for the so-called *Solfeggio Scale* healing frequencies first brought forward by Dr. Leonard Horowitz (a retired dentist not a musicologist, or someone at all familiar with music theory) who is also a very prominent presence on the Internet as someone who is opposed to 440 HZ tuning. While he is opposed to 440 and maybe rightly so, instead of advocating for 432 tuning, he advises 444 which is truly unharmonious and possibly worse than 440. Sadly, many sound therapists have fallen for the Solfeggio scale nonsense and use these very unharmonious/dissonant frequencies in their practices.

Try organizing and/or participating in a **drum circle.** I used to join in a regular drum circle in my old neighborhood in Washington, DC. The energy of 50 or more people beating on drums in a small space was incredible. Sadly, misguided pandemic policies moved that physical drum-circle space into a virtual one for quite a long time (I am not sure what is happening with it currently, maybe they are now only allowing "vaccinated" participants to join).

Experiencing a drum circle through a screen and tiny computer microphone hardly equals to the real in-person experience and should never be viewed as a replacement for it—neither should any online connection ever replace in-person ones. What started as augmented virtual connectivity has begun to shift into surrogacy—this is something we must not allow to happen. It is not something we should accept as the "new normal"—this stripping away of our cultures and *essential* human connection, nor should anyone be disallowed from participating in society and within social groups based on medical status or willingness or unwillingness to be injected with potentially harmful or lethal experimental drugs.

But as so many of us have been driven into isolation out of necessity (long before the lockdowns and anti-social distancing madness) in the process of fleeing artificial EMF-environments, we have had to accept some substitutes for in-person socialization[lxxxvii] and many of us may not

---

[lxxxvii] But we need to hold onto hope that this situation will not last forever, and in the meantime it is important to take care not to live all of one's social life online. It is better to have fewer real, in-person friends than dozens or hundreds of virtual ones. Many online connections of course, turn out to be very beneficial and sustaining, but we all

be in a position to buy a musical instrument. So the next best thing to playing instruments or hitting turning forks, is to listen to recorded music. Again, choose what most appeals to you personally but also note how the music makes you feel. Maybe you always loved heavy metal or discord or hard rock—maybe it really "pumps you up," but is that a good thing? Do you feel worn out or anxious afterward? The point here is to find frequencies with which to resonate that will bring about healing, not add further stress to an overburdened body.

My partner Sam happens to be a very gifted musician and I am certain that his regular acoustic guitar playing has aided me in my own healing. Sam has also recently begun what he calls his ***Intention Project***,[lxxxviii]for which he is creating personalized healing acoustic guitar recordings. While creating these pieces he focuses in on the intended recipient and so far, those recipients have felt a personal connection with and direct benefit from the music. This is another way we can contribute to healing ourselves and our world, through sound coupled with intention.

## *22- Work with the Energy Body*

Since ancient times, Chinese medicine has regarded the body as an energy system possessing clearly defined energy pathways known as **meridians**. Several charts detailing these meridians can be readily found on the Internet. Disease, amongst traditional healers of the Orient, was and still is, understood to be caused by imbalances in these meridians.

Healing then, it is understood, can come about by working with the energy pathways to restore balance. Adding pressure to specific points along the pathways is known as **acupressure** and is used to treat a variety of illnesses.

---

need in-person physical connections, so long-distance virtual ones should never be viewed as replacements for these. Even if you have only one friend, it is better to spend more quality time with this person than to spread yourself thin virtually.

[lxxxviii] Anyone interested in becoming a client can contact Sam directly by emailing to samj@fastmail.com

Ruth and Gary have used acupressure self-massage quite successfully in reducing and eliminating painful symptoms due to EMF exposures. They even fashioned their own acupressure aids out of knobby sticks, helping to reach harder-to-reach parts of the back. One can alternatively also buy such devices. My personal favorite is called ***The Backnobber II*** by Pressure Positive, which can be purchased on Amazon.com.

Others use **acupuncture**, which works on the same principle, but makes use of tiny needles poked in the skin at key meridian points instead. However, many of us who are sensitized to EMFs also seem to have an aversion to metals in general and cannot tolerate acupuncture. I am one such canary and have only experienced pain during acupuncture sessions and no noticeable benefit. But I have made extensive use of acu*pressure* both self-applied and professionally administered, and have experienced wonderful results from this approach.

Other ancient eastern healing practices also see the body as comprised of specific measurable energy systems. In India, **Ayurvedic medicine** teaches that the body has at least 7 **chakras** (energy centers), while some claim the existence of 12 chakras. These energy centers run from the top of the head (crown chakra) down to the base of the spine (root chakra) and are thought to rotate in a specific direction and at a specific rate. Disruptions to the rotation of these centers are understood, as in the Chinese meridian system, to cause disease. The approach to righting these chakra rotations takes many forms and can involve use of Indian Ayurvedic herbal medicines, dietary changes, meditation or hands-on energy healing.

**Hands-on energy healing** has been used for thousands of years in different forms in practically all cultures of the Earth. Reiki is a form of hands-on energy healing out of Japan that has gained popularity in the West in recent times. And others like **Jin Shin** can be performed on one's own but it really helps to have the guidance of a trained therapist like Jeanice Barcelo, especially at the beginning. And other forms stem from Christian and Shamanic healing traditions. All involve laying-on of hands over ailing bodies and directing renewing energy into the recipient from God, or the Universe, or Love—depending on one's belief system. Many EHS sufferers have reported benefits from different types of energy healing. I personally like the Jin Shin method, one that is also practiced and taught by Jeanice.

## 23- *Color Yourself Healthy!*

As a teenager, my church youth group leaders organized a special event at which we young women were gathered to learn about how to “color ourselves beautiful,” by choosing the right colors of clothing and makeup for our own particular complexions. I learned that with my auburn hair and brown eyes I was considered to be an “autumn” and as such, did best with autumn, earth-toned colors. (This was not actually bad advice and I think is true on the whole.)

But color therapy can extend well beyond helping one dress one’s best, it can have actual healing effects, especially when coupled with light frequencies.

Some alternative historians believe it possible that ancient cathedrals were once used not as churches but as healing centers. Proponents of this theory speculate that the specific geometric design, along with the types of materials chosen, used for building the cathedrals, may have been done with the express intention of creating a space that through its own particular shape along with the electromagnetic properties inherent to the minerals in the building materials used, could bring about spontaneous healing for those who entered without any other interventions.

It is theorized that use of **stained-glass windows** in these spaces was also intended for healing by juxtaposing specific colors together and allowing light to pass through, so that frequencies were created in order to affect the persons seated within its healing rays at profound physiological levels.

I found further evidence to back this latter hypothesis when I happened to encounter an alternative healing massage therapist on a trip to Mexico. She reported using **colored heat-tolerant plastic gels** over top of incandescent theater lights with great success in healing many of her own ailments, including partial deafness.

In order to use colors for healing in this way, she referenced a now out-of-print (but still available on eBay and Amazon.com) book called *Let There Be Light,* by Darius Dinshah. I have not tried this yet personally but find the idea intriguing. Another way to use color and light for healing is to buy **colored glass bottles**, fill them with water and place them in the sun for several hours. Therapeutic frequencies are then allegedly imbued within the water that, once ingested, can also supposedly bring about healing. For best results the glass should be impregnated with the color and not added as an after-coating.

## *24- Nurture and Sustain Positive Healthy Relationships*

Avoid unsupportive so-called *friends* and even family (if abusive) and surround yourself with loving and supportive people.

It can be challenging enough to try to cope and stay physically, emotionally and mentally healthy while carrying the stigma of EHS and/or MCS or espousing a different viewpoint from the mainstream, while also trying to avoid exposures to EMFs (and chemicals) during the rapidly encroaching onslaught permeating our world at present.

A supportive, understanding non-judgmental friend in these *uncertain times* can be a true lifesaver.

## *25- Reexamine Priorities and Beliefs*

Those of us whom have been injured by EMFs and consequently lost our former positions in society, connections with disbelieving friends and family, careers, and sense of purpose, have been forced to abandon our once beloved digital devices and other EMR-emitting technologies in order to preserve our health and escape daily physical torment.

This intensive life upheaval often drives us to a deep spiritual bottom and forces us to reexamine beliefs, things, and relationships we once held dear and paramount to our lives. But this kind of drastic lifestyle change can be a blessing in disguise if we allow ourselves to benefit from the hidden lessons with which it presents us.

After reprioritizing what is important, connection with the natural world—once considered a luxury only possible on weekends away or on annual vacations—becomes a daily event and serves as an opportunity to enrich our lives in unnamable and innumerable ways.

By reading, writing, talking with friends *in person* (giving more attention to a select few loyal friends) and taking the opportunity to develop closer bonds with loved ones, we end up spending more quality time together instead of exchanging quality for quantity and approval as we vie to garner the most "likes" and "friends" on social media. Once addicted to the accompanying dopamine hit of pursuing online "connections," we lose sight of what is truly important in human relationships. This is an opportunity to regain what was once lost—take it!

## 26- Preserve Humanity

While many of us found ourselves suddenly under house arrest in 2020-2021 (and sadly for some again in early 2022), resultant of extremely misguided efforts to contain a phantom, unproven contagion, we were pushed into replacing physical human contact in favor of virtual connection.

Virtual human connections, while sometimes necessary or convenient, should never become a permanent replacement for real-world in-person ones. We need human touch and face-to-face interactions in order to preserve our humanity and wellbeing.

We should not let this temporary situation in which we have found ourselves become a model for our future societies or we stand to lose irreplaceable, priceless, enriching human connection, which up until now has defined our collective human experience. Consenting even once to tyrannical public-health "orders" and so-called mandates (but legally only "guidelines"), sets us up for future abuse, whenever those in authority decide to cry "pandemic." But even if in our fear and ignorance we once submitted, it does not mean we have to do it again. It is time to take back our own authority and preserve humanity, now targeted for genetic altering and genocide via alleged prophylactic "vaccines."

## 27- Take a Holistic Approach

*"None of our textbooks could tell us the how and why of healing. They explained the basics of scientific medicine—anatomy, biochemistry, bacteriology, pathology, and physiology—each dealing with one aspect of the human body and its disconnects. Within each subject the body was further subdivided into systems. The chemistry of muscle and bone, for example was taught separately from that of the digestive and nervous*

*systems. The same approach is used today, for fragmentation is the only way to deal with a complexity that would otherwise be overwhelming. The strategy works perfectly for understanding spaceships, computers, or other complicated machines, and it's very useful in biology. However, it leads to the reductionist assumption that once you understand the parts, you understand the whole. That approach ultimately fails in the study of living things—hence the widespread demand for an alternative, holistic medicine—for life is like no machine humans have ever built: It's always more than the sum of its parts... Except in the simplest species removal of anything more than a few cells always destroys the organization, and hence the organism."*

—Robert O. Becker, M.D.—*The Body Electric*

We live in a war-based culture. The mechanistic view of the world and its inhabitants has led to this culture, which divides living organisms into its parts and everything into good and bad, and hero and enemy.

But nature's intelligent creation actually demonstrates health in terms of balance and imbalance, harmony and disharmony. Relationships between organisms and species in the natural world are symbiotic. There is no *good* or *bad* bacterium, "virus", fungus, or parasite in nature. When equilibrium is established all of these microscopic parts benefit the whole of creation.

Fighting a phantom "evil" virus, especially when seeking to destroy and eradicate it, does nothing but to leave us with a poisoned environment (upon which all life depends) and poisoned people, and hardly works to achieve a goal of health as proclaimed. It is not until we change our attitudes towards each other and our world that we will know true health, peace and prosperity.

We need to ask ourselves how successful the "war on cancer," "war on terrorism," "war on drugs," "war on crime," and "war on infectious disease" has been? And now we are being asked to come together to wage war on an unseen common enemy—the phantom SARS-CoV2 *virus.*

Disease conditions create disease, not exposure to specific microorganisms. Disease conditions include; starvation, malnourishment, fatigue, exhaustion, poisoning, exposure to extreme heat or cold (especially for prolonged periods), lack of basic sanitation[lxxxix], and interference with our Earth's natural electromagnetic field by way of artificial EMF exposures. These conditions can also be viewed as imbalances in the natural order. Disease conditions as listed, were the cause of *all* so-called infectious, communicable disease at all times in human history.[271] In each case of *plague outbreak,* one or more of these conditions can be found. Blaming this suffering and death on invisible, and often unproven, pathogens, works to divert our attentions from the real culprits, especially where profitable industries are permitted to continue in their polluting practices as a result of our being blindsided.

Understanding and uncovering the *real* cause of disease and finding truly effective ways to heal and restore balance to our bodies and our world is the only way we will ever end suffering and disease.

## 28- Be Gentle and Kind to Your Body and Your World

Try to focus on the bigger picture instead of dissecting it to its smallest invisible parts—parts, which we then attempt to label, isolate and contain, to the point where we are only left with a theory of reality while fully in denial of our direct observation and experience of it.

Educator and EMF activist, Max Igan[272] during an interview with Josh Del Sol at the 2019 5G Crisis Awareness and Accountability summit, gives this amusing, yet disturbing analogy: "You don't tear a kitten to pieces in order to love it."

Likewise, we do not have to dissect and label "love" in order to benefit from and experience its healing life-giving effects.

---

[lxxxix] When speaking on the issue of sanitation, I do not mean to say that it is necessary to "sanitize", as in chemically eradicate microbes, and one's environment. I am speaking merely of the need for access to clean unpolluted water and air. Drinking water from cesspools in which decaying animals and human waste abounded, was a major cause of disease at the turn of the 20th century in cities all over the world, and is still a problem in less developed nations today.

## *29- Remember Who You Are*

You are a powerful creator with unlimited, untapped potential.

There is a reason the "powers that be" (but clearly *should not* be) are doing all they can to keep us from life-affirming and sustaining contact with our natural world and with each other.

Our power threatens their control positions and the best way for them to maintain obscene wealth, and political and industrial power, is to keep the people sick, overworked, and afraid.

So, remember who you are and do not believe for a minute that any artificial technology or so-called artificial "intelligence" is better or more intelligent than you. Machines cannot and will never be able to posses or experience the traits or feelings of love, compassion, intuition and wisdom. True intelligence can only be exhibited and measured through the inclusion of these uniquely human traits.

## *30- Lastly, Maintain a Positive Outlook and Sense of Humor*

...As reason and logic are in about as short supply as toilet paper was at the start of the pandemic panic—effectively leaving the collective shelves of the human psyche empty—as it is constantly bombarded with messages of fear.

Humor lightens our mood, reduces stress and can help us to step outside ourselves and see the absurdity and folly of our hasty actions and emotional reactions. It may be exactly what we need now more than ever to see us through the hard times ahead—especially if you happen to be a little yellow canary.

APPENDIX B

# Resources/Further Reading

## BOOKS

(Arranged under topic, although some books cover multiple topics and could cross over into other categories, however I still listed these under one only and also chronologically with most recent publications listed first.)

### Alternative Medicine

*Earth Medicine: A Field Guide—Healing in seasons and cycles*, Merrill Page, 2021

*Cancer and the New Biology of Water,* Thomas Cowan, MD, 2019

*Human Heart, Cosmic Heart—A Doctor's Quest to Understand, Treat, and Prevent Cardiovascular Disease*, Thomas Cowan, MD, 2016

*The Lethal Dose; Why Your Doctor is Prescribing It*, Dr. Jennifer Daniels, MD/MBA, 2013

*Death by Modern Medicine,* Dr. Carolyn Dean, MD ND, 2005

### EMF Biological Effects and as Weapons

*Welcome to the Masquerade; Prelude to the Coming Reset,* Shannon Rowan, John Hamer 2022

*Are Wireless Devices Really Safe? Or is Wireless Radiation Harmful?* Jeanice Barcelo, 2020

*The Dark Side of Prenatal Ultrasound and the Dangers of Non-Ionizing Radiation,* Jeanice Barcelo, 2019

*The Invisible Rainbow; A History of Electricity and Life,* Arthur Firstenberg, 2017

*EMF off! –A call to consciousness in our misguidedly microwaved world,* Olga Sheean, 2017

*Irradiated: A comprehensive compliance and analysis of the literature on radiofrequency fields and the negative impacts of non-ionizing electromagnetic fields (particularly radiofrequency fields) on biological organisms,* Wireless Action, 2017

*An Electronic Silent Spring; Facing the Dangers and Creating Safe Limits*, Katie Singer, 2014

*Cross Currents—The Perils of Electropollution, the Promise of Electromedicine*, Dr. Robert O. Becker, 2014

*Disconnect; The Truth About Cell Phone Radiation*, Devra Davis, 2013

*Overpowered; What Science Tells Us About The Dangers of Cell Phones and Other Wi-Fi-Age Devices,* Martin Blank, PhD, 2013

*Dirty Electricity—Electrification and the Diseases of Civilization*, Sam Milham, 2010

*Public Health SOS: The Shadow Side of the Wireless Revolution*, Magda Havas and Camilla Reese

*Controlling the Human Mind; The Technologies of Political Control OR Tools for Peak Performance*, Dr. Nick Begich, 2006

*Would You Put Your Head in a Microwave Oven?* Gerald Goldberg, MD, 2007

*Cell Phones; Invisible Hazards in the Wireless Age,* Dr. George Carlo and Martin Schram, 2001

*The Body Electric; Electromagnetism and the Foundation of Life,* Robert O. Becker M.D., 1985

*Operation Mind Control,* Walter Bowart, 1977

*Silent Weapons for Quiet Wars, Operations Research Technical Manual TW-SW7905.1*, 1979, free download–

http://www.coalitionoftheobvious.com/SILENT_20WEAPONS_20for_20QUIET_20WARS.pdf

## Germ Theory/Virology

*The Achilles' Heel of the Biomedical Paradigm*, James McCumiskey, 2021

*The Contagion Myth: Why Viruses (including "Coronavirus") Are Not the Cause of Disease*, Thomas S. Cowan, MD, and Sally Fallon Morell, 2020

*Béchamp or Pasteur?: A Lost Chapter in the History of Biology,* Ethel D. Hume, 2011

*The Ultimate Conspiracy: The Biomedical Paradigm,* James McCumiskey, 2012

*Good-Bye Germ Theory–Ending a Century of Medical Fraud and How to Protect Your Family*, Dr. William P. Trebling, 2006

*Virus Mania- How the Medical Industry Continually Invents Epidemics, Making Billion Dollar Profits at Our Expense*, Torsten Engelbrecht and Claus Kohnlein, 2007

## Technological/Digital Addiction

*Glow Kids,* Nicholas Kardaras, Ph.D., 2016

*The Glass Cage,* Nicholas Carr, 2014

*The Shallows; What the Internet is Doing to Our Brains*, Nicholas Carr, 2011

*Alone Together–Why We Expect More from Technology and Less from Each Other*, Sherry Turkle 2011

### *Vaccination*

*The Autism Vaccine—The Story of Modern Medicine's Greatest Tragedy,* Forest Maready, 2019

*Dissolving Illusions; Disease, Vaccines, and the Forgotten History,* Suzanne Humphries, MD, and Roman Bystrianyk, 2015

*Anyone Who Tells You Vaccines Are Safe and Effective is Lying,* Vernon Coleman, 2014

*Vaccination Voodoo—What You Didn't Know About Vaccines,* Catherine J. Frompovich, 2013

*The Poisoned Needle; Suppressed Facts about Vaccination,* Eleanor McBean, 1957

## DOCUMENTARY FILMS

*Take Back Your Power*

*Resonance—Beings of Frequency, 2012*

*WiFi Refugees* RT Documentary, 2017

*Generation Zapped*
www.generationzapped.com

*Lo and Behold; Reveries of a Connected World,* Werner Herzog, 2017

*Microwave, Science, and Lies*

## LINKS

Rally together, petition, and especially make an effort to block 5G installations in your communities. Get involved with the growing number of people waking up to the issue of EMF harm. Help us reach critical-mass awareness.

Here are links to a list of action groups that can help you get started:

https://keepyourpower.org/

http://www.electrosmogprevention.org/stop-5g-action-plan/10-actions-to-help-stop-5g/

https://wearetheevidence.org/

http://www.friendsofmerrymeetingbay.org/fombnew/pages/what_we_do/events/Speaker%20Series%202021-2022/Nusiance/Nuisance.htm

Massachusetts for safe technology
https://mailchi.mp/06e5ccdf490a/action-please-send-testimony-to-nh-legislators-13425057?fbclid=IwAR3BU4mdWmPv0eMkUtDSHYMQYIFUbFL6YKiB_SrSi26WsZrRX5PPBZg0M_E

***Other links related to EMF education listed below:***

International scientist appeal to the UN: https://emfscientist.org/

Database of over 6,000 scientific studies linking EMR and biological harm: www.Emf-portal.org

Microwave News: http://www.microwavenews.com

Physicians for Safe Technology: https://mdsafetech.org

Living with EHS: http://www.weepinitiative.org/livingwithEHS.html

Environmental Health Trust (EMF-related science, policies, facts, resources): https://ehtrust.org

The Bioiniative Report reviews the rationale for EMF and RF exposure standards: www.bioinitiative.org

Dr. Magda Havas biological studies on EMFs for both harm and healing: www.Magdahavas.com

Less EMF Inc. offers shielding products and EMF meters: www.lessemf.com

Claire Edwards, EMF awareness activist, writer, researcher
https://forlifeonearth.weebly.com/about.html

EMF Safety Zone—EMF Protection, Products, Consulting, Education, and Resources
https://emf-protection.us

Stop 5G action plan, No SMART meters campaign
https://www.electrosmogprevention.org

https://scientists4wiredtech.com/about/

The EMF Community—News, Information, Discussion and Online Consulting with EMF Experts
https://emfcommunity.com

Resources for EHS and EHS stories
www.radiationrefuge.com

Jeanice Barcelo, EMF educator, author, researcher
www.radiationdangers.com

Steven Magee, EMF researcher
https://www.environmentalradiation.com

Please consider donating to Professor Olle Johansson's work at the following links:

https://honeywire.org/research

https://www.emfsa.co.za/news/fundraiser-to-support-associate-professor-olle-johanssons-ongoing-research/

***Links related specifically to children's health and EMF exposures:***

https://www.techsafeschools.org

https://www.mothersforsafetech.co.uk

http://www.parentsforsafetechnology.org

https://mieuxprevenir.blogspot.com/2018/10/united-states-sidelined-childrens.html

http://wiredchild.org/health-effects.html

# ACKNOWLEDGMENTS

I often feel I would not be here today at all were it not for the unconditional support I have received from a few choice friends and family members. Living with a condition that sets oneself against the world can be very lonely. It has been the support I received from these true comrades that has seen me through some very tough times.

My very deep appreciation goes out to my partner Sam, who has, against the odds, stuck by my side throughout eight long years of this ordeal. So many canaries lament the loss of partners whose own strength failed to meet the challenge and left them to face impossible-seeming situations alone. Sam, you have been my rock and my best friend, taking care of me in so many ways when I needed it most, words are not adequate to express my gratitude.

A few of my oldest friends, Sara Kelleher, Katherine Cottle, Britt Halter, Jane Goss, Solveig Mortensen and Christine Turner have also stuck by my side through good and bad times, thank you so much!

And friends I have made along the way, like the other canaries whose stories I presented in this book and others still such as, John Maier, Dave Shreeve, Birgit DeGregario, Tim Heiierman, Ricky Berger, and James McCumiskey. Thank you for helping shoulder my burden and providing safe spaces in which I can express any opinion without judgment.

And special thanks to my good friend, fellow researcher, and colleague, John Hamer, who has been both an invaluable mentor professionally and emotionally supportive on a personal level. Without your help I would not have had the confidence or know-how to write a book of my own. You have also helped me keep sane through extremely difficult times brought on by recent world events. Thank you for your unending bottomless generosity to both myself and Sam and your wonderful friendship.

Thanks to other newfound friends, especially to Paul Senyszyn, whose generosity in setting up a makeshift office for me at his home, without which I would not have completed this book for quite a long time to come.

I also want to thank the EMF experts who directly contributed to this work through graciously offering their time to me, (in many cases) a stranger, and from whose own publications I have learned so much, and with a special thank you to Claire Edwards for contributing to this book with her excellent foreword and for sharing her invaluable expertise, guidance and friendship.

# ABOUT THE AUTHOR

Shannon Rowan, BFA, is a "WiFi refugee," social critic, freethinking fine artist, writer, geopolitical author and researcher, photographer, dancer, children's book author/illustrator and EMF-awareness activist. She lives in the wilds of Northern California with her partner and their two cats and is an outdoor enthusiast, avid surfer and sea kayaker. She had a career as a photojournalist and contributed to major national and international news and travel magazines as photographer and writer. Later careers related to dog walking and training, and hula-hoop performance art.

She is the author and illustrator of: *The Goldfish of Brandywine Farms*, illustrator of: *Baltimore Sideshow*, a poetry book collaboration with Katherine Cottle, (both available on Amazon.com), and is co-author and illustrator of *The Great Masquerade; Prelude to the Coming Reset* with geopolitical author John Hamer.

For more information visit: www.wifi-refugee.com

# ENDNOTES

---

[1] The **EUROPAEM EMF Guideline 2016** states that EHS develops when people are "continuously exposed in their daily life" to increasing levels of EMFs, and that "reduction and prevention of EMF exposure" is necessary to restore these patients to health.[105] EHS should no longer be considered a disease, but an injury by a toxic environment that affects an increasingly large portion of the population, estimated already at 100 million people worldwide,[106][107] and that may soon affect everyone[108] if the worldwide rollout of 5G is permitted. International Appeal to Stop 5G on Earth and in Space. 18 September 2018. https://www.5gspaceappeal.org/the-appeal

[2] Glaser, Z.R. 1971/72. Bibliography of reported biological phenomena ('effects') and clinical manifestations attributed to microwave and radio-frequency radiation. Naval Medical Research Institute MF12.54.015-004B, Report No. 2, revised. 106 pp. [N.B.: this document was reduced to 25 pages so that it could be posted online and does not contain all of the 2,311 references.] https://magdahavas.com/from-zorys-archive/pick-of-the-week-1-more-than-2000-documents-prior-to-1972-on-bioeffects-of-radio-frequency-radiation/

[3] World Health Organization. "Biologic Effects and Health Hazards of Microwave Radiation." This 350-page document is the result of a symposium (convened by WHO, among others), held in Warsaw on 15-18 October 1973, with the participation of 60 researchers specializing in the biological effects of microwaves on humans. This document describes primarily the adverse effects of microwave radiation on neurological, vascular and cardiac systems, as well as on the thyroid and thrombocytes; it also reports that microwaves can cause Type II diabetes, sleep issues, cataracts, opacification of the ocular lens, and behavioural disorders, among others. EMF Off! https://www.emfoff.com/symposium/.

[4] The EMF-Portal is a project of the *femu* working group of the Institute for Occupational, Social and Environmental Medicine of the Uniklinik RWTH Aachen University. The internet information platform EMF-Portal of the RWTH Aachen University summarizes systematically scientific research data on the effects of electromagnetic fields (EMF). The core of the EMF-Portal is an extensive literature database with an inventory of 36,091 publications and 6,968 summaries of individual scientific studies on the effects of electromagnetic fields. https://www.emf-portal.org/en.

[5] For a more complete list of all the ways you are exposed to electromagnetic radiation, see the list given at the end of the article entitled "The 5G Dementors Meet the 4G Zombie Apocalypse": *"Technicians have been given free rein to dream up ways of attacking populations: from under manhole covers; from cabinets on the street; from lamp posts that blast blue light with no more diffusers, as well as 5G EMR in laser-like beams; from adhesive strips of tiny but powerful antennas hidden under carpets; on the street; in trains; in planes; in cars; in buses; blasting through the walls of our homes; from our television sets; from fridges, hairdryers, milk cartons, babies' diapers, baby monitors, "smart" phones, "smart" meters, and soon from the billions of devices that are planned to be connected to the Internet of Things. This is not to mention the plethora of "wearables" and cell phone apps that purport to help you monitor your health status while seriously undermining it. The plan is to beam 5G down to Earth from satellites in the Earth orbits, from networked civil aircraft, from pseudosatellites in the stratosphere ... In other words, from everywhere. The stated plan and the word trumpeted in the 5G literature is to "blanket" every inch of the Earth, with no escape for any of the approximately 100 million people worldwide already made sick by the toxic environment supplied courtesy of the first to fourth generations of WiFi, to which 5G will be additional."* https://www.globalresearch.ca/5g-electromagnetic-mad-zone-poised-self-destruct-5g-dementors-meet-4g-zombie-apocalypse/5689176.

[6] "The Radiation Poisoning of America." Amy Worthington. Global Research. 9 October 2007. https://emf-protection.us/2015/02/the-radiation-poisoning-of-america/.

[7] International Appeal to Stop 5G on Earth and in Space. 18 September 2018. "Biological effects occur even at near-zero power levels. Effects that have been found at 0.02 picowatts (trillionths of a watt) per square centimetre or less include altered genetic structure in E. coli[95] and in rats, altered EEG in humans, growth stimulation in bean plants, and stimulation of ovulation in chickens." https://www.5gspaceappeal.org/the-appeal.

[8] European Union: "Parliament Briefing: Effects of 5G wireless communication on human health", February 2020 https://forlifeonearth.weebly.com/eu-briefing-effects-of-5g-on-human-health.html.

[9] "For Life on Earth: Worldwide Web of Life." Claire Edwards. Radiation Dangers. 25 November 2020. https://radiationdangers.com/2020/11/25/for-life-on-earth-worldwide-web-of-life/.

[10] Arthur Firstenberg, Newsletter of the 5G Space Appeal: "441,449 Low Earth Orbit Satellites Operating, Approved and Proposed." 5 January 2022.

[11] 5G is a Weapon System: 5GW Fifth Generation Warfare. Video. Mark Steele. 8 July 2022. https://www.bitchute.com/video/2lUtZvLeHReO/.

[12] "You Are Not Mostly Empty Space." Ethan Siegel. Forbes. 16 April 2020. https://www.forbes.com/sites/startswithabang/2020/04/16/you-are-not-mostly-empty-space/.

[13] "The Schumann Effect Part 1: How the Earth Influences Your Brain Waves" (video and transcript). https://subtle.energy/the-schumann-effect-how-the-earth-influences-your-brain/.

[14] "Frequency and Our Body Organs." https://shepherdsheart.life/pages/body-organ-frequencies. The Organ Frequencies. https://www.taijiworld.com/organ-frequencies.html.

[15] "High electrical field effects on cell membranes." U. Pliquetta and others. *Biochemistry*. Vol. 70, Issue 2, May 2007, pp. 275-282.https://www.sciencedirect.com/science/article/abs/pii/S1567539406001459

[16] "DNA Molecules Can 'Teleport', Nobel Winner Says". John E. Dunn. Techworld.com. 16 January 2011. http://www.rexresearch.com/montagnier/montagnier.htm.

[17] Why Children Are Especially at Risk of Damage from EMFs." Dr. Joseph Mercola. 8 August 2021. https://lifeforce-vitality.com/blogs/news/why-children-are-especially-at-risk-of-damage-from-emfs.

[18] "WiFi: A Thalidomide in the Making — Who Cares?" Barrie Trower. http://stopsmartmetersbc.com/wifi-a-thalidomide-in-the-making-barrie-trower/.

[19] Barrie Trower on EMFs and Reproductive Devastation. Video. https://www.youtube.com/watch?v=y5Vyz2iS_PM

[20] Writing by Hand Boosts Brain Activity and Fine Motor Skills, Study Shows." Claire Gillespie. 24 October 2020. https://www.verywellmind.com/how-the-dying-art-of-handwriting-boosts-brain-activity-and-fine-motor-skills-5083814.

[21] 5G: Gigantic Health Hazard. Video. 25 February 2019. Barrie Trower and Sir Julian Rose. https://www.youtube.com/watch?v=TZ3t6n2SkUY.

[22] "Bird Feathers as Dielectric Receptors of RF Fields." J. Bigu del Blanco, C. Romero-Sierra & J. A. Tanner. http://thegreentimes.co.za/wp-content/uploads/2019/09/Tanner-Birds-as-Microwave-Sensors.pdf.

[23] "Bees, Birds and Mankind: Destroying Nature by 'Electrosmog'." Ulrich Warnke. https://kompetenzinitiative.com/wp-content/uploads/2019/08/ki_beesbirdsandmankind_screen.pdf.
All bees dead after cell tower activated in Idaho, US. 11 June 2021. https://radiationdangers.com/2021/06/11/all-bees-dead-after-cell-tower-activated-in-eagle-idaho-video/.

[24] "Insects have declined by as much as 80%, say entomologists." Garden Drum. 19 May 2017. https://gardendrum.com/2017/05/19/insects-declined-much-80-say-entomologists/.

[25] Radiofrequency Radiation from Wireless Technology Damages and Kills Plants and Trees." https://www.weebly.com/uploads/1/3/7/4/137404875/wireless_kills_trees.pdf.

[26] "Waves Upon the Oceans — and the Internet of Underwater Things." Shannon Rowan. 24 January 2022. https://wifi-refugee.com/blog%2Fessays/f/waves-upon-the-oceans%E2%80%94and-the-internet-of-underwater-things.

[27] "MIT Researcher's New Warning: At Today's Rate, Half of All U.S. Children Will be Autistic by 2025." 11 June 2014. https://althealthworks.com/mit-researchers-new-warning-at-todays-rate-1-in-2-children-will-be-autistic-by-2025/.
"Prevalence of autism in U.S. children increased by 119.4 percent from 2000 (1 in 150) to 2010 (1 in 68). (CDC, 2014). Autism Society. https://www.autism-society.org/what-is/facts-and-statistics/.

[28] "Brain, other CNS and intracranial tumours incidence statistics." Cancer Research UK. https://www.cancerresearchuk.org/health-professional/cancer-statistics/statistics-by-cancer-type/brain-other-cns-and-intracranial-tumours/incidence.

[29] "Indeed Study Shows that Worker Burnout is at Frighteningly High Levels." Jack Kelly. 6 April 2021. https://www.forbes.com/sites/jackkelly/2021/04/05/indeed-study-shows-that-worker-burnout-is-at-frighteningly-high-levels-here-is-what-you-need-to-do-now/.

[30] "Why is Arthur Firstenberg Not Telling You That 5G is a Weapon?" Claire Edwards. 20 March 2021. https://forlifeonearth.weebly.com/why-is-arthur-firstenberg-not-telling-you-that-5g-is-a-weapon.html.

[31] "Our Children are Now in Grave Danger." Claire Edwards. 10 October 2020. https://stateofthenation.co/?p=31293

[32] *Berlin Electropolis: Shock, Nerves, and German Modernity*. Andreas Killen. 16 January 2006. https://www.amazon.com/Berlin-Electropolis-Modernity-Cultural-Criticism/dp/0520243625.

[33] *The Invisible Rainbow: A History of Electricity and Life*. Arthur Firstenberg. 9 March 2020. Chapter 5: Chronic Electrical Illness. https://www.amazon.com/Invisible-Rainbow-History-Electricity-Life/dp/1645020096/ref=tmm_pap_swatch_0?_encoding=UTF8&qid=1657733242&sr=1-1.

[34] Shell Shock." BBC Inside Out. The story of Dr. Arthur Hurst's maverick approach to curing "shell shock": sending men out to dig the fields. https://www.bbc.co.uk/insideout/extra/series-1/shell_shocked.shtml.

[35] *The Invisible Rainbow: A History of Electricity and Life*. Arthur Firstenberg. 9 March 2020. Chapter 8: Mystery on the Isle of Wight. https://www.amazon.com/Invisible-Rainbow-History-Electricity-Life/dp/1645020096/ref=tmm_pap_swatch_0?_encoding=UTF8&qid=1657733242&sr=1-1.

36 Electricity and flu connection may explain viruses. Video. Claire Edwards. 15 September 2019. https://www.brighteon.com/d18e2f53-d3ee-4277-9ebb-ade2caca30da.

37 [11]

38 Cell Phone Industry 'War-game' Memo: The Disinformation Campaign to Confuse the Public." Environmental Health Trust. 14 May 2019. https://ehtrust.org/cell-phone-industry-wargame-memo-the-disinformation-campaign-to-confuse-the-public/.

39 "BBC Fake News on 5G Decoded: Health Impacts Denied Despite Overwhelming Scientific Evidence." Claire Edwards. Global Research. 25 August 2019. https://www.globalresearch.ca/online-bbc-fake-news-5g-decoded/5687055.

40 Guidelines for limiting exposure to time-varying electric, magnetic, and electromagnetic fields (up to 300 GHz). Page 506. International Commission on Non-Ionizing Radiation Protection. 1998. https://www.icnirp.org/cms/upload/publications/ICNIRPemfgdl.pdf.

41 Agenda 21 & 2030 Explained Within 15 Minutes by Rosa Koire. 20 July 2020. https://odysee.com/Agenda-21-Agenda-2030-Rosa-Koire:e.

42 "China: Social Crediting System Has 'Restored Morality' by Blacklisting Over 13 Million People." Breitbart. 14 May 2019. https://www.breitbart.com/asia/2019/05/14/china-social-credit-system-has-restored-morality-by-blacklisting-over-13-million-people/.

43 Information on graphene oxide. For Life on Earth. https://forlifeonearth.weebly.com/graphene-oxide-information.html.

44 What Graphene Oxide and Nanotechnology Have to Do With 5G. Video interview with Ricardo Delgado Martin of La Quinta Columna. 25 June 2022. https://australiaoneparty.com/what-graphene-oxide-and-nanotechnology-have-to-do-with-5g-interview-with-ricardo-delgado-martin-25-jun-2022/.

[45] "Graphene Oxide in All Vaccines Makes the Vaccinated Walking Time Bombs." Claire Edwards. 26 July 2021.
https://forlifeonearth.weebly.com/graphene-oxide-in-all-vaccines-makes-the-vaccinated-walking-time-bombs.html.

[46] [11], The Covid-19 Genocide of 2020. Video. Claire Edwards. 23 October 2020. https://forlifeonearth.weebly.com/the-covid-19-genocide-of-2020.html.

[47] "Dr. Martin Pall Prophesied Societal Breakdown 'Within Months, Not Years' of the Rollout of 5G." The Frank Report. 7 September 2020.
https://frankreport.com/2020/09/07/dr-martin-pall-prophesied-societal-breakdown-within-months-not-years-of-the-rollout-of-5g/.

[48] The Great Global Warming Swindle. Video documentary.
https://www.youtube.com/watch?v=oYhCQv5tNsQ.

[49] [30]

[50] Arthur Firstenberg, *The Invisible Rainbow,* 2017, AGB Press, Santa Fe, New Mexico, p. 76-77

[51] [50] p. 30

[52] https://www.bemri.org/publications/natural-fields/427-natural-and-human-activity-generated-electromagnetic-fields-on-earth/file.html

[53] *Resonance, Beings of Frequency*, documentary film, 2013

[54] [53]

[55] [53]

[56]Jeanice Barcelo, *Are Wireless Devices Really Safe?,* 2021

[57] Jagadis Chunder Bose, *Physiology of the Ascent of Sap*, 1923

[58] Peter Tompkins and Christopher Bird, *The Secret Life of Plants,* 1973, p. 95

[59] [58] p. 195

[60] [58], p. 195

[61] https://www.dailymail.co.uk/sciencetech/article-2524598/Experiment-finds-plants-die-placed-internet-Wi-Fi-routers.html

[62] Dr. Gerald Goldberg, MD, *Would You Put Your Head in a Microwave Oven*, 2007, Foreword

[63] [62] Chapter 1

[64] http://www.brooks.af.mil/AFRL/HED/hedr

[65] Dr. Gerald Goldberg, MD, *Would You Put Your Head in a Microwave Oven*, 2007, Chapter 2

[66] [65], Chapter 2

[67] [65], Chapter 2

[68] https://scientists4wiredtech.com/2019/09/tuning-in-to-microwave-sickness-from-wireless-radiation/

[69] https://phiremedical.org/

[70] Dr. Erica Mallery-Blythe, 'EHS- A Summary' https://emfacademy.com/wp-content/uploads/2017/12/Mallery-Blythe-v1-EESC.pdf

[71] These statistics were relayed to me by Dr. Olle Johansson, some of which can be found in *The Electronic Silent Spring*, by Katie Singer.

[72] Katie Singer, *An Electronic Silent Spring* p.85

[73] The microwave syndrome or electro-hypersensitivity: historical background, PubMed.gov, US National Library of Medicine, National Institutes of Health, 2015, https://www.ncbi.nlm.nih.gov/pubmed/26556835

[74] Dimitris J. Panagopoulos, 'Polarization: A Key Difference between Man-made and Natural Electromagnetic Fields, in regard to Biological Activity'

[75] 2019 5G Summit Crisis Talks with Josh Del Sol

[76] Arthur Firstenberg, *The Invisible Rainbow,* 2017, AGB Press, Santa Fe, New Mexico, p.368

[77] [76] p.367

[78] https://www.activistpost.com/2018/01/flu-microwave-sickness-share-many-symptoms.html

[79] You can support Olle and his research by sending donations at the following link:
https://www.emfsa.co.za/news/fundraiser-to-support-associate-professor-olle-johanssons-ongoing-research/

[80] Arthur Firstenberg, *The Invisible Rainbow,* 2017, AGB Press, Santa Fe, New Mexico, p.375

[81] Devra Davis, PhD, MPH, *Disconnect—The Truth about Cell Phone Radiation,* 2013

[82] [80] p 378 (Ollie's interview, chapter 'Land of the Blind')

[83] [80] p.74

[84] [80] p.31

[85] [80] p.32

[86] [80] p.32-33

[87] [80] p.62

[88] [80] p.62

[89] Dr. Elaine Aron PhD, *The Highly Sensitive Person,* 1997

[90] [89]

[91] [89]

[92] [89]

[93] Arthur Firstenberg, *The Invisible Rainbow,* 2017, AGB Press, Santa Fe, New Mexico, p.138

[94] https://www.ncbi.nlm.nih.gov/pubmed/26613326

[95] Liz Barris, www.ThePeoplesInitiative.org

96 Liz Barris' video created to raise awareness for the Children's Wireless Protection Act https://www.youtube.com/watch?v=TJg9_22nVMo

97 http://americanassociationforcellphonesafety.org/

98 The legislators Guide to Warning Labels and Cell Phones and the Layman's Guide to Non Thermal Effects from Wireless Devices and Infrastructures"...
http://www.americanassociationforcellphonesafety.org/uploads/Non_Thermal_Paper_10-10_AAA.pdf

99 Here is a video interview on the lawsuit. The video had nearly 20,000 views prior to being pulled from YouTube but Liz posted a copy to one of her YouTube Channels—
https://www.youtube.com/watch?v=M3gQOW2sfRE&feature=youtu.be

100 Liz Barris' Congressional White Paper presented to Congress when lobbying against this bill...
http://stopsmartgrid.org/wp-content/uploads/2013/10/Legal-Constitutional-and- Human-Rights-Violations-of-Smart-Grid-and-Smart-Meters.pdf

101 Liz Barris' video on the Federal Cell Tower Roll Out. https://www.youtube.com/watch?v=solvrkfKg-I

102https://www.prnewswire.com/news-releases/pilot-study-shows-dramatic-difference-in- brain-activity-with-ehs-electrohypersensitive-cases-as-compared-to-controls-non- ehs-300566854.html?tc=eml_cleartime

103 [101]

104 The study: https://peerj.com/articles/2727/

105 Source: https://www.ncbi.nlm.nih.gov/pmc/articles/PMC4773875/

106 Olga Sheean, EMF Off! A Call to Consciousness in a Misguidedly Microwaved World, 2017, p.52

107 [106] p.52-53

108 [106] p.53

109 [106] p.54

110 [106] p.99 - 106

[111] See https://olgasheean.com/ and https://www.emfoff.com/ for more on Olga's work, books, videos, courses and blogs.

[112] https://olgasheean.com/2020-vision/

[113] You can find all of the included posts and more at this link:

https://www.radiationrefuge.com/stories-from-people-with-EHS-news_1_1_2_1

Also Bruce's story, as he tells it on his site can be found here: *https://www.radiationrefuge.com/Bruce-Evans-my-EHS-story-news_1_1_2_1_44*

And my own as I wrote it in 2015 and shared to his site is here: https://www.radiationrefuge.com/Shannnons-EHS-story-news_1_1_2_1_26

[114] https://www.emfacts.com/2018/09/implications-for-5g-property-owner-can-be-forced-to-cut-trees-if-they-interfere-with-wi-fi/

[115] https://www.wisegeek.com/what-does-it-mean-to-be-a-canary-in-a-coal-mine.htm

[116] https://www.mirror.co.uk/news/uk-news/school-wifi-networks-could-put-12602188

[117] [116]

[118] [116] Also see: https://www.mirror.co.uk/news/uk-news/schoolgirl-found-hanging-tree-after-6926586

https://www.dailymail.co.uk/news/article-3339511/Schoolgirl-15-hanged-developing-allergic-reaction-WiFi-school.html

[119] Debra Fry, "Why Die for WiFi? My Did – Will Yours?" Toward Better Health, May, 3, 2016.

[120] [119]

[121] [119]

[122] [119]

[123] Olga Sheean, *EMF Off!; A call to consciousness in our misguidedly micro-waved world*, 2017, p.183-5

[124] GMI Reporter, Is Wi-Fi in Schools Harming Our Children? The Testimony of Rodney Palmer, GreenMedinfo.com, Oct. 4, 2018.
http://www.greenmedinfo.com/blog/can-wi-fi-schools-cause-sudden-cardiac-arrest-testimony-rodney-palmer

[125] *https://timesofindia.indiatimes.com/home/education/news/is-wifi-a-danger-to-our-kids/articleshow/64341468.cms*

[126] Jeanice Barcelo, *The Dark Side of Prenatal Ultrasound and the Dangers of Non-ionizing Radiation*, p 240, citing Barry Trower

For more on the link between increased risk of suicide and EMR exposures, see also Jeanice's website: https://radiationdangers.com/2022/02/11/increased-suicides-linked-to-cell-towers/

[127] Jeanice Barcelo, *The Dark Side of Prenatal Ultrasound and the Dangers of Non-ionizing Radiation*, p.247

[128] Olga Sheean, *EMF Off!—A call to consciousness in our misguidedly micro-waved world*, 2017, InsideOut Media , p. 182

[129] [128] p.182

[130] [128] p.183-184

[131] https://www.cnn.com/2018/07/31/europe/france-smartphones-school-ban-intl/index.html
https://parenting.firstcry.com/articles/the-shocking-reason-france-banned-wi-fi-in-schools/

https://www.powerwatch.org.uk/news/2015-02-05-france-wifi-restrictions.asp

https://www.lemonde.fr/planete/article/2015/01/29/une-loi-pour-encadrer-l-exposition-aux-ondes_4565339_3244.html

[132] https://www.jrseco.com/lloyds-insurance-company-does-not-cover-health-damage-caused-by-electromagnetic-radiation/

[133] https://5gfreecalifornia.org/stories/canary-dies/

[134] https://www.rfglobalnet.com/doc/fixed-wireless-communications-at-60ghz-unique-0001

[135] https://www.aip.org/fyi/2019/noaa-warns-5g-spectrum-interference-presents-major-threat-weather-forecasts

[136] https://www.drrobertyoung.com/post/the-connection-between-5g-and-the-corona-virus

[137] Jeanice Barcelo, *The Dark Side of Prenatal Ultrasound and the Dangers of Non-ionizing Radiation*, p. 83-84

[138] https://www.emfanalysis.com/wp-content/uploads/2015/06/EMF-Effects-via-Voltage-Gated-Calcium-Channels-Dr-Martin-Pall.pdf

[139] https://americanpressassociation.com/news-and-events/dr-cameron-kyle-ny-its-not-pneumonia-but-high-altitude-pulmonary-edema/

[140] https://radiopaedia.org/articles/radiation-pneumonitis?lang=us and https://radiopaedia.org/cases/covid-19-pneumonia-66?lang=us

[141] 'Mental health issues persist for many COVID survivors,' Aidin Vaziri, San Francisco Chronicle, December 6, 2020

[142] Evidence linking Covid-19 and 5G installations/exposures: https://pubmed.ncbi.nlm.nih.gov/34778597/

https://esmed.org/MRA/mra/article/view/2371

https://www.drrobertyoung.com/post/the-connection-between-5g-and-the-corona-virus

[143] 5G being covertly installed in schools during lockdowns https://www.takebackyourpower.net/5g-biometrics-covertly-installed-during-lockdown/

[144] Robert J. Burrowes, 'Deadly Rainbow; Will 5G Precipitate the Extinction of All Life on Earth'

[145] https://www.bbc.com/news/business-50258287

[146] https://radiationdangers.com/2020/04/24/study-shows-direct-correlation-between-5g-networks-and-coronavirus-outbreaks-2/

[147] See Claire Edward's excellent article for a juxtaposed outbreak map next to a cell phone coverage map for Italy: https://forlifeonearth.weebly.com/graphene-oxide-in-all-vaccines-makes-the-vaccinated-walking-time-bombs.html

[148] https://www.congress.gov/bill/116th-congress/senate-bill/893/text

[149] https://www.healthleadersmedia.com/innovation/president-trump-signs-2-trillion-stimulus-package-telehealth-gets-another-boost

[150] You can find out more about Bruce's fight on his YouTube and Bitchute channels, here:
https://www.youtube.com/channel/UCU109fUcF6HYgnJ9s6lvKDw
https://www.bitchute.com/channel/JiPYYzHG0x6t/

[151] https://thepeoplesinitiative.org/wp-content/uploads/2020/02/FCC-Lawsuit-2020-RF-Standards.pd

[152] See the fundraising video here—https://www.youtube.com/watch?v=jNpGDUbWZFo

[153] Fundraising trailer for Liz Barris' documentary film project https://vimeo.com/216441333

[154] https://olgasheean.com/crisis-to-freedom/

[155] Increases in asthma have been shown to correlate directly with increases in EMF exposures. https://www.irjet.net/archives/V2/i3/Irjet-v2i379.pdf

[156] Walter Bowart, *Operation Mind Control*, 1978

[157] https://www.foxnews.com/story/report-metrolink-engineer-texting-with-teen-moments-before-killer-commuter-crash

[158] Autonomous vehicles run down and kill pedestrians
https://www.foxnews.com/auto/autonomous-uber-ignored-pedestrian-it-fatally-hit-report-says

https://www.msn.com/en-us/news/us/waymo-automatic-vehicle-hits-pedestrian-in-lower-haight/ar-AARTqfT

https://news.yahoo.com/pedestrian-killed-accident-autonomous-uber-133400019.html

https://apnews.com/article/north-america-us-news-ap-top-news-dara-khosrowshahi-az-state-wire-fc96c4ec74fc4f68bfda092304bea82e

[159] https://www.nbcnews.com/tech/tech-news/self-driving-uber-car-hit-killed-woman-did-not-recognize-n1079281

[160] Nicholas Carr, *The Glass Cage*, 2014, p.11

[161] [160] p.119

[162] [160] p.186

[163] [160], p.7

[164] [160] p.186

[165] 'Strategic Aspects of the Asymmetric War on the US, Including Scalar Electromagnetic Weapons and their Terrorist Uses,' Tom E. Bearden, First edition October 13, 2004.

[166] https://www.trtworld.com/magazine/elon-musk-s-astonishing-mission-to-colonise-mars-here-s-how-he-ll-do-it-42246

[167] https://www.trtworld.com/magazine/elon-musk-s-astonishing-mission-to-colonise-mars-here-s-how-he-ll-do-it-42246

[168] https://5galert.com/2021-05-31-astronomers-call-on-world-governments-5g-satellites.html

https://www.globalresearch.ca/more-than-50000-5g-satellites-encircling-the-earth-appeal-by-astronomers-safeguarding-the-astronomical-sky-it/5700687

https://safetechinternational.org/independence-5g-vs-astronomy-losing-the-sky/

[169] see Ray Kurzweil, *The Singularity is Near; When Humans Transcend Biology*, Penguin Books, 2005

[170] [169] p.306

[171] Sherry Turkle, *Alone Together*, 2011

[172] https://mashable.com/article/autonomous-vehicle-perception-coronavirus

[173] Links to Dr. Stephan Lanka's work on viruses including recent studies disproving the viral theory linked to the Covid-19 pandemic: https://abruptearthchanges.com/2017/11/17/dr-stefan-lanka-the-history-of-the-infection-theory/

https://www.weblyf.com/2020/05/microbiologist-and-virologist-dr-stefan-lanka-viruses-do-not-cause-diseases-and-vaccines-are-not-effective/

https://greatreject.org/dr-stefan-lanka-claims-about-viruses-are-false/

https://projekt-immanuel.de/en/projekt-immanuel/

[174] www.cnet.com/news/how-elon-musk-pronounces-x-ae-a-12-his-new-sons-name/

[175] https://www.businessinsider.com/countries-tracking-citizens-phones-coronavirus-2020-3#italy-has-created-movement-maps-12

[176] https://japantoday.com/category/tech/japanese-government-proposes-cyborgs-and-robotic-avatars-for-all-by-2050

[177] Nicholas Carr, *The Glass Cage*, 2014, p.121

[178] [177], p.67

[179] [177] p.80

[180] Things That Make Us Smart: Defending Human Attributes in the Age of the Machine (New York: Perseus, 1993)

[181] Nicholas Carr, *The Glass Cage*, 2014, p.7, 131

[182] Werner Herzog, 2017 documentary film, *Lo and Behold; Reveries of a Connected World*

[183] Mammals, Birds, Insects and Plants Harmed by Radiation Emanating from Wi-Fi, Cellphone Towers, Microwave Transmitters, etc., https://radiationdangers.com/effects-on-nature/mammals-birds-insects-and-plants-harmed-by-radiation-emanating-from-wi-fi-cellphone-towers-microwave-transmitters-etc/

[184] Video - Animals are Dying Near Wind Turbines and Cell Towers - Farmers are Organizing in France, https://radiationdangers.com/effects-on-nature/video-animals-are-dying-near-wind-turbines-and-cell-towers-farmers-are-organizing-in-france/

[185] Hundreds of Thousands of Birds Drop Dead Across American South, https://radiationdangers.com/effects-on-nature/hundreds-of-thousands-of-birds-drop-dead-across-american-south/

[186] Birds Dying, Going Blind in D.C. — Experts "Puzzled" About the Cause, https://radiationdangers.com/effects-on-nature/5g-news-birds-dying-going-blind-in-d-c-experts-puzzled-about-the-cause/

[187] Dying birds fall from the sky 'screaming and bleeding from their eyes' in horrific incident in Australia, https://radiationdangers.com/effects-on-nature/dying-birds-fall-from-the-sky-screaming-and-bleeding-from-their-eyes-in-horrific-incident-in-australia/

[188] Arthur Firstenberg, *The Invisible Rainbow*, 2017, AGB Press, Santa Fe, New Mexico, Chapter 16, citing *Bienenvater,* Issue No. 9, 2003 and F. Ruzicka, Shaden Durch Elektrosmog, *Bienenwelt*, 2003, Vol 10: 34-35.

[189] 40% of Honeybee Colonies GONE In Less Than One Year, https://radiationdangers.com/effects-on-nature/40-of-honeybee-colonies-gone-in-less-than-one-year/

[190] All Bees DEAD After Cell Tower Activated in Eagle, Idaho - VIDEO, https://radiationdangers.com/effects-on-nature/all-bees-dead-after-cell-tower-activated-in-eagle-idaho-video/

[191] Wi-Fi Makes Trees Sick, Study Says, https://radiationdangers.com/effects-on-nature/wi-fi-makes-trees-sick-study-says/

[192] 5G Causing Illnesses, Stabbing Feeling, Migraines, Dizziness, Vertigo, Skin Rashes, Blurred Vision And More, https://radiationdangers.com/microwave-sickness/5g-causing-illnesses-stabbing-feeling-migraines-dizziness-vertigo-skin-rashes-blurred-vision-and-more/

[193] Electro-Sensitivity and Microwave Illness — The Epidemic of our Time (VIDEO), https://radiationdangers.com/microwave-sickness/electro-sensitivity-and-microwave-illness-the-epidemic-of-our-time-video/

[194] Smart Meters Responsible For Tidal Wave Of Mysterious Illnesses Due To EMFs, https://healingthebody.ca/smart-meters-causing-tidal-wave-of-mysterious-illnesses/

[195] Please see: The Covid vaccines are human-to-transhuman conversion kits to enslave & eliminate humanity, https://rumble.com/vri3uf-thumbnail-233-nanotechnology-identified-in-vaccines-composition-and-purpose.html and/or Dr. Ricardo Delgado of La Quinta Columna Connects the "Vaccine" Graphene Oxide Nanotechnology to the Great Reset/Transhumanism Agenda, https://www.expandingawarenessrelations.com/dr-ricardo-delgado-of-la-quinta-columna-connects-the-vaccine-graphene-oxide-nanotechnology-to-the-great-reset-transhumanism-agenda/

[196] Indiana life insurance CEO says deaths are up 40% among people ages 18-64, https://www.thecentersquare.com/indiana/indiana-life-insurance-ceo-says-deaths-are-up-40-among-people-ages-18-64/article_71473b12-6b1e-11ec-8641-5b2c06725e2c.html

[197] Killed via a phone in a nail salon, https://www.bitchute.com/video/1WjUWWDXZEbT/

[198] Tell Me Again How the Vaccine is Safe, https://www.brighteon.com/50f80fe0-dd42-44a9-a78e-0a72368d7660

[199] Brain Short Circuit, https://t.me/drchristianenorthrup/3500

[200] 405 Athlete Cardiac Arrests, Serious Issues, 237 Dead, After COVID Shot, https://goodsciencing.com/covid/athletes-suffer-cardiac-arrest-die-after-covid-shot/

[201] UK sees 44% increase in child deaths after jab rollout for young teens, data shows, https://www.lifesitenews.com/news/uk-sees-44-increase-in-child-deaths-after-jab-rollout-for-young-teens-data-shows/

[202] Baby of 'fully vaccinated' mom dies after born bleeding from mouth, nose: VAERS report, https://www.lifesitenews.com/news/baby-of-fully-vaccinated-mom-dies-after-being-born-bleeding-from-mouth-nose-vaers-report/

[203] See Arthur Firstenberg's book, *The Invisible Rainbow*, for more details about this.

[204] See Jeanice's article on the topic here: *https://radiationdangers.com/2020/01/31/is-covid-19-a-cover-up-for-radiation-sickness/*

[205] The "flu" and microwave sickness share the same symptoms: https://radiationdangers.com/microwave-sickness/the-flu-and-microwave-sickness-share-many-of-the-same-symptoms/

[206] You can read more about Jeanice's experience of first dealing with radiation poisoning at the following: https://radiationdangers.com/my-story/my-health-has-taken-a-downturn/ hhttps://radiationdangers.com/my-story/microwave-radiation-poisoning-my-symptoms-are-worsening/, https://radiationdangers.com/my-story/digital-opt-out-meter-has-been-removed-from-our-home/

[207] Jeanice's book can be found here: (https://birthofanewearth.com/2019/04/are-wireless-devices-really-safe/

[208] Jeanice's website can be accessed here: https://radiationdangers.com/

[209] See Jeanice's article for more info: *https://radiationdangers.com/microwave-hearing-and-tinnitus/is-it-tinnitus-or-is-it-microwave-auditory-effect/*.

[210] Nipul Parekh, Ultrasonic Waves Are Invisibly Harassing People All Over the World, Gadgetvibe.com, May 10, 2018. http://www.gadgetvibe.com/2018/05/10/ultrasonic-waves-are-invisibly-harassing-people-all-over-the-world citing E.M. Grigor, Effect of ultrasonic vibrations on personnel working with ultrasonic equipment, *Sov. Phys. Acoust*, 1966, and T. Kamakura, et al., Survey of Ultrasound Exposure: Through topic of parametric loudspeaker, *J. Acoust. Soc. Jpn. (J)*, 2011, Vol 67(5): 200-203, https://ci.nii.ac.jp/naid/10029477564/ and H.O. Parrack, Effect of Air-Borne Ultrasound on Humans, *International Audiology*, 1966, Vol 5(3), https://www.tandfonline.com/doi/abs/10.3109/05384916609074198

[211] Ultra Hearing Fetus, ACFNewsource.org, aired on Sep 30, 2002 and Jan 1, 2003. http://acfnewsource.org.s60463.gridserver.com/science/ultra_hearing_fetus.html

[212] High-Pitched Screaming and Sleep Disorders in Children Disappear when EMF/Microwave Frequencies are Removed (VIDEO), https://radiationdangers.com/babies-and-children/high-pitched-screaming-and-sleep-disorders-in-children-disappear-when-emf-microwave-frequencies-are-removed-video/

[213] Smart Meter Noise and "The Hum" – Ultrasonic Noise that People and Animals Can Hear, https://radiationdangers.com/smart-meters/smart-meter-noise-and-the-hum-ultrasonic-noise-that-people-and-animals-can-hear/

[214] Jennifer McGraw, Parents Blame Elementary School's Cell Tower After 4th Student Diagnosed With Cancer, CBS Sacramento, Mar 12, 2019. https://sacramento.cbslocal.com/2019/03/12/school-cell-tower-causing-cancer/

[215] GMI Reporter, Is Wi-Fi in Schools Harming Our Children? The Testimony of Rodney Palmer, GreenMedInfo.com, Oct 4, 2018. http://www.greenmedinfo.com/blog/can-wi-fi-schools-cause-sudden-cardiac-arrest-testimony-rodney-palmer

[216] Pauline Anderson, Long-term Cell Phone Use Linked to Brain Tumor Risk, Medscape.com, November 13, 2014. http://www.medscape.com/viewarticle/834888

[217] Dina Fine Maron, Major Cell Phone Radiation Study Reignites Cancer Questions, ScientificAmerican.com, May 27, 2016. https://www.scientificamerican.com/article/major-cell-phone-radiation-study-reignites-cancer-questions/

[218] S. Lehrer, et al., Association between number of cell phone contracts and brain tumor incidence in nineteen U.S. States, *Journal of Neuro-Oncology,* Feb 2011, Vol 101(3): 505-7. https://www.researchgate.net/publication/44853472_Association_between_num-ber_of_cell_phone_contracts_and_brain_tumor_incidence_in_nineteen_US_States

[219] A. Ciccone, Brain Tumor Histology and Incidence in Adolescents and Young Adults, NeurologyAdvisor.com, Feb 24, 2018. https://www.neurologyadvisor.com/brain-tumors/brain-tumor-malignant-cancer-incidence-young-adults/article/478958/

[220] US Brain Tumor Association, Brain Cancer #1 for Children 15-19 yrs While FDA Minimizes Risk From Conclusive NIH Study - US Brain Tumor Association.com, PR Newswire, Nov 5, 2018, https://www.prnewswire.com/news-releases/brain-cancer-1-for-children-15-19-yrs-while-fda-minimizes-risk-from-conclusive-nih-study—us-brain-tumor-associationcom-300743742.html

[221] Q.T. Ostrom, et al., American Brain Tumor Association Adolescent and Young Adult Primary Brain and Central Nervous System Tumors Diagnosed in the

United States in 2008-2012, *Neuro Oncol.*, Jan 2016, Vol 18(Suppl 1): i1–i50. https://www.ncbi.nlm.nih.gov/pmc/articles/PMC4690545/

[222] Professor Tom Butler, On the Clear Evidence of the Risks to Children from Smartphone and WiFi Radio Frequency Radiation, University College Cork, Ireland citing H.M. Mialon, & E.T. Nesson, Mobile Phones and the Risk of Brain Cancer Mortality: A Twenty-Five Year Cross-Country Analysis, 2018. https://ssrn.com/abstract=2674296 or http://dx.doi.org/10.2139/ssrn.2674296

[223] For the online version of this story with hyperlinks visit: *https://forlifeonearth.weebly.com/ehs-christmas-diary.html*

[224] Walter Bowart, *Operation Mind Control*, 1977

[225] *The Minds of Men* documentary film

[226] Dr. Jose Delgado, *Physical Control of the Mind; Toward a Psychocivilized Society*, 1969

[227] https://www.globalresearch.ca/study-electromagnetism-vaccinated-persons-luxembourg/5749516

https://www.orwell.city/2021/06/dr-jose-luis-sevillano-variable-pseudo-magnetism.html

[228] https://forlifeonearth.weebly.com/graphene-oxide-in-all-vaccines-makes-the-vaccinated-walking-time-bombs.html

[229] https://www.expandingawarenessrelations.com/dr-ricardo-delgado-of-la-quinta-columna-connects-the-vaccine-graphene-oxide-nanotechnology-to-the-great-reset-transhumanism-agenda/

[230] *Lo and Behold; Reveries of a Connected World,* Werner Herzog, 2017 and Nicholas Kardaras, *Glow Kids*, p. 171

[231] *Glow Kids*, Nicholas Kardaras, Ph.D, 2016, preface

[232] [231] p. 8

[233] *Lo and Behold; Reveries of a Connected World,* Werner Herzog, 2017

234 https://japantoday.com/category/tech/japanese-government-proposes-cyborgs-and-robotic-avatars-for-all-by-2050

235 [234]

236 [234]

237 Videos/info related to trees being cut down to make way for 5G and how 5G affects trees
http://stop5g.whynotnews.eu/?p=635

238 British MP calls cutting down trees to make way for 5G environmentally 'ridiculous' https://www.youtube.com/watch?v=M–2HsWtpfc

239 http://stop5g.whynotnews.eu/?p=635

240 https://www.alfavedic.com/alfacast-episode/countering-the-3d-control-grid-w-sofia-smallstorm/

241 Nicholas Carr, *The Glass Cage*, 2014, p. 132

242 [241], p. 128

243 [241], p. 133

244 [241] p. 132

245 [241] p. 131

246 'Cell Towers on the Ocean Floor' January 2022 article, Arthur Firstenberg

247 [246]

248 See, *The Lethal Dose*, Dr. Jennifer Daniels, and, *The Ultimate Conspiracy—the Biomedical Paradigm*, James McCumiskey

249 https://michaelsavage.com/birx-says-government-is-classifying-all-deaths-of-patients-with-coronavirus-as-covid-19-deaths-regardless-of-cause/

250 https://www.organiclifestylemagazine.com/cdc-admits-finacial-hospital-incentives-drove-up-covid-19-death-rates

251 Dr. Jennifer Daniels M.D., *The Lethal Dose; Why Your Doctor is Prescribing it, 2013, p.44*

[252] Alex Berenson, *Unreported Truths about COVID-19 and Lockdowns, Part 4:Vaccines,* see section 2

[253] Here is just one of the many links available for tracking information about Covid-19 vaccine deaths-https://www.precisionvaccinations.com/covid-19-vaccine-related-fatalities-updated
These vaccines are also being linked to thousands of cases of blood clots, stroke, aneurysm and other blood disorders.
https://conejoguardian.org/2021/12/14/more-vc-nurses-blow-whistle-on-overwhelming-numbers-of-heart-attacks-clotting-strokes/

[254] https://dianawest.net/Home/tabid/36/EntryId/4415/Report-Consequently-death-after-vaccination-is-considered-suicide-by-the-French-courts.aspx

[255] See this article about privacy invasion and cell phones:
https://www.electricsense.com/cell-phone-spying/

[256] Health dangers of LED and CFL light bulbs with Dr. Robert Hanson video
https://www.youtube.com/watch?v=P4gvBePS7qk

https://www.findhealthtips.com/dangerous-health-effect-of-led-clf-bulbs/

[257] https://radiationdangers.com/2019/11/30/the-dangers-of-led-lightbulbs/

[258] Links to articles about blue light dangers:
https://www.salzburg.gv.at/gesundheit_/Documents/feb_mystery_in_the_skin.pdf

https://www.foodsmatter.com/es/computers_wifi_bluetooth/articles/johansson_tv_healthy_volunteers.pdf

https://www.aoa.org/news/clinical-eye-care/diseases-and-conditions/blue-lights-link-to-prostate-and-breast-cancers?sso=y

https://www.hindustantimes.com/health/these-lights-are-harmful-for-health-may-trigger-breast-and-prostate-cancer/story-9loGV6AVpsfSAEhwWGNpzI.html

https://pubmed.ncbi.nlm.nih.gov/8867387/

[259] https://justinhealth.com/functional-medicine-mistakes-emf-sunlight-and-your-mitochondria-podcast-135/

[260] https://wellnessmama.com/5600/earthing-grounding/

[261] www.stetzerelectric.com the official source for Stetzer filters

[262] Sam Milham, *Dirty Electricity; Electrification and the Diseases of Civilization*, 2010, Chapter 8

[263] https://www.masterresource.org/wind-power-health-effects/infrasound-growing-liability-windpower/

[264] https://emfsol.com/mold-produces-600-times-more-bio-toxins-with-emf/

[265] *We Want to Live*, Aajonus Vonderplanitz, p. 136-7

[266] https://microwavenews.com/news-center/prenatal-exposure-weak-magnetic-fields-leads-childhood-asthma

[267] https://www.news-medical.net/news/20200630/COVID-19-triggers-changes-in-blood-platelets.aspx

[268] https://www.bmj.com/content/373/bmj.n958/rr-2

[269] https://www.armorguardsmg.com

[270] https://hexagonalwater.com/jerald_pollack.html

[271] See the excellent book, *Dissolving Illusions* by Suzanne Humphries for more information.

[272] Thecrowhouse.com

www.ingramcontent.com/pod-product-compliance
Ingram Content Group UK Ltd.
Pitfield, Milton Keynes, MK11 3LW, UK
UKHW021711190726
13853UKWH00001B/487